Context-aware Systems and Applications

Context-aware Systems and Applications

Edited by
Gannon Curran

Willford Press
www.willfordpress.com

Published by Willford Press,
118-35 Queens Blvd., Suite 400,
Forest Hills, NY 11375, USA

ISBN: 978-1-68285-342-9

Cataloging-in-Publication Data

Context-aware systems and applications / edited by Gannon Curran.
p. cm.
Includes bibliographical references and index.
ISBN 978-1-68285-342-9
1. Context-aware computing. 2. Mobile computing. 3. Ubiquitous computing. 4. Computer networks.
5. Telecommunication systems. I. Curran, Gannon.
QA76.5915 .C65 2017
004--dc23

For information on all Willford Press publications
visit our website at www.willfordpress.com

WILLFORD PRESS

Printed in the United States of America.

Contents

Preface

Context-aware systems are a group of systems that are self-aware with regard to their physical environment and behave according to the conditions that are imposed by their physical setting. Context-awareness, or ubiquitous computing is a budding field of computer science and seeks to create highly personalized, intelligent systems that can closely follow the user in his/her day to day activities. This book on context-aware systems and applications puts together the latest research that has taken place in this field. It presents the complex subject of context-aware systems in the most comprehensible and easy to understand language. It will help the readers in keeping pace with the rapid changes in this field. This book is a resource guide for experts as well as students.

This book has been the outcome of endless efforts put in by authors and researchers on various issues and topics within the field. The book is a comprehensive collection of significant researches that are addressed in a variety of chapters. It will surely enhance the knowledge of the field among readers across the globe.

It gives us an immense pleasure to thank our researchers and authors for their efforts to submit their piece of writing before the deadlines. Finally in the end, I would like to thank my family and colleagues who have been a great source of inspiration and support.

Editor

1

Towards a Context-Aware Framework for Improving Collaboration of Users in Groupware Systems

Luis G. Montané-Jiménez, Edgard Benítez-Guerrero, Carmen Mezura-Godoy

Facultad de Estadística e Informática, Universidad Veracruzana, Xalapa, Mexico 91020

Abstract

A Context-Aware Groupware System (CAGS) enables the members of a team to communicate, cooperate and coordinate their activities to achieve a common goal, by providing them tools that are aware of their current execution context and adapt accordingly. CAGS can be found in several domains such as entertainment, particularly Collaborative First-Person-Shooter (FPS) video games. In CAGS, the means of collaboration traditionally provided to users (e.g. text and audio messaging) are not necessarily adequate: for instance, in a FPS, messages can distract the gamer due to the speed of the game. This paper reports a study that, for Collaborative FPS, identifies advantages/disadvantages of current means of collaboration and social behaviors that arise when players interact face-to-face or remotely. Based on this study, a context-aware conceptual model and architecture is proposed for CAGS aimed to improve user collaboration.

Keywords: Context-Awareness, Groupware Systems, Video Games

1. Introduction

A Groupware System (GS) enables the members of a team to communicate, collaborate and coordinate their activities in order to achieve a common goal, in a loosely-coupled manner, by using a combination of mobile and fixed technologies. With advances on wireless networks and the emergence of new electronic devices (e.g. tablets, smart phones) the demand for GS has increased, causing these systems to become an essential element in several application domains, such as commerce, health, business, and entertainment. It is in this last domain where we are witnessing the proliferation of On-Line Multi-Player Video Games (particularly those of type First-Person Shooter - FPS), where players of a team collaborate to defeat other teams.

Traditional GS provides their users with information and services in predetermined ways: the same data items are presented in the same way to all the users, and the same services are invoked in the same sequence. Although this behavior is enough for some applications, a GS should satisfy more efficiently the information/services needs of its users. A GS can be enhanced if the context of its users is considered in the execution of activities, in order to provide adequate and relevant data and/or services; i.e. it is context-aware. Context variables related to a single user have been traditionally used in context-aware groupware systems (CAGS); for instance, in a FPS video game, contextual data of relevance for a player can be his/her energy level, the number of available munitions, and his/her current virtual location. However, context variables related to social aspects have been less studied. Awareness, defined as an understanding of the activities of others [10], is a classical concept that has been explored in CAGS; for instance, a player of a FPS video game can know the current virtual location of his/her teammates and make decisions based on that knowledge. Social Presence (SP), another relevant concept defined as the degree of relevance of the person performing an interaction [34], is also being investigated.

Now, the means of collaboration traditionally offered by current CAGS, particularly FPS videogames (e.g. text and audio messaging) at first sight seem limited. For instance, sending messages requires the user to deviate his/her attention from the game, which can affect his/her performance. It is then necessary to analyze the traditional means of collaboration offered in CAGS, particularly in Video Games of type First-Person-Shooter, in order to identify elements and conditions necessary to improve them. This paper presents a review of the main aspects related to Context-Aware Groupware Systems, and report an exploratory study considering as case study a Video Game of type FPS, which identifies advantages/disadvantages of current means of collaboration. Based on the findings of

*Corresponding author. Email: lmontane@uv.mx

the exploratory study, a conceptual model and an architecture for CAGS is proposed.

The remainder of this paper is organized as follows. Section 2 introduces Context-Aware Groupware Systems and describes the trends in Context-Aware Collaborative Video Game Systems. Section 3 describes the exploratory study and Section 4 defines the context-aware conceptual model and architecture for improving users collaboration. Section 5 presents a summary and introduces future works. This paper is an extended version of [25]. We have rewritten the "Context-Aware Groupware Systems and Video Games" and "Model and Architecture for CAGS" Sections in order to accommodate ideas we collected at the conference. Finally, we have added a new section (Section 4.3) describing related works about social presence.

2. Context-Aware Groupware Systems and Video Games

2.1. Context-Awareness and Groupware

Context-Aware Computing focuses on the creation of systems that use context to provide relevant information and/or services to the user, where relevancy depends on the user's task [8]. Let us note that there is no consensual definition of context: Dey [8] defines it as "any information that can be used to characterize the situation of an entity. An entity is a person, place or object that is considered relevant to the interaction between a user and an application, including the user and applications themselves", while for Chen [6] is "the set of states and values of the environment that either determine the behavior of an application or in which an application event happens and is of interest to the user".

Nowadays, most of the work related to modeling, management and use of context are intended for systems that do not support collaborative work. These works have traditionally considered context variables such as temperature or location (see Table 1). However, social context variables such as roles, social rules, status of the group activity, or status of other users, have been not considered in detail [26]. Therefore, it is relevant to explore the use of contextual information in collaborative activities in order to provide tools and mechanisms to make the collaborative process more efficient.

It is possible to define a Context-Aware Groupware System (CAGS) as software that enables the members of a team to communicate, cooperate and coordinate their activities in order to achieve a common goal by providing tools which are sensitive to any contextual information that can enhance collaboration. These tools acquire, manage and use context information produced by individual and group activities (in the form of user preferences, states of activities, physical/virtual locations, and sequences of activities, for instance) in order to support users to achieve their common goals more effectively and efficiently than in a traditional Groupware System.

Building a CAGS is a challenging task, as it requires considering the traditional aspects of software development (requirements analysis, design, implementation, and evaluation), but also it is necessary to consider the aspects related to the incorporation of context awareness: modeling, acquisition and use of context. With respect to models, some proposals are *activity-centered* (e.g. [2, 16–18]), while others are *user-centered* (e.g. [14, 24, 27, 32]). For instance, [3] presents a model for context-aware mobile collaborative systems based on the notions of user and group spaces, where users execute activities under a certain context either in isolation or within a working group, respectively. Let us note that most models do not allow dynamic changes in the rules and conditions that impact the performance of users in a collaborative activity; however, in real world scenarios these variables tend to constantly change. Thus, models considering dynamic changes in protocols and social rules defined and adapted by end users are needed.

With respect to development tools, one can find (a) *architectures* (e.g. [11, 23, 31]), (b) *middlewares* (e.g. [12, 13, 38, 39]), and (c) *frameworks* (e.g. [19–21, 36]). All the aforementioned works are important because they identify variables of collaboration and context awareness to support users to conduct their collaborative activities more efficiently.

2.2. Context-Aware Collaborative Video Games

Current Collaborative Video Games can be considered as basic context-aware systems, as they already use information about the performance of their users (points scored or game levels completed) in order to re-adapt the game to offer a more attractive user experience (changing dynamically the difficulty level, for instance). Table 2 shows the traditional variables on Collaborative Videogames. This can be fun but it is limited, as it has already been noticed by researchers whose works are aimed to take collaborative games to the next level. [28] presents a qualitative study to discuss the impact of awareness tools in Quake (a FPS Video Game), while [29] shows that users play

Table 1. Traditional variables on Context-Aware Systems [35].

Name	Description
Physical Context	Ambient temperature, lighting, noise.
User Context	Location, preferences, physical activity (run, walk, stop).
Time Context	Time of day, date.

better when they use awareness tools. Other related works are [9] which presents an analysis for improving communication tools through the use of user and group information, [40] which proposes a method to design games that support collaborative learning, and [22] which presents a study aimed to analyze collaborative activities performed by students and teachers in a game-based learning environment. So, it is desirable to propose mechanisms to change the behavior of the game depending on the current social context.

The aforementioned works are important because they represent the first efforts aimed to have context-aware collaborative video games. However, they are limited:

(i) User preferences about the Graphical User Interface (GUI) are not considered, so the same GUI is presented to all the players. This can have a negative effect in the performance of a player (for instance a player can get easily distracted when messages are displayed at the upper part of the screen instead of at the bottom part), so user preferences need to be taken into account (see Fig. 1).

(ii) Those works consider only awareness as the social aspect (for instance, reporting the presence of nearby enemies in radar); however social aspects are much more than just awareness and, for this reason, it seems attractive to propose models considering roles, group rules, goals, unusual behaviors, objects manipulated, social protocols, types of communication (e.g. visual, physical), actions (e.g. write, speak, etc.) to enable the representation of complex situations.

(iii) It is not clear if the traditional means of collaboration (e.g. voice/text messaging) are sufficient or appropriate to meet the needs of users in a distributed setting. It seems then interesting to investigate if the information generated when players are face-to-face and that is lost when they are geographically distributed can be detected and used in alternative means of collaboration.

Table 2. Traditional players variables on Collaborative Videogames.

Name	Description
User Level	Levels achieved.
Active Hours	Hours of training.
Score	Historical score obtained.
Achievements	Achievements unlocked.

Figure 1. Video Game screen. It presents energy level, bonuses, score, munitions.

(iv) The speed and efficiency to collaborate and communicate in a FPS Video Game are important and need to be considered, otherwise there is a high possibility that the players cannot generate a strategy and then can be easily defeated. Thus, solutions to this issue need to be proposed.

The study presented in Section 3 was conducted to explore and analyze these issues.

3. Exploratory Study

An exploratory and analytical study was designed and executed in order to identify whether there are additional elements that traditional means of collaboration do not consider. The following describes the design of the experiments considered in the study as well as the materials and methods used to perform them, and finally discusses the obtained results.

3.1. Experimental design

For this study, two different experiments were designed and executed. In the first experiment (e_1) a group of players were divided in two teams of equal size, and they played a FPS game in a face-to-face setting: they were in the same room and each player knew who were his/her teammates (see Fig. 2). For the second experiment (e_2), players were also divided in two teams of equal size, but this time they played in different physical locations (e.g. home, school, etc.). The intention of defining e_1 and e_2 was to detect behaviors that give us indications of the existence of other variables that arise from a face-to-face interaction between players, and these variables are not covered by traditional tools offered in the game (e.g. messages, audio messages predefined). Experiment e_1 was performed in a different day than experiment e_2.

The first experiment e_1 was conducted with 8 players, all of them students, 25% are women and 75% are men,

Figure 2. Group of players in a face-to-face setting.

Figure 3. Video Game screen with location map and the flag.

all between 19 and 26 years old and with at least one year of experience in playing FPS video games. The second experiment e_2 was performed with 8 players, students and workers, 25% are women and 75% are men, all between 20 and 29 years old and also with at least one year of experience in FPS Video Games.

The following variables were measured per player: number of eliminated enemies, player deaths, and the score generated individually and in group.

3.2. Materials

AssaultCube (AC) [1] is an open source First Person Shooter (FPS) Collaborative Video Game that can be executed in desktop environments and has a server to be executed online. AC provides collaborative scenarios with tools that users can use to coordinate and communicate (e.g. private and group messaging, predetermined audio messaging). The chosen collaborative scenario was the mini-game Flag Capture. In this game, the members of a team must keep a yellow flag as long as possible (see Fig. 3). Each time a team member holds the flag for a 15-second period, he earns points for his team and for his personal account; therefore, teammates need to protect each other, and they must also eliminate enemies that attempt to recover the flag. However, when the player carrying the flag is eliminated, the flag is available again so the enemy team can recover it to try to earn points.

For both experiments, e_1 and e_2, an instrument (a survey) to measure user satisfaction using a Likert scale was prepared. The instrument consists of 12 questions that refer to 6 dimensions related to user satisfaction (see Table 3), such as (D1) Coordination (1-2), (D2) Communication (3-4), (D3) Cooperation (5-6), (D4) Entertainment (7-8), (D5) Ease (9-10) and (D6) Awareness (11-12). The possible answers to the questions go from 1 to 5, where "1" stands for Strongly Disagree, "2" for Disagree, "3" Neither agree nor disagree, "4" Agree and "5" Strongly Agree. The consistency of this instrument was validated by applying a Cronbach Alpha coefficient to the answers obtained from the players in e_1. The Cronbach Alpha coefficient was applied to each dimension of the scale, and the results were as follows: D1 = 0.7, D2 = 0.5, D3 = 0.6, D4 = 0.8, D5 = 0.5 and D6 = 0.5. As can be seen, values of dimensions D1, D3 and D4 are equal to or greater than 0.6, therefore the questions in these dimensions have a moderate/acceptable reliability. In the case of dimensions D2, D5 and D6, they did not reach a value of 0.6 or higher; however, the obtained data were not significant to discard these items. In addition, the average of all the dimensions was 0.6, which is considered moderately acceptable. For future applications, it is expected to improve the items (D2, D5 and D6) and increase the sample size.

3.3. Procedure

The procedure for observation and evaluation of the video game was the following:

1. The AC Video Game was installed on computers that were involved in the experimental study.
2. During 15 minutes, the participants were provided with a manual of the AC video game with the connection steps in order to connect to the server and operate successfully the means of collaboration.
3. During 10 minutes, a test mission (Team Deathmatch) was launched in order to familiarize participants with the game.
4. During 15 minutes, the activity "Team Keep the flag" (a_1) was executed.
5. The questionary was applied to the participants.

3.4. Results

The results obtained during the game at e_1 can be seen in Fig. 4. In this case the winner was the blue team. The total score in the game is based on: i) accumulated points for every 15 seconds a player keeps the flag, ii) accumulated points for every eliminated enemy, iii) decreased points for every eliminated teammate and iv) decreased points for every dead player of the team.

In the AC video game experiment 1 (e_1), the players did not use the communication tools (e.g. instant messenger, predefined audio messages). This occurred because they were in the same place; therefore, they communicated with their own voice and gestures. For example, some players expressed explicit phrases as: "Protect me, I have the flag!". So roles (e.g. guards, flag bearer) were created on-the-fly, and constantly changed during the game. Also, in many cases, group behavior depended on reactions of the bearer of the flag. For instance, if his teammates heard and saw that he was in a state of frustration, they protected him even when the bearer did not request help explicitly.

Table 3. Relation of statements in the survey.

No.	Dimention	Statement
1	D1	Text messages offered by the game were helpful to organize you with your teammates.
2	D1	Location map in the game helped you generate an estrategy with your other teammates.
3	D2	Predefined audio messages were enough to transmit your emotions in the game.
4	D2	Text messages were useful for transmiting information to your teammates during the collaborative activity.
5	D3	The map locating your teammates is useful to support the group activity.
6	D3	The video game offers suitable collaboration tools to help your team.
7	D4	The video game kept you entertained during the development of the game.
8	D4	Information (e.g. messages, alerts, etc.) shown on the game interface possively affect your entertainment.
9	D5	The video game makes it easy to collaborate with your teammates.
10	D5	The video game makes it easy to eliminate players of the opposite team.
11	D6	Knowing the location of your teammates on the map helped you to execute game activities
12	D6	Knowing the score of all the players positively changed your behavior during the collaborative activity.

Red Team					
Player	Deaths	Killed enemies	Messages	Flags	Total score
1	13	28	0	0	193
2	15	8	0	1	47
3	5	7	0	0	54
4	13	17	0	9	329
Total	46	60	0	10	623

Blue Team					
Player	Deaths	Killed enemies	Messages	Flags	Total score
1	5	0	0	0	-20
2	16	33	0	11	585
3	10	2	0	0	-30
4	17	8	0	7	192
Total	48	43	0	18	727

Figure 4. Results of experiment e_1.

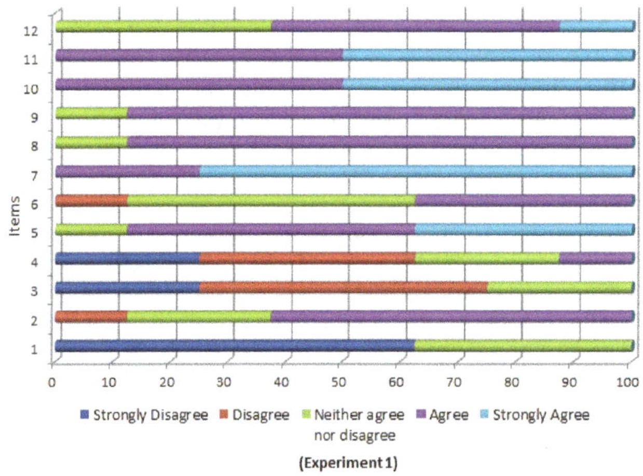

Figure 5. Results of the survey applied to the participants in e_1.

Fig. 5 presents the charts summarizing the results of the survey applied to players in e_1. The results related to dimension (D1) show an adverse opinion regarding the use of text messaging for coordination (item 1), but a positive attitude regarding the use of the map location (item 2). Dimension 2 (D2) shows the adverse opinion regarding for communication of players with AC tools. Dimension 3 (D3), related to cooperation, shows a positive trend in item 5, which is related to location map; however, in item 6, the trend of neutrality and dissatisfaction predominated. In dimension 4 (D4) and dimension 5 (D5), regarding ease of use and entertainment, the trend is strongly in favor. Finally, in dimension 6 (D6), related to awareness, the trend is to accept it positively.

For the results of e_2, see Fig. 6. In this case, the winner was the red team; however, they had an unfavorable opinion regarding the collaboration mechanisms provided by AC. Most messages sent by the teams were not to coordinate and organize their actions, but to joke or threat their enemies. More points were generated in e_2 than in e_1; this can be explained if one considers that a more competitive environment is created when players are competing face-to-face, so eliminating enemies or capturing the flag is more difficult.

Fig. 7 presents the charts with the results obtained by the application of the survey to players in e_2. As can be seen, D1 and D2 show an attitude of neutrality and dissatisfaction, in D3 only the item 5 is predominant with a positive attitude, in item 6 showing a slight trend in favor. Finally, in D4, D5 and D6 also the point of view was in favor.

Let us note that the group behavior of helping the bearer of the flag without any explicit request was not present in experiment 2 (e_2). This was because the tools currently offer by the video game are not sufficient to transmit states of frustration, so the bearer needed

Red Team					
Player	Deaths	Killed enemies	Messages	Flags	Total score
1	9	11	2	1	105
2	10	11	3	2	142
3	12	10	3	1	83
4	10	34	10	8	558
Total	41	66	18	12	888

Blue Team					
Player	Deaths	Killed enemies	Messages	Flags	Total score
1	18	7	2	1	26
2	0	21	0	4	258
3	14	3	2	0	-21
4	15	8	15	-1	0
Total	47	39	19	4	263

Figure 6. Results of experiment e_2.

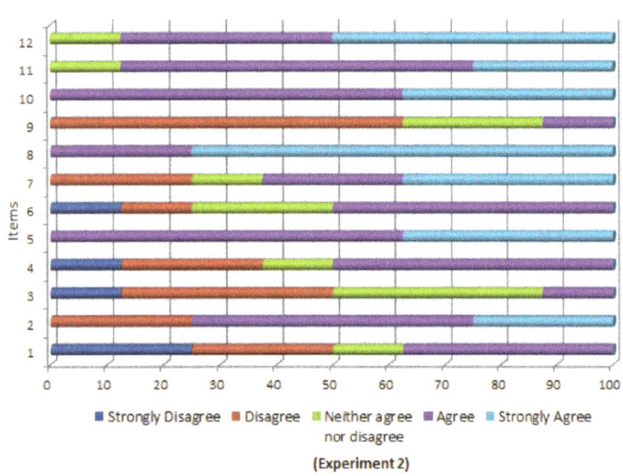

Figure 7. Results of the survey applied to the participants in e_2

to request help explicitly through text messages or predefined audio messages.

3.5. Discussion

The qualitative results obtained in experiments e_1 and e_2 were not in favor of collaboration mechanisms (D1, D2, D3). This give us evidence that the traditional mechanisms offered in the FPS videogame are not sufficient and are not presented in the best way; therefore, it is interesting to propose tools that include contextual information to enhace collaborations. The evaluation also helped us identify social behaviors that could be studied in greater depth, for example, identification of moods, detection of group patterns, and recognition of social interactions, to name few. This information can be considered as *awareness*, which helps the users to know the state of the activity undertaken. Therefore, the awareness and social behaviors identified in these experiments can be used to measure the relevance of users (i.e. Social Presence) within a Groupware System, which would allow to have new systems sensitive to such information.

4. Context-Aware Model and Architecture

Based on the findings of the exploratory study, this Section proposes a conceptual model and a generic architecture for CAGS.

4.1. Model

A conceptual model for Context-Aware Groupware Systems is presented in Table 4. This model comprises a set of social context variables that we consider of interest for a FPS-like Collaborative Video Game. The variables are categorized in three classes (interactive, cohesive, and affective), and for each of them a description and some examples are given. For instance, a group is a cohesive variable specifying a set of users that executes a collaborative activity, such as a team in a video game.

As complement to these variables, and in order to support the expected proactive behavior of a CAGS, we propose *social triggers*. They are rules of the form Event-Condition-Action, where Event refers specifically to a social event of relevance in the system. For example, Luis (p_1) can set a trigger (d_1) for the activity *Teem keeps the flag* (a_1), so that when he takes the flag (ac_1) some actions are automatically triggered, such as sending a predefined message to his closest teammate requesting for assistance (ac_2), transmitting a predefined audio message to all team members to inform them the place to where he is going (ac_3), or requesting information about where to find the closest place to recharge power (ac_4). Social triggers are intended to speed up the

Table 4. Social Context Variables

Category	Elements	Description	Examples
Interactive variables	Objects	Artifacts used in an activity.	Messages, bonus refills.
	Tasks	Individual actions to achieve goals.	Capturing the flag.
	Events	Occurred events in interactions.	Captured flag.
	Users	Entities that perform tasks and belong to a community	Players: Hugo and Luis.
	Locations	Physical or virtual locations of actors	Luis is participating in the activity "Team Keep the Flag" at the university.
Cohesive variables	Groups	A collection of users executing an activity.	Red team, blue team.
	Roles	Functions expected by a user.	Leader, protector, bearer.
	Goals	Community objectives.	The team must retrieve the flag.
	Alliances	Subset of users in the group.	Users with the aim of achieving a record score.
	Activities	Activities performed by the group.	Participate in the battle to keep the flag.
	Rules	Behaviors defined by the group.	If Luis has a leader role then he will receive all the messages from the team.
Affective variables	Moods	Moods of users executing an activity.	Frustation, happiness.
	Gestures	Gestures of the users.	Laugh or cry of a player.

participation of a player in the game and to avoid possible distractions.

4.2. Architecture

Fig. 8 presents a context-aware architecture based on the proposed model. For incorporating context-awareness into Groupware Systems, it is not only necessary to capture individual and social contextual data from different sources, but it is also important to consider a reasoning mechanism (logical or probabilistic) being able to infer high-level contextual information, but also to help the users of Groupware Systems to make decisions based on the current context or to proactively take actions. This architecture has four layers that are described below.

Groupware System Layer. This layer has direct communication with the users through the user interface and the means of collaboration provided by the Groupware System. The Groupware System send events occurred in the group to the Contextual Data Acquisition Layer.

Context-Usage Layer. This layer uses information provided by the Context-Management layer in order to infer and reason with social and individual contexts. Once this process is complete, the GS is adapted and manipulated depending on the result of the reasoning process.

Context-Management Layer. The primary functions of the Context-Management Layer are store, update and retrieve the context information acquired from the Contextual Data Acquisition Layer. Therefore, in this layer there is a data store that is accessed in order to process information.

Contextual Data Acquisition Layer. It is located at the lowest level; it acquires information directly from internal (wrt to the GS) and external data providers

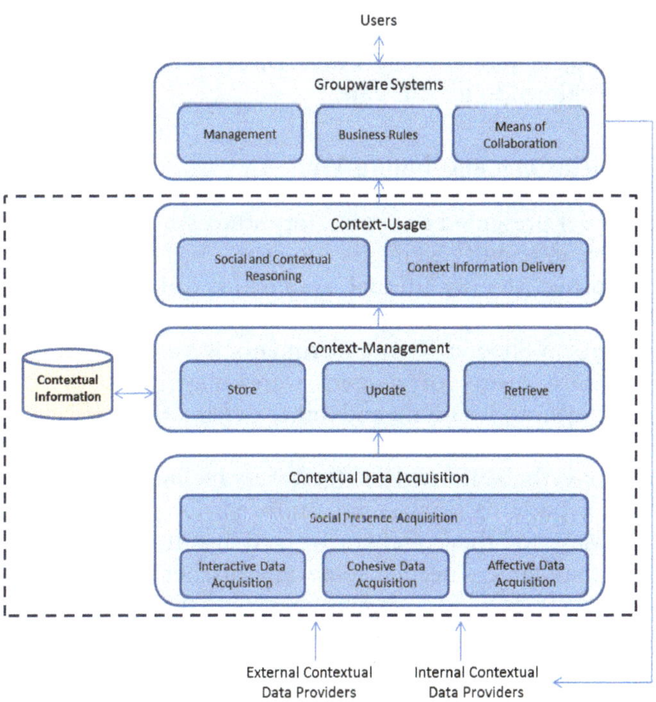

Figure 8. Conceptual Architecture.

(such as web services, video capture from video cameras and audio capture from microphones). In this layer, we propose the incorporation of a sensor of Social Presence to determine the level of relevance of users involved in a collaborative activity. The relevance of users will be a composite categorical variable built from basic interactive, cohesive and affective data obtained from the game. For future stages, the data acquisition layer be able to communicate with the groupware through the information delivery via a service platform that provides web services. For information retrieval, we propose retrieve game data for lapses of 15 seconds for each player, therefore, considering that the game of "Team Keep the Flag" lasts 15 minutes, at the end of the game each player will have 60 readings of cohesive and interactive variables.These lapses will be analyzed in order to see the behavior and variability information.

4.3. Discussion

Let us note that several works [4, 5, 34, 37] have explored social presence in educational settings, but few works (as ours) have studied it in gaming scenarios [7]. In the educational domain, [15] presents a critique of how students perceive the social means to support their study activities, and [33] indicates that technological progress has led to the creation of new models for different types of communication, using synchronous and asynchronous methods for online teaching. In game scenarios, [30] explores social presence experimenting with collaborative mobile games. Few works address this problem for different types (face-to-face and computer-mediated) of interactions: the work presented in this paper is aimed to fill this gap.

5. Conclusions and Future Work

This paper presented an exploratory study which had as objective to identify elements and conditions necessary to improve the traditional means of collaboration offered in CAGS, particularly in Video Games of type First-Person-Shooter. Two experiments were executed, one where teams of gamers were playing at the same physical location, and other where they were geographically distributed. As result, it was found that nonverbal information (gestures, feelings, group behavior patterns) that is not captured nor processed is relevant for the collaborative activity. It was also found that gamers needed to deviate his/her attention of the game to send text messages, and that most of the time text messaging were not used to collaborate. Then, faster and more efficient collaboration tools are needed. With this evidence, a model and an architecture for CAGS that takes into account these findings were proposed.

In future works, we will modify AssaultCube to integrate elements of context-awareness in order to support users to achieve their goals more efficiently. It is expected that the game will automatically provide appropriate means of collaboration (e.g. messages, maps) depending on the executed interaction (e.g. player captures the flag on the battlefield), without requiring the user to invoke such tools in an explicit manner. In order to accomplish this, we will implement a service layer based on the conceptual architecture. Finally, we will design and execute an experimental study to compare both the regular version of the game and its context-aware version.

Acknowledgements. The authors thank the reviewers of this paper for their useful comments. The first author gratefully acknowledges to CONACYT for scholarship No. 350124 for graduate studies.

References

[1] AssaultCube, P. (2013) What is assaultcube?, http://assault.cubers.net/.

[2] Bardram, J.E. and Christensen, H.B. (2007) Pervasive computing support for hospitals: An overview of the activity-based computing project. *Pervasive Computing, IEEE* **6**(1): 44–51.

[3] Benítez, E., Mezura, C. and Montané, L. (2012) Context-aware mobile collaborative systems: Conceptual modeling and case study. *Sensors* **12**(10): 13491–13507.

[4] Biocca, F., Harms, C. and Burgoon, J.K. (2003) Toward a more robust theory and measure of social presence: review and suggested criteria. *Presence: Teleoperators and Virtual Environments* **12**(5): 456–480.

[5] Bulu, S.T. (2012) Place presence, social presence, co-presence, and satisfaction in virtual worlds. *Computers and Education* **58**(1): 154–161.

[6] Chen, G. and Kotz, D. (2000) *A Survey of Context-Aware Mobile Computing Research*. Technical Report TR2000-381, Department of Computer Science, Darthmouth College.

[7] de Kort, Y.A.W., IJsselsteijn, W.A. and Poels, K. (2007) Digital games as social presence technology: Development of the social presence in gaming questionnaire (spgq). In *Proc. of the 10th International Workshop on Presence (PRESENCE'07)* (Barcelona, Spain): 195–203.

[8] Dey, A.K. (2001) Understanding and using context. *Personal Ubiquitous Computing* **5**(1): 4–7.

[9] Diamantaki, K., Rizopoulos, C., Charitos, D. and Tsianos, N. (2011) Theoretical and methodological implications of designing and implementing multiuser location-based games. *Personal and Ubiquitous Computing* **15**(1): 37–49.

[10] Dourish, P. and Bellotti, V. (1992) Awareness and coordination in shared workspaces. In *Proc. of the ACM conference on Computer-supported cooperative work (CSCW '92)* (New York, NY, USA): 107–114.

[11] Duran-Limon, H.A., Blair, G.S., Friday, A., Grace, P., Samartzidis, G., Sivaharan, T. and Wu, M. (2003)

Context-aware middleware for pervasive and ad hoc environments. Tech. rep.

[12] Favela, J., Alba, M. and Rodríguez, M. (2002) Autonomous agents for ubiquitous collaborative environments. In Coello Coello, C., de Albornoz, A., Sucar, L. and Battistutti, O. [eds.] *Proc. of the Second Mexican International Conference on Artificial Intelligence: Advances in Artificial Intelligence (MICAI'02)* (Springer Berlin-Heidelberg), *Lecture Notes in Computer Science* **2313**, 351–360.

[13] Gu, T., Pung, H.K. and Zhang, D.Q. (2005) A service-oriented middleware for building context-aware services. *Network and Computer Applications* **28**(1): 1–18.

[14] Hervás, R., Bravo, J. and Fontecha, J. (2010) A context model based on ontological languages: A proposal for information visualization. *Universal Computer Science* **16**(12): 1539–1555.

[15] Hrastinski, S. and Aghaee, N.M. (2012) How are campus students using social media to support their studies? an explorative interview study. *Education and Information Technologies* **17**(4): 451–464.

[16] Hung, L.X., Lee, S., Lee, Y.K. and Lee, H. (2007) Activity-based access control model to hospital information. In *Proc. IEEE Conference on Embedded and Real-Time Computing Systems and Applications (RTCSA'07)* (Daegu): 488–496.

[17] Kaenampornpan, M. and O'Neill, E. (2004) An integrated context model: Bringing activity to context. In *Proc. of the Workshop On Advanced Context Modelling, Reasoning and Management (UBICOMP'04)* (Nottingham, UK).

[18] Kofod-Petersen, A. and Cassens, J. (2006) Using activity theory to model context awareness. In *Modeling and Retrieval of Context: Second International Workshop (MRC'05)* (Springer-Verlag), *Lecture Notes in Computer Science* **3946**: 1–17.

[19] Kulkarni, D., Ahmed, T. and Tripathi, A. (2012) A generative programming framework for context-aware cscw applications. *ACM Transactions on Software Engineering and Methodology* **21**(2): 11:1–11:35.

[20] László Kovács, Péter Mátételki, B.P. (2009) Service-oriented context-aware framework. In *Proc. of the Fourth European Young Researchers Workshop on Service Oriented Computing (YR-SOC'09)* (Pisa, Italy).

[21] Lu, C., Chang, M., Kinshuk, Huang, E. and of Context-Aware Mobile Educational Game, U. (2011) Usability of context-aware mobile educational game. *Knowledge Management & E-Learning* **3**.

[22] Marty, J.C. and Carron, T. (2011) Observation of collaborative activities in a game-based learning platform. *IEEE Transactions on Learning Technologies* **4**(1): 98–110.

[23] Mayrhofer, R. (2004) An architecture for context prediction. In *Advances in Pervasive Computing* (Austrian Computer Society (OCG)), **176**, 65–72. Part of the Second International Conference on Pervasive Computing (PERVASIVE'04).

[24] Messeguer, Medina, R. and J.Fave (2010) Supporting context-aware collaborative learning through automatic group formation. In *Proc. of the 4th International Symposium on Ubiquitous Computing and Ambient Intelligence (UCAmI'10)* (Valencia, España).

[25] Montane-Jimenez, L., Benitez-Guerrero, E. and Mezura-Godoy, C. (2013) A context-aware architecture for improving collaboration of users in groupware systems. In *Proc. of the 9th International Conference Conference on Collaborative Computing: Networking, Applications and Worksharing (Collaboratecom' 2013)*: 70–76.

[26] Montane-Jimenez, L.G., Benitez-Guerrero, E. and Mezura-Godoy, C. (2013) Context-aware groupware systems and video games: State of the art. *Proc. of the Mexican International Conference on Computer Science (ENC'13)* : 55–59.

[27] Mowafi, Y. and Zhang, D. (2007) A user-centered approach to context-awareness in mobile computing. In *Proc. of the 2007 Fourth Annual International Conference on Mobile and Ubiquitous Systems: Networking&Services (MobiQuitous'07)* (DC, USA: IEEE Computer Society): 1–3.

[28] Nova, N. (2001) Awareness tools: lessons from quake-like. In *Proc. of the Playing with the Future Conference.*

[29] Nova, N., Wehrle, T., Goslin, J., Bourquin, Y. and Dillenbourg, P. (2007) Collaboration in a multi-user game: impacts of an awareness tool on mutual modeling. *Multimedia Tools Applications* **32**(2): 161–183.

[30] Nova, N., Girardin, F. and Dillenbourg, P. (2010) The effects of mutual location-awareness on group coordination. *Human-Computer Studies* **68**(7): 451–467.

[31] Olivares, A., Mendoza, S. and de Luca, A. (2011) An architecture to support context of use in groupware systems. In *Proc. 8th International Conference on Electrical Engineering Computing Science and Automatic Control (CCE'11)* (Merida City): 1–6.

[32] Romsaiyud, W. and Premchaiswadi, W. (2010) Soa context-aware mobile data model for emergency situation. In *Proc. IEEE Conference on ICT and Knowledge Engineering'10* (Bangkok): 93–97.

[33] Roseth, C., Akcaoglu, M. and Zellner, A. (2013) Blending synchronous face-to-face and computer-supported cooperative learning in a hybrid doctoral seminar. *TechTrends* **57**(3): 54–59.

[34] Rourke, L., Anderson, T., Garrison, D.R. and Archer, W. (1999) Assessing social presence in asynchronous text-based computer conferencing. *Distance Education* **14**(2): 50–71.

[35] Schilit, B. and Theimer, M.M. (1994) Disseminating active map information to mobile hosts. *IEEE Network* **8**(5): 22–32.

[36] Schroyen, J., Luyten, K., Gabriels, K., Robert, K., Teunkens, D., Coninx, K., Flerackers, E. *et al.* (2009) The design of context-specific educational mobile games. In Trant, J. and Bearman, D. [eds.] *Proc. of the Workshop on Museums and the Web: The International Conference for Culture and Heritage Online* (Indianapolis, USA: Toronto: Archives and Museum Informatics).

[37] Shen, K.N. and Khalifa, M. (2008) Design for social presence in online communities: a multi-dimensional approach. *AIS Transactions on Human-Computer Interaction (1)* : 33–54.

[38] Siegemund, F. (2004) A context-aware communication platform for smart objects. In *Proc. of the Second International Conference on Pervasive Computing (PERVASIVE'04)* (Vienna, Austria: Springer-Verlag Berlin Heidelberg), **3001**, 69–86.

[39] Yau, S.S. and Karim, F. (2004) An adaptive middleware for context-sensitive communications for real-time applications in ubiquitous computing environments. *Real-Time Systems* **26**: 29–61.

[40] Zea, N.P., Sanchez, J.L.G. and Gutierrez, F.L. (2009) Collaborative learning by means of video games: An entertainment system in the learning processes. In *Proc. IEEE International Conference on Advanced Learning Technologies (ICALT'09)* (USA: IEEE Computer Society): 215–217.

2

Context analysis approach for context aware applications deployed on pervasive environments

Nesrine Khabou[1,*], Ismael Bouassida Rodriguez[1,2], Mohamed Jmaiel[1,3]

[1]ReDCAD laboratory, University of Sfax, National School of Engineers of Sfax, Tunisia

[2]Univ de Toulouse, LAAS, Toulouse, France

[3]Research Center for Computer Science, Multimedia and Digital Data Processing of Sfax, Sfax, Tunisia

Abstract

The increase of mobile and interconnected devices leads to the growth of demands for context aware applications. These applications deployed on top of pervasive environments must adapt themselves to context changes. Context aware applications have to continuously sense their physical environment, and adapt their behavior accordingly. These applications must perform four phases starting by collecting and monitoring context, then analyzing context, deciding adaptation actions and finally executing the planned adaptation actions to deal with the context changes. In this paper, we focus on the second phase.We propose a context analysis approach that relies on different thresholds defined according to the user needs to detect context changes and raise notifications when changes occur. The analysis approach is performed according to three different steps. A context storage step, a context classification step, and a threshold calculation step.

Keywords: Pervasive computing, context awareness, adaptation, analysis techniques, threshold calculation, mathematical models.

1. Introduction

Recently, we have been witnessing how various applications are being integrated more deeply into daily users [6]. These technological advances are providing the hardware infrastructure necessary to achieve the pervasive computing paradigm [14]. Pervasive computing is defined by Mark Weiser [16] as: *"The most profound technologies are those that disappear. They weave themselves into the fabric of everyday life until they are indistinguishable from it"*. One of the important properties of pervasive computing is context awareness. Moreover, pervasive computing and ubiquitous applications need to be aware of their environment in order to adapt to changing contexts and provide correct services [1]. Context awareness refers to the ability of an application to acquire and reason about context and adapt its behavior accordingly.

Therefore, applications on top of pervasive environments need to be context aware so that they can adapt themselves to the changing context and the execution environments. The applications that feature this property should perform four phases illustrated in Figure. 1 starting by collecting and monitoring context from different sources like sensors for example. These monitored information are exploited and analyzed during the second phase (Analyze). The analysis phase is responsible for detecting context changes and raising notifications when changes are detected. These notifications are used in the third phase (Decide) to decide the adaptation actions related to the changes. In the last phase (Act), the adaptation actions are executed to react to the changes and adapt the application behavior accordingly. Our interest is focused on the second phase (Analyze). In fact, detecting context changes in a timely manner is crucial to avoid undesirable situations such as device crashes, etc. In this paper, we propose

* Corresponding author. Email: nesrine.khabou@gmail.com

and we detail a context analysis approach in pervasive environments. Our proposed approach is performed according to three steps. First, context parameters are stored in context databases for example for further use. The context storage step is out of the scope of this paper. Second, the stored context parameters are retrieved from the databases then classified to facilitate the context use. Finally, the context parameters are analyzed and changes are detected by comparing the context parameter values and the computed threshold values. The threshold values are computed and updated via different mathematical models. Notifications are triggered whenever the context values cross the threshold value.

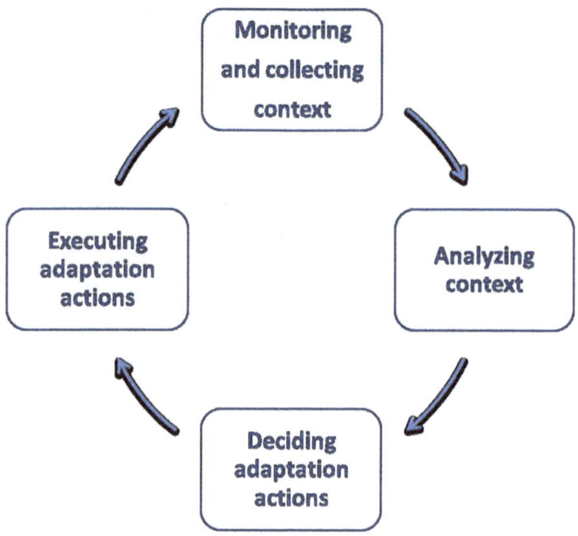

Figure 1. Context aware application cycle

The rest of the paper is organized as follows. In section 2, we present some studies related to context change detection (Context analysis). In section 3, we present a case study named "Smart Campus System". Its purpose is to provide suitable services to users according to context. The analysis approach is detailed in section 4. First, we define a context parameter classification step which takes into account the context parameter evolution. Second, we detail the threshold calculation step that allows to analyze context and detect changes. In section 5, we illustrate the usefulness of our approach through an illustrative scenario related to the Smart Campus case study. The last section concludes the paper and gives some directions for future work.

2. Related work

In the context changes detection (context analysis) research direction, different techniques are proposed to detect the context changes.

Cioara et al. [4] propose to use the context entropy concept to detect the context changes and determining the degree of fulfilling a predefined set of policies. Furthermore, context situation entropy measures the level of the system's self and execution environment disorder. So, if the system context situation entropy value exceeds a fixed threshold value, then the system is in a critical state and it must execute adaptation actions to control itself and to keep the system entropy below the threshold.

Although this approach allows for a self adaptation in context aware systems, it is restricted to external context parameters such as temperature, humidity and light etc. Besides, using fixed threshold values is inappropriate in dynamic environments and can cause false alarms.

In other studies, context changes are detected by comparing a context value saved in a repository with a new context value. In fact, Zheng et al. [18] have addressed the issue of context change detection by proposing a context-aware middleware which conforms to the CORBA component model. The proposed middleware is composed of context aware services such as a context collector, a context interpreter, a context repository and a context analyzer. The latter is in charge of filtering and analyzing context parameters to determine relevant context changes and notify the application afterwards. Context filtering is based on a comparison of the context values stored in the context repository with the new context value in order to detect context changes.

The proposed middleware enables to save the scare resources. In fact, the component deployment is performed "just-in-time". However, this middleware does not specify context parameters to take into account.

Another approach for dynamic context management is proposed by Taconet et al. in [15]. They present CA3M, a context aware middleware, which enables applications to adapt their behavior by dynamically taking into account context changes. The authors model the application by "entities", which represent a physical or a logical phenomenon (person, concept, etc.) and "observable", which defines something to observe. For instance, a mobile device state is an example of an observable which may take a finite number of values (e.g low battery, almost low battery or normal battery). They consider that the change of an "observable" state leads to a different application behavior.

Other approaches are used to analyze context. In fact, Bouassida et al. [3] proposed a model driven approach for collaborative pervasive systems. The authors have identified different levels for such systems. Then, models for each level have been proposed. They have implemented a refinement procedure, that allows the transition between levels and they have detailed a selection procedure, that allows the adaptation to context changes according to given policies. The selection procedure is based essentially on fixed thresholds. Then, once a node state (identified by battery level, memory consumption, CPU usage, etc.) remain below/under the threshold values, a notification is raised. Although this approach enables to detect context changes, it may cause false detections by using fixed thresholds.

In the work of Hussein et al., [7], the authors proposed an architecture based approach for developing context aware adaptive systems. The proposed architecture has three layers that partition the system in a way that allows the context aware adaptive system requirements to be handled using the different layers. The context is first modelled at layer two (The system and its context representation), processed and managed at layer three (The change management). This layer is responsible for adapting the system to cope with context. It is based on adaptation rules that rely on fixed thresholds to trigger adaptation actions. The above layered approach allows for context aware system development. The used technique for context change detection is based on fixed threshold which can lead to false detections or missing alarms.

Birje et al. [2] propose a multiagent model to monitor the resource availability and control the device state in wireless grid. The mobile agent models aim to provide not only a resource monitoring scheme to keep track of all devices and their resource utilization at any instant but also a device state control such that avoiding an overloaded state. For that reason, the authors consider three resource states: an overload/poor state, an underload/excellent state and a normal state. To evaluate the resource state, the authors define two thresholds. A minimum threshold $threshold_{min}$ and a maximum threshold $threshold_{max}$. The proposed multi agent model monitors and controls the resource utilization, the resource availability, the device mobility and the device state by using the defined thresholds. Although the proposed approach takes into account various parameters, the use of fixed thresholds can lead to false or missing detections.

In [10], the authors provide an experimental study about the detection capabilities provided by different monitoring techniques used in web based applications. They study four monitoring techniques. The system level monitoring that consists in gathering different system parameters like CPU usage, memory consumption, etc.These parameter values are forwarded to a central system that compares the collected parameter values against threshold values that are set so that when the thresholds are reached, alert messages are triggered. In the container level monitoring, application containers that include monitoring services gather details related to the application server parameters for example the JVM memory, the instanced objects, etc. The gathered data is used by the information technology staff to reason about the application server thumbprint and to react when necessary in order to adjust the application and improve its performance. The end-to-end monitoring aims to identify if users are facing problems while accessing the application. End-to-end monitoring tools are based on monitoring agents that are geographically dispersed that gather details such as availability, response time, etc. These data are used to detect anomalies and identify their origin. Finally, log analysis which consists in log files that record data about the system and application usage to detect anomalies. The authors propose their own monitoring technique. They present a framework used to detect and isolate performance anomalies in web based applications. The proposed framework is composed of different building blocks: A monitoring module, a performance analyzer, an anomaly detector, a root cause failure analysis, a recovery module, a workload variation module and a system dimensioning module. The monitoring module is responsible for collecting, online, a set of system, application and application server parameters. These parameters are transmitted to the performance analyzer to detect the performance anomalies. When anomalies are detected, the recovery module is responsible for executing the reconfiguration action. For the workload variation and the system dimensioning modules, they are in charge of controlling the workload demands and the resource consumption. The authors use AOP (Aspect Oriented Programming) to monitor the system, the application and the application server parameters. In order to detect anomalies, the authors measure the correlation degree between the response time and the number of user transactions processed. Alert messages are triggered when the correlation degree suddenly decays. An additional analysis based on the Dynamic-Time Wrapping algorithm is performed to detect performance anomalies. The proposed approach is designated to web based applications, so that it is restricted to few parameters.

Exposito et al. [5], have presented a context aware framework for alert generation. The framework is responsible for identifying alarm events identifying the monitoring entities being able to manage the alarm events when it is generated and forwarding data to the monitoring entity when requested. The authors divided the alarm events into three categories: instantaneous, interval based and multiparametric. For the instantaneous alarms, they are triggered when the parameter value exceeds a pre-established range as well as when they have an excessive deviation from a certain value. For the interval based alarms, they are associated to situations lasting for a span of time, they are based on the knowledge of the dynamics of parameters and the reasoning over their temporal evolution. Finally, for the multiparametric alarms, they are build by considering simultaneously several heterogeneous parameters. Summarizing, the diverse alarm categories are characterized by thresholds in order to detect the event. However, specifying pre-established thresholds is not appropriate in dynamic environments. Furthermore, only data related to patients are considered, as the proposed framework is evaluated through a health care domain.

Miao et al. [11], have proposed an online light weight failure detection approach that aims at detecting nodes failure in a wireless sensor network. The authors design a correlation graph that characterizes internal correlations inside a node. Then, failures are detected by identifying the changes and anomalies of the correlation graph. The proposed approach aims at analyzing node metrics. The authors classified node metrics into four categories such as timing metrics, traffic metrics, task metrics and other metrics. Then, the correlation graph nodes represent the node metrics and the edges represent the pairwise correlations of the metrics. Each edge has a weight which denotes the correlation score between the corresponding metrics. The correlation graph is maintained periodically for each node in the wireless sensor network. Then, a node is assumed to be faulty if there is a missing correlation between two metrics. The proposed approach allows to detect node failure in wireless sensor network by constructing a correlation graph of a node. Although the proposed approach relies on minimal a-priori knowledge and can be generalized to a wide variety of wireless sensor network applications, it is restricted to some context parameters related to nodes. Furthermore, the proposed approach does not treat for example link failure.

Mohamed et al. [12], have proposed an approach based on the modeling of physical phenomena, allowing to use available resources to predict the values that are supposed to be read by sensors. Then, a comparison of the predicted values and the real readings allows to detect failures. Predicting is performed through a prediction engine that is responsible for deducing the values expected to be read by sensors based on the law sets defined in the environment model. If the predicted values are outside the sensor's tolerance margin, then, a failure is detected. The proposed approach allows to detect failures using a comparison between predicted values and values actually read from sensors. However, the technique used to detect failures can cause false alarms since the predicted engine could produce inaccurate values.

3. Case study: Smart campus system

To motivate the use of our approach, we introduce in the following an example of a smart campus system illustrated in Figure. 2. Smart campuses, with the ability to collect and analyze data, are built in order to benefit the institutions, the actors -the students and the teachers- by providing services which facilitate interaction between them. For instance, in the smart campus, every actor is equipped with a personal device such as a Personal Digital Assistant (PDA), a smart phone, a tablet, etc. and the smart campus system provides different services to actors based on their current situations. To maintain the collaboration between students and teachers, these devices need to be aware of their environment and their execution context, etc. Consequently, smart campuses contain an infrastructure that allows devices and systems to be monitored and adapted autonomously according to the changing context.

In this paper, our work deals with several context parameters such as temperature, pressure, position and light which need to be monitored. Furthermore, a number of context resources such as the memory consumption, the energy, the CPU load and the available bandwidth should be monitored to assess both the devices and the communication state especially in pervasive environments. The campus architecture depicted in the Figure. 2 involves different kinds of participants. On the one side, two controlling servers called Smart Campus Servers (SCS) namely SCS_1 and SCS_2 which are Ethernet-connected and equipped with important storage and high computational capabilities. On the other side, fixed (Presence sensors, cameras, lamps, air conditioners) and mobile (laptop, PDA, phones) devices are placed inside of the campus. The mobile devices are usually resource constrained in terms of memory, bandwidth and energy for example. Hence, a periodic monitoring by the controlling servers is needed in order to check their state which changes according to context. Furthermore, two gateways called Smart Campus Gateways (SCG), SCG_1 and SCG_2 implementing software interfaces are used to connect devices to the corresponding controlling servers to transmit information between them. The campus space is composed of separate rooms (Research laboratories, Meeting rooms, classrooms (Classroom1 and Classroom2)). Each room is equipped with fixed devices as well as mobile devices that are carried by the different actors.

Because of the complexity of the interaction between the different entities, we propose to focus our study on a part of the Smart Campus. It consists in the gateway SCG_2 connected to both the SCS_2 and the devices located in the Classroom2.

After presenting the case study, we detail in the following the proposed approach.

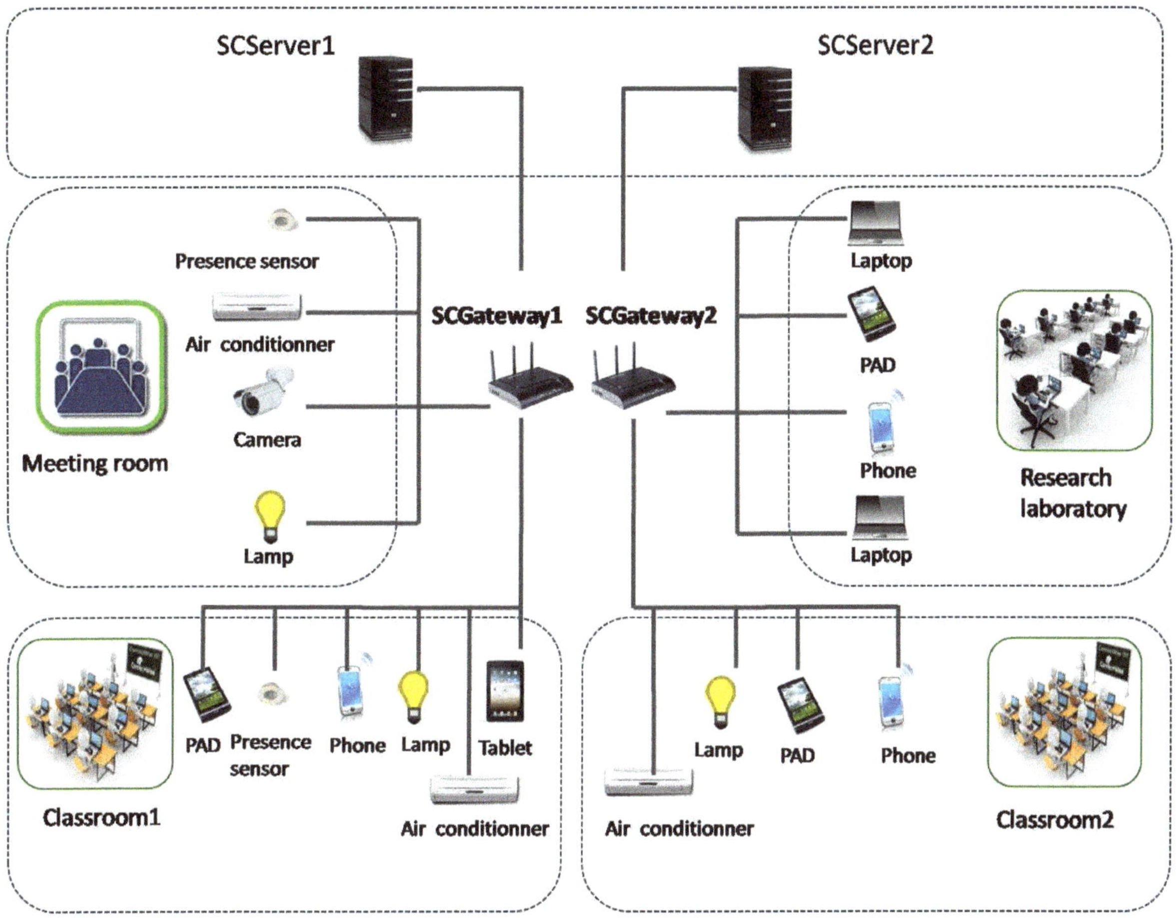

Figure 2. Case study: Smart campus

4. The proposed approach

Our context analysis approach is performed according to three steps: A context storage step, a context processing and classification step and a threshold calculation step. The context storage step is out of the scope of this paper.

In the following we present respectively the context classification step and the threshold calculation step.

4.1. Context parameter classification step

With a wide range of context parameters, context parameters need to be classified into different categories in order to use context easily. Since context parameters evolution is dynamic especially in pervasive environments, we propose to divide the context parameters into three categories according to the context parameter evolution. We have identified three categories. Parameters whose evolution is characterized by a tendency, parameters whose evolution is characterized by peaks and parameters whose evolution is characterized by bursts. Therefore, this classification covers almost all context parameter types. Indeed, each context parameter evolves over time, so, it belongs inevitably to one of the defined context categories.

The tendency category. A tendency is defined as a line in a graph which shows the general direction that a set of points seem to be heading as illustrated in Figure. 3(a). For example, each mobile device such as PDA, mobile phone, etc. mentioned in section 3 periodically monitors its resource state namely the battery level, the memory consumption that belong to the tendency category.

The peak category. Peaks are defined as high values with sharp rise followed quickly by sharp fall implying a narrow period width [13]. We define a peak *"Peak"* by its amplitude "A_p" and its period "T" as depicted in Figure. 3(b). In fact, the peak is a very narrow period of high values- That is, its amplitude "A_p" exceeds for n times the average amplitude "A_{av}" of the time series formed by the context parameter evolution. We define a peak by the following formulas:

$$\begin{cases} A_p & = & n \times A_{av} \\ T & \leq & m \times TimeUnit \end{cases}$$

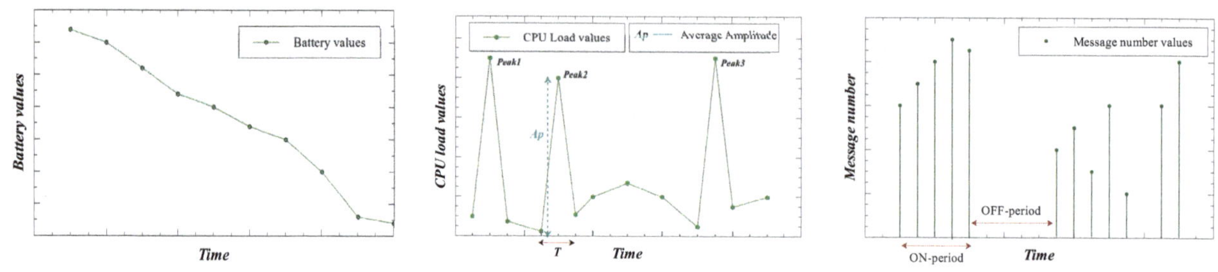

(a) Tendency category (b) Peak category (c) Burst category

Figure 3. The proposed context categories

Where A_p defines the peak's amplitude, A_{av} defines the average amplitude of the context parameter evolution, n and m are constants fixed by the application designers. The $SCSs$ receive the monitored data from the mobile devices in order to analyze them. For that reason, the $SCSs$ CPU load can rise suddenly reaching high values after a high rate of requests. Hence, the CPU load especially of the $SCSs$ can be modeled by a peaked function. The link load and the available bandwidth belong also to this context category.

The burst category. A burst consists on a relatively wide contiguous region of values. Otherwise, a burst is defined as a large number of occurring events [8]. As depicted in the Figure. 3(c), we model the wide region by an ON period and the other by an OFF behavior [17]. The ON-period models a single flow such as the transfer of a single web page, and the OFF-period models the user's thinking time. The ON-period and the OFF-period are strictly alternating. The message number received by the $SCSs$ during an ON-period is modeled as a burst.

4.2. Threshold calculation step

In the following, we detail some elements about threshold calculation for each context category defined previously.

Threshold calculation for the tendency category. For this category, the context parameter evolution is described by a tendency. In order to avoid false detections as well as missing alarms, we need to define thresholds which are uncorrelated with the context parameter evolution. The notification raised when the context parameter behavior crosses the threshold is illustrated in Figure. 4.

Uncorrelated thresholds can be fixed thresholds, uncorrelated adaptive thresholds and step function thresholds. For instance, these thresholds are defined by the application designers according to their needs in terms of Quality of Service (QoS).

For the adaptive threshold denoted in the Figure. 4(b), mathematical methods can be applied in order to update threshold values at runtime. However, for this kind of context parameter characterized by a tendency, adaptive threshold must be uncorrelated with the context parameter evolution in order to avoid false detections and missing alarms.

For the step function threshold described in the Figure. 4(c), thresholds are defined per period according to the application designers needs.

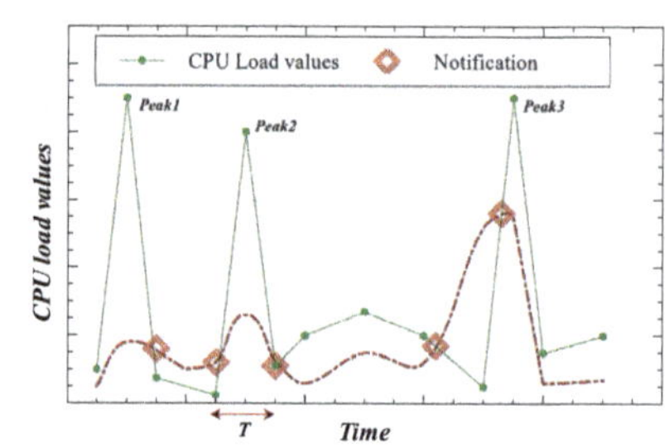

Figure 5. Threshold calculation for the peak category

Threshold calculation for the peak category. In this context category, the idea consists in specifying adaptive thresholds that are correlated with the context parameter evolution. Adaptive thresholds are calculated via different mathematical models such as Exponential Weighted Moving Average (EWMA) technique used by Lahyani et al. [9], Simple Moving Average (SMA). Mathematical functions defined by the application user are used to update thresholds.

Threshold calculation for the burst category. In this context category, our idea consists on transforming a bursty model into a tendency/peak model. So we propose to apply an aggregate function G in each ON-period. The obtained model coincides with a tendency function as illustrated in Figure. 6(b). A notification is raised when the context parameter crosses the threshold calculated in an ON-period.

For the smart campus case study (section 3), the gateway SCG_2, analyzes the received data from the

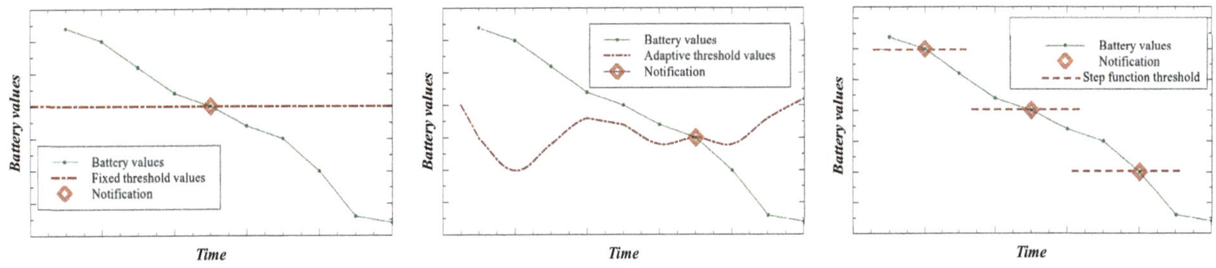

(a) Fixed Threshold (b) Adaptive Threshold (c) Step Function Threshold

Figure 4. Threshold calculation for the tendency category

devices. The maximum queue size for this device is set by the application designer. So that, if the traffic received by this gateway in a period T_i exceeds a maximum, then a burst is identified in T_i. To detect bursty periods, applying the aggregate function G consists in calculating the slope of the scatter diagram obtained in each ON-period. Second, in each ON-period, we compute the intensity of each slope formed in each ON-period. We obtain the Figure. 4(b). Consequently, if the slope intensity exceeds the threshold, then a burst is detected and appropriate adaptation actions are triggered.

5. Illustrative scenario

To illustrate the usefulness of our approach, we elaborate the following scenario which is conducted in the smart campus system presented in Figure. 2. It highlights the ability of an application to react accordingly to the changing context. We focus on the interaction of the actors of the Classroom2 and SCS_2 through SCG_2. Classroom2 is used by the researchers, the teachers and the students. Different context parameters are considered in this scenario.

- The actors' position: It belongs to the peak category. Adaptive thresholds are applied to this context parameter.
- The memory consumption: It belongs to the tendency category. Fixed, adaptive or step function thresholds are applied to this context parameter.

The scenario starts as follows: At the beginning of each course session, the presence sensor captures and localizes the mobile actors. Their positions are then forwarded to SCS_1. SCS_1 runs the analysis algorithm based on thresholds on the context parameter (mobile actors' position) and takes the appropriate decision. Each student participating to a course uses a tablet device to which the course will be dispatched through bluetooth. The tablet device holds the analysis algorithm in order to detect context changes. During the course, the students' tablet display appropriate slides and they follow their courses. Furthermore, the students can write annotations on their tablets and publish their comments to share knowledge between all the group members to enrich the course and enhance the collaboration. A student holding a tablet is participating to the course by exchanging information and slides.

For each amount of data received, the tablet device retrieves the memory consumption from the operating system using probes. Since the memory consumption belongs to the tendency category, fixed threshold(Threshold3), step function threshold (Threshold1) and uncorrelated adaptive threshold (Threshold2) are applied for this context parameter as illustrated in the Figure. 7.

- $Threshold_3$: A fixed threshold (value=90). This threshold is specified by the application designer. It corresponds to a critical memory. For instance, if the memory values crosses this fixed threshold, the analysis module detects a device crash.
- $Threshold_2$: An adaptive threshold. This threshold is updated using the mathematical formula defined by the user. The mathematical formula is described as follows:

$$Threshold_t = \lambda \times t + \beta$$

Where λ and β are two constants defined by the user. λ represents the slope and β represents the intercept. This threshold level corresponds to a normal memory state. As shown in Figure. 7, we do not need to trigger adaptation actions when memory values oscillate around the $Threshold_2$. However, we must react when the memory not only goes past the $Threshold_2$ but comes closer to the critical threshold ($Threshold_3$).

- $Threshold_1$: A step function threshold. The step values are defined by the user according to his needs.

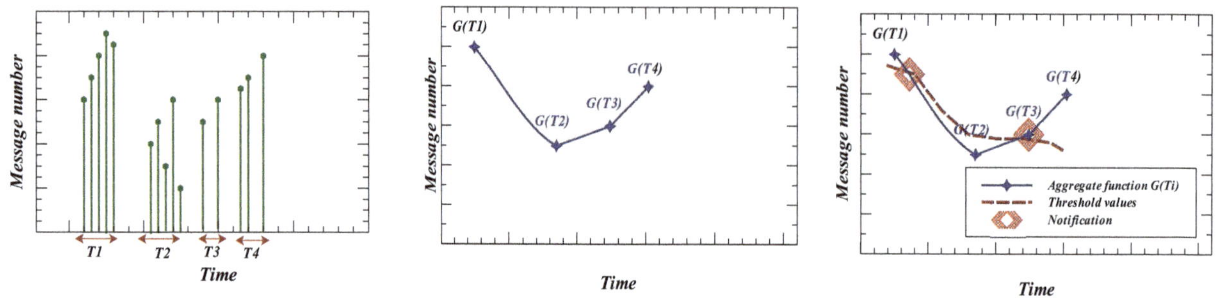

(a) Burst model (b) Aggregation function application (c) Threshold calculation

Figure 6. Threshold calculation for the burst category

- For the first step (threshold value=30), we can consider that the course to which the student is participating does not require exchanging information.
- For the second step (threshold value=50.5), we can consider that the student is participating to a course that requires interaction with the teacher, displaying and exchanging information.
- For the third and the fourth step (threshold value=20, threshold value=20.5) respectively, we can consider that the student is in a pause time or he is not doing any activity.
- For the last step (threshold value=70.5), it means that we must react once the memory values cross this threshold and increase rapidly.

6. Conclusion

The challenges for context aware applications design and implementation are to perform four phases. Collecting context from different sources such as sensors, widgets, etc., analyzing context to detect changes, deciding adaptation actions and executing the planned action to react dynamically to context changes. In this paper, we have considered context analysis. we have proposed an analysis approach for context change detection in pervasive environments. The proposed analysis approach is divided into three steps. A context storage step, a context classification step and a threshold calculation step. The context storage step is out of the scope of this paper. We have proposed a context classification based on context parameter evolution. This context classification takes as input the collected context parameter and attributes for each context parameter one or many categories. Three categories have been identified. Second, we have detailed the threshold calculation step that aims at analyzing context and identifying context changes. Threshold calculation allows to analyze context parameters, identify context changes and notify the context aware application. The thresholds can be fixed, step function and adaptive ones. For the adaptive thresholds, mathematical models are applied in order to update the threshold used to detect context changes. Further, when the context parameter crosses the threshold, a notification is triggered to react to the context changes. As future work, we plan first to stretch the context parameter classification. Second, our aim consists not only in detecting context changes but also predicting context behavior using mathematical models such as autoregression.

7. Acknowledgements

This research is supported by the Itea2's A2NETS (Autonomic Services in M2M Networks) project[1].

References

[1] Alagar, V., Mohammad, M., Wan, K. and Hnaide, S.A. (2014) A framework for developing context-aware systems. *EAI Endorsed Transactions on Context-aware Systems and Applications* **14**(1). doi:10.4108/casa.1.1.e2.

[2] Birje, M. and Manvi, S. (2011) Multiagent model for device state control in the wireless grid. In *Electronics Computer Technology (ICECT), 3rd International Conference on,* **3**: 456 –460. doi:10.1109/ICECTECH.2011.5941834.

[3] Bouassida Rodriguez, I., Sancho, G., Villemur, T., Tazi, S. and Drira, K. (2009) A model-driven adaptive approach for collaborative ubiquitous systems. In *Proceedings of the 3rd workshop on Agent-oriented software engineering challenges for ubiquitous and pervasive computing*, AUPC 09 (New York, NY, USA: ACM): 15–20.

[4] Cioara, T., Anghel, I., Salomie, I., Dinsoreanu, M., Copil, G. and Moldovan, D. (2010) A self-adapting algorithm for context aware systems. In *Roedunet International Conference (RoEduNet), 2010 9th*: 374 –379.

[5] Esposito, A., Tarricone, L. and Zappatore, M. (2010) A versatile context-aware pervasive monitoring system:

[1] https://a2nets.erve.vtt.fi/

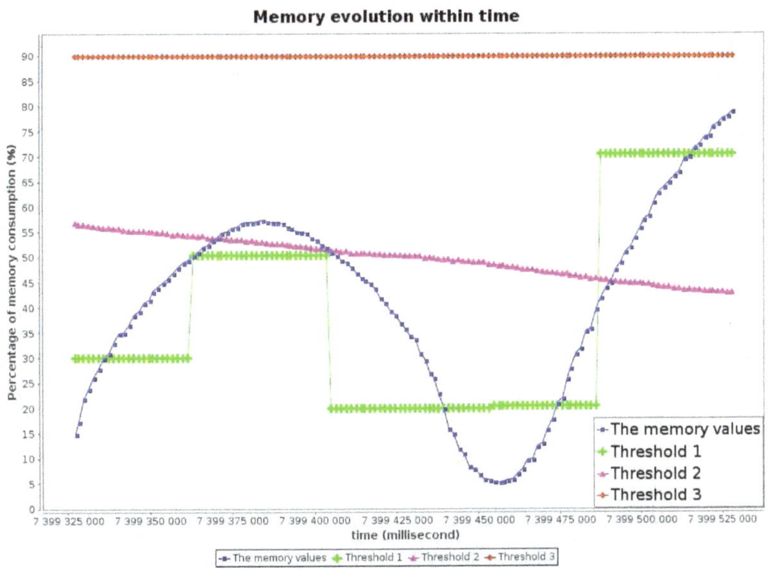

Figure 7. Application of threshold on the memory context parameter

Validation and characterization in the health-care domain. In *Industrial Electronics (ISIE), 2010 IEEE International Symposium on*: 2791–2796.

[6] Gilman, E., Davidyuk, O., Su, X. and Riekki, J. (2013) Towards interactive smart spaces. *JAISE* **5**(1): 5–22.

[7] Hussein, M., Han, J., Colman, A. and Yu, J. (2012) An architecture-based approach to developing context-aware adaptive systems. In *Engineering of Computer Based Systems (ECBS), 2012 IEEE 19th International Conference and Workshops on*: 154–163. doi:10.1109/ECBS.2012.13.

[8] Klan, D., Karnstedt, M., Politz, C. and Sattler, K. (2008) Towards burst detection for non-stationary stream data. In *KDML*: 57–60.

[9] Lahyani, I., Khabou, N. and Jmaiel, M. (2012) Qos monitoring and analysis approach for publish/subscribe systems deployed on manet. In *Parallel, Distributed, and Network-Based Processing, Euromicro Conference on* (Los Alamitos, CA, USA: IEEE Computer Society): 120–124.

[10] Magalhaes, J. and Silva, L. (2012) Anomaly detection techniques for web-based applications: An experimental study. In *Network Computing and Applications (NCA), 2012 11th IEEE International Symposium on*: 181–190. doi:10.1109/NCA.2012.27.

[11] Miao, X., Liu, K., He, Y., Liu, Y. and Papadias, D. (2011) Agnostic diagnosis: Discovering silent failures in wireless sensor networks. In *INFOCOM, 2011 Proceedings IEEE*: 1548–1556.

[12] Mohamed, A., Jacquet, C. and Bellik, Y. (2012) A fault detection and diagnosis framework for ambient intelligent systems. In *Ubiquitous Intelligence Computing and 9th International Conference on Autonomic Trusted Computing (UIC/ATC), 2012 9th International Conference on*: 394–401.

[13] Palshikar, G.K. (2009) Simple Algorithms for Peak Detection in Time-Series. In *Proc. 1st Int. Conf. Advanced Data Analysis, Business Analytics and Intelligence*.

[14] San Martín, L.A., Peláez, V.M., González, R., Campos, A. and Lobato, V. (2010) Environmental user-preference learning for smart homes: An autonomous approach. *J. Ambient Intell. Smart Environ.* **2**(3): 327–342.

[15] Taconet, C., Kazi-Aoul, Z., Zaier, M. and Conan, D. (2009) Ca3m: A runtime model and a middleware for dynamic context management. In *Proceedings of the Confederated International Conferences, CoopIS, DOA, IS, and ODBASE 2009 on On the Move to Meaningful Internet Systems: Part I*, OTM '09 (Berlin, Heidelberg: Springer-Verlag): 513–530.

[16] Weiser, M. (1995) Human-computer interaction (San Francisco, CA, USA: Morgan Kaufmann Publishers Inc.), chap. The Computer for the 21st Century, 933–940.

[17] Yang, X. (2002) Designing traffic profiles for bursty internet traffic. In *In Proceedings of IEEE Global Internet*.

[18] Zheng, D., Wang, J., Han, W., Jia, Y. and Zou, P. (2006) Towards a context-aware middleware for deploying component-based applications in pervasive computing. In *Proceedings of the Fifth International Conference on Grid and Cooperative Computing*, GCC '06 (Washington, DC, USA: IEEE Computer Society): 454–457.

3

PaisleyTrees: A Size-Invariant Tree Visualization

Katayoon Etemad [1,*], Dominikus Baur[1], John Brosz[1], Sheelagh Carpendale[1] and Faramarz F. Samavati[1]

[1]University of Calgary

Abstract

Squeezing large tree structures into suitable visualizations has been a perennial problem. In response to this challenge, we present PaisleyTrees, a size-invariant tree visualization. PaisleyTrees integrate node-of-interest focus with tree-cut presentations to support rapid tree navigation without resorting to zooming and panning. This visualization offers the ability to work with trees of arbitrary depth and breadth, and maintains legibility for displayed elements. These advantages are achieved by using a hybrid layout, inspired by traditional Paisley patterns, that combines node-link, nested and adjacency-based tree layout techniques, and offers both depth and breadth elision.

Keywords: Information Visualization, Tree Layout, Hybrid Layout, Mobile, Graphs.

1. Introduction

Modern computers are increasingly efficient at storing and processing large data-sets. Despite this, visualization of large data-sets is still a challenging task. This is apparent with visualizations of tree data structures, as can be seen with linguistic and genealogical hierarchies. As information must eventually be read, extracted and understood by a human, an attractive and aesthetic visualization of large trees can prove vital in performing this challenging task.

While node-link diagrams (Battista [1]) are a valuable method for tree visualization, an informative and aesthetically pleasing arrangement of nodes and edges is required. In addition, the scalability of these types of diagrams to large trees remains a challenge. As a tree grows, node-link diagrams do not necessarily provide the best solution. At some point, zooming and panning operations start to take up a large share of all interaction. As an alternative, treemaps [2] usually come with fixed screen size requirements, but they adjust to growing trees by shrinking their internal components. The problem of scalability is particularly acute on touch-enabled displays and high-resolution, small-screen devices. Keeping the problem of increasingly large trees on small-screen devices in mind, we explore the creation of a tree visualization with several attributes: (1) an aesthetically pleasing structure using a layout inspired by Paisley patterns, (2) support for trees of arbitrary breadth and depth by using an appropriate tree-cut, (3) a resulting visualization that is size-invariant and adjustable for various screen sizes from mobile to desktop, and (4) support for the common task of tree traversal through quick navigation. In this paper we introduce a novel tree visualization called PaisleyTree (see Figure 1).[1] To keep PaisleyTrees size-invariant while supporting arbitrarily large trees, we use a tree-cut based on a hybrid layout that combines aspects of node-link, adjacency and nesting layouts.

The remainder of this paper is structured as follows. An overview of relevant previous research is provided in Section 2. The design of PaisleyTrees, including the layout, visualization, interaction and implementation is explained in Section 3. In Section 4, we discuss our design decisions and rationales. In Section 5 we provide some examples, and Section 6 contains a discussion. We conclude the paper in Section 7.

2. Related Work

Tree visualizations usually fall into one of two categories: they show the connections between nodes either *explicitly* by drawing edges or *implicitly* by using the position of node elements [4] (see Figure 2).

Explicit approaches include the most common tree layouts: the node-link diagram and its variations. The main methods used to reduce the display footprint involve changing the layout algorithm (e.g., [5, 6]), adjusting the representation of edges (e.g., *Cheops* [7]), or using the third dimension (*ConeTrees* [8]).

*Corresponding author. Email: ketemad@ucalgary.ca

[1]An early concept of PaisleyTree was mentioned in GRAND NCE poster by our group presented at GRAND [3].

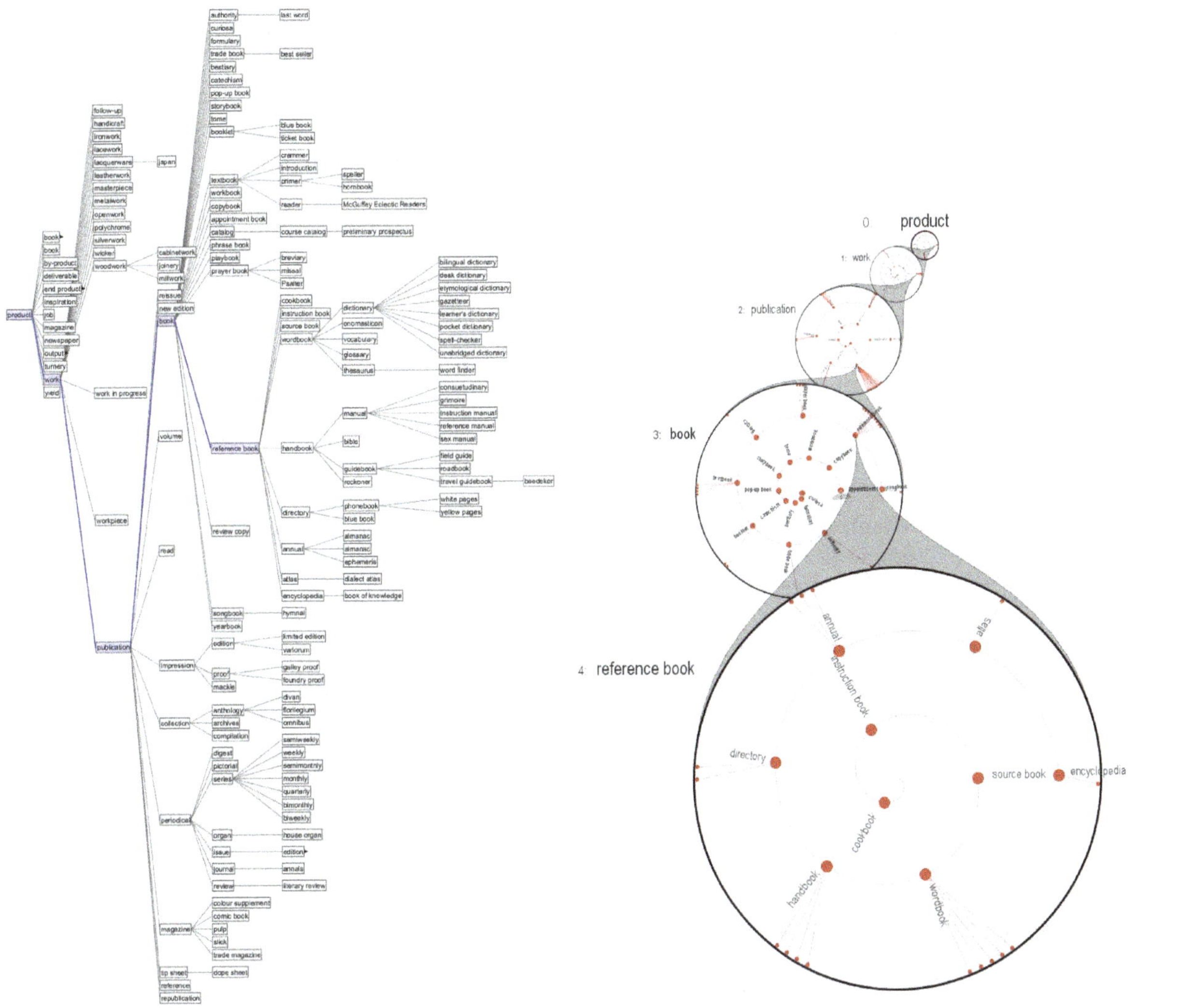

Figure 1. Left: SpaceTree, Right: PaisleyTree. Both trees show the subtree of Wordnet rooted at the word "Product" with 197 nodes and 9 levels.

Neumann et al. increased possible layout size by borrowing optimal node configurations from nature in *Phyllotrees* [9]. Still, all these adjustments cannot prevent the increases in tree size causing a resulting increases in the visualization's space requirements.

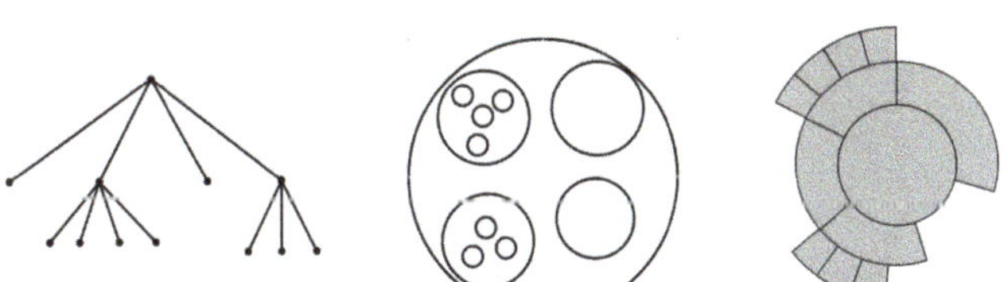

Figure 2. *Tree layouts, Node-link (explicit), Nested (implicit) and Adjacency (implicit).*

Implicit approaches to encoding connections eliminate the need for explicit visual edges. The spatial relationship between two elements sufficiently illustrates in which way they are related. Treemaps [2] and their derivatives (e.g. *Cascaded Treemaps* [10] and *Bubble Trees* [11]) use *nesting* to encode parent-child-relationships. In ShamsehTree [12] a sequence of concentric circles are used to present the ancestral path for a tree-cut. The underlying layout of this tree visualization is designed based on symmetric arrangement of nodes, inspired by Persian floral patterns. Another implicit approach relies on *adjacency* (e.g., *Docuburst* [13], *Icicle Tree* [14], *Sunburst* [15] and *Information Slices* [16]). Implicit techniques are often used when scale is an issue since they save screen space by skipping edge drawing.

The scaling problem of tree visualizations can be partially overcome by other methods. The family of *SpaceTree* [17] and *Degree-Of-Interest Trees* [18, 19] visualize a tree-cut (a connected subset of a the tree), thus using the screen space more efficiently by focusing only on nodes-of-interest. This idea of tree-cut originates in linguistics where thesaurus trees require scale reduction [20]. In Degree-Of-Interest Trees, a specific DOI value is assigned to each node and the size of the node in visualization is a function of that value. DOI value of each node is the importance and its distance from the node of interest. In PaisleyTree we also use a tree-cut which is defined in the proximity of the node of interest (NOI). However, we use a cut from NOI's ancestor nodes up to the root and its descendant nodes down to a specified level. This specific cut is defined to fit to our hybrid aesthetic layout inspired by Paisley patterns and to ease navigation.

Displaying trees in a non-uniform fashion using hybrid layouts provides an excellent opportunity to address the scale challenge. In addition, it can help us to create more pleasing layouts. Examples include *Elastic Hierarchies* [21] and *Linked Treemaps* [22], that use both node-links and treemaps. In Elastic Hierarchies some nodes are presented in treemap layout and connected with links. Linked Treemaps take advantage of the third dimension, presenting the treemap with links added in the 3D space. PaisleyTrees is a 2D layout that uses all three of these approaches — node-link, nesting and adjacency — in one visualization to address both the large scale and aesthetical design challenges.

Finally, the problem of growing visualizations becomes especially obvious on mobile devices with small screens. Mobile visualizations therefore have to take more creative approaches for displaying trees. *Tablorer* [23] converts the underlying tree into a nested table, while *Magic Eye View/Rectangular View* [24] combines distortion with dropping branches of the tree. The *Radial Edgeless Tree* [25] uses a variation of a Voronoi diagram and adjacency. PaisleyTree is a size invariant visualization that works for both large and small screens.

3. PaisleyTrees

In this section, we describe the basics of the PaisleyTrees visualization. The design rationales and possible variations are discussed in Section 4. First, in Section 3.1, we discuss the layout of PaisleyTrees, followed by an overview of the visualization of PaisleyTrees in Section 3.2. We then present a discussion about the interaction ideas in Section 3.3, and give an overview of the implementation in Section 3.4.

3.1. PaisleyTree Layout

According to Roger Scruton, "Styles may change, details may come and go, but the broad demands of aesthetic judgement are permanent." [26] One of the objectives of our visualization is to use an aesthetically pleasant layout. However, the challenge is how to design such a layout whilst meeting our other visualization objectives (see Section 1). In fact "Aesthetics can be considered an added bonus, or a by-product when striving for readability and effectiveness." [27]

Combinations of the node-link, nested and adjacency layouts can be used for designing an aesthetically pleasing layout (see Figure 2). Node-link layout relies on drawing edges in an explicit fashion, which tends to lead to clutter. In contrast, nested and adjacency layouts avoid this by implicitly representing edges. One method for handling a large tree visualization is to consider a node-of-interest (NOI) to be the center of attention. This could be accomplished by shrinking ancestor and descendant nodes as they get further from the NOI (see Figure 3a). By definition, one characteristic of nested layouts is that the largest nodes are the further ancestor nodes. In our design, we use adjacency layout for the NOI and its ancestral nodes to the root, and nested and node-link layouts for descendant subtrees of the NOI.

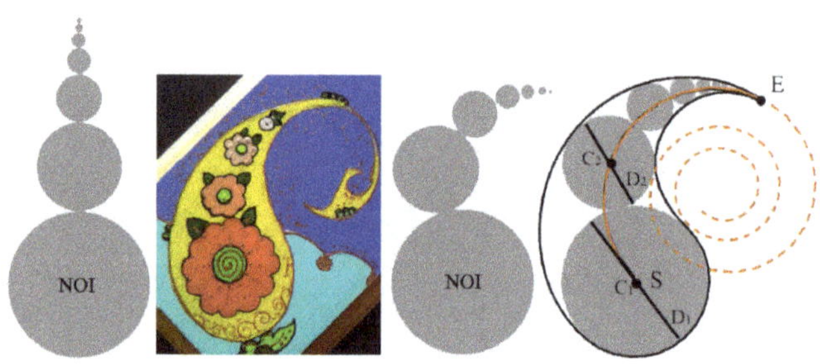

Figure 3. *(a) Shrinking ancestral chain. (b) A traditional Paisley pattern. (c) The ancestral chain arranged on a spiral curve. (d) The node circles filling the Paisley outline.*

As a strategy to create an aesthetic design, we have looked at many patterns used in the traditional arts from different cultures. We chose Paisley patterns used in traditional Persian and Indian designs (see Figure 3b). The main motivation behind our use of this leaf-like pattern is that, as they shrink, nodes further away from the NOI fit nicely within the Paisley outline, which follows a spiral curve (see Figure 3c). For our implementation, we use Fermat's spiral [28] in polar coordinates

$$r = k\sqrt{\theta}, \quad \theta_0 \leq \theta \leq \theta_1$$

where k is a constant controlling the curvature of the spiral. As shown in the Figure 3d, we denote the start and the end points of the spiral by S and E. To arrange the nodes on the spiral, we represent each node with a circle, denoted by $n_1, n_2, n_3, \ldots$. In this arrangement the node n_{i+1} is tangent to the node n_i and has a smaller size. Each circle n_i is specified by (C_i, D_i) where

C_i is the center and D_i is the diameter of the circle. The basic idea is that the sum of diameters roughly approximate the arc-length L of the spiral from S to E. We use a numerical approximation for finding L using the sum of small lines formed by a discrete sample of the curve's points. The diameter of the first circle ($D = D_0$) is usually known, and we need to find the diameter of the other circles. Since the circles are shrinking, we need to calculate their contraction factor α. Thus, the sum of the diameters (except the NOI circle which is 1/2 of D_0, as shown in Fig 3d) is approximately the same as L:

$$\frac{1}{2}D + \alpha D + \alpha^2 D + \ldots + \alpha^{n-1} D = L. \quad (1)$$

To determine α we find a numeric solution to the non-linear Equation 1. Notice that an approximate value for α is sufficient.

Now we can determine the distribution of circles on the spiral. To do this, we need to know the position of each circle. Note that the first circle is positioned at $C_1 = S$ and its radius is $R_1 = \frac{D}{2}$. The second circle has contraction factor α and must be tangent to the first circle as shown in Figure 3d. Thus, the arc-length of the spiral from C_1 to C_2 (the center of the second circle) must be $R_1 + R_2$ where $R_2 = \alpha R_1$. Consequently, we find C_2 by traveling the distance of $R_1 + \alpha R_1$ along the spiral from C_1. The centers of the other circles are determined similarly.

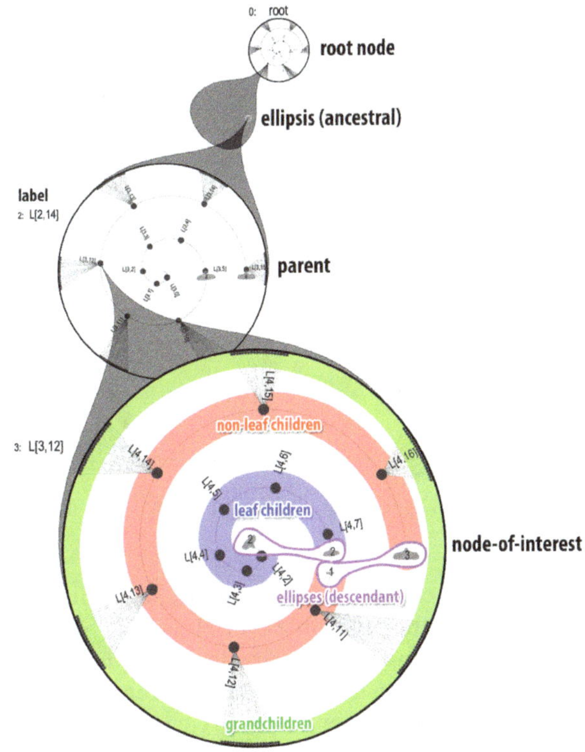

Figure 4. *Visual organization of a PaisleyTree. In each node grandchildren are positioned on the perimeter of the circle. Non-leaf children on the outer ring of the spiral and leaf children are positioned on the inner ring of the spiral.*

3.2. PaisleyTree Visualization

PaisleyTree (see Figure 1 right) is our constructed tree visualization based on the above layout and on the requirements outlined in Section 1. To address the challenge of large tree visualization, we use tree-cuts, hybrid layout and elision to make the amount of visual elements manageable.

The main approach of the PaisleyTree is to use a specific tree-cut (i.e. a connected subset of the tree) for achieving our requirements. This tree-cut presents the NOI along with its hierarchically proxemic nodes, including its ancestor nodes up to the root and its descendant nodes down to a specified level. Our implementation includes two levels of descendants from the NOI. The main representation of each node is a circle (as discussed in Section 3.1). Node circles are arranged from bottom to top, with the node-of-interest at the bottom, the root node at the top and intermediate nodes in between. Circles and their contents further up the spiral are diminished in size, reflecting their distance from the NOI.

As shown in the Figure 4, each node circle displays two levels of its descendant sub-tree: all of the node's direct children are placed as dots (nested layout) on a spiral within the node circle while grandchildren

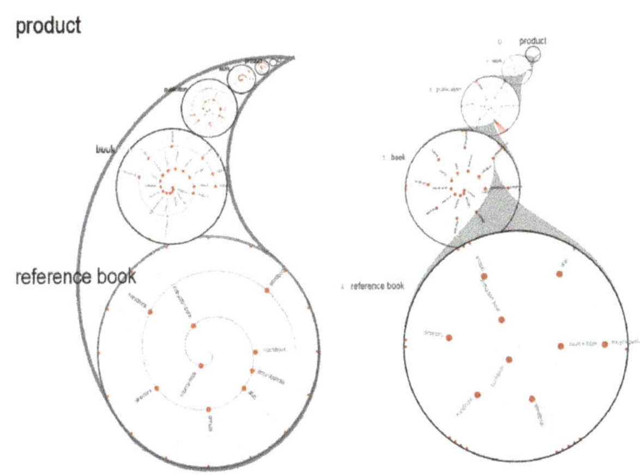

Figure 5. *Left: PaisleyTree with the paisley outline. Right: PaisleyTree without the paisley outline.*

are placed on the perimeter of the node circle (this region is highlighted with green) using node-link layout. To reduce cluttering, the children are divided into two groups, leaf children and non-leaf children. Leaf children are placed on the spiral's inner ring (highlighted with blue), non-leaf children on the outer (highlighted with red). The children are

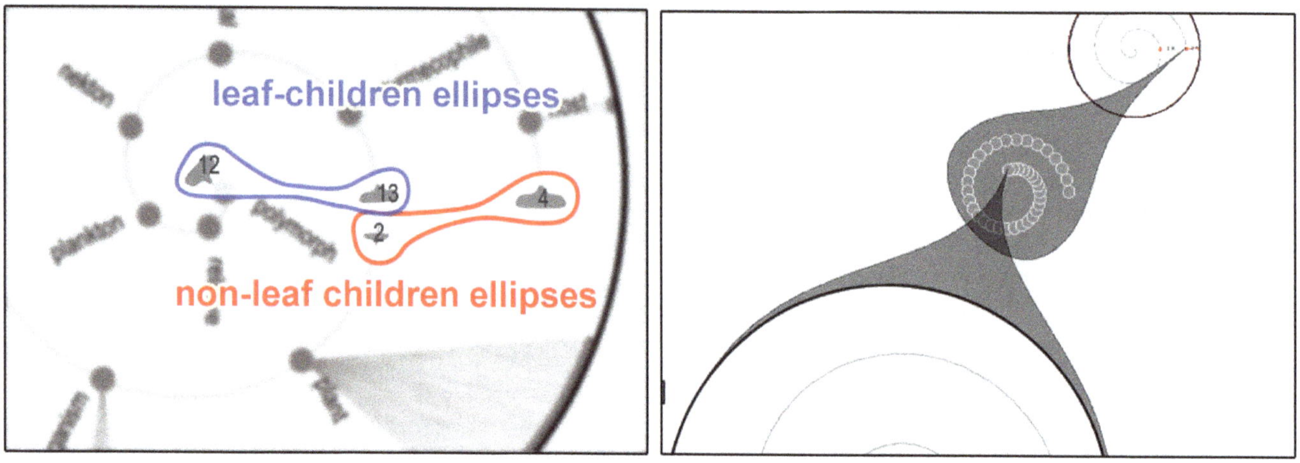

Figure 6. Types of elision in PaisleyTrees: descendant ellipses (left), an ancestral ellipsis (right).

distributed at identical distances to support legibility and interactivity in the available spiral space. All grandchildren are shown on the perimeter of the node circle and arranged close to their respective parent (i.e., a non-leaf child). The perimeter of the circle is divided into equal sections for each set of grandchildren.

Within each node circle the children's labels are displayed in a radial angular orientation. The angle of orientation for each label is perpendicular to the spiral's tangent. Depending on a label's angle, its text direction can change between left-to-right and right-to-left to make it easier to read. Additionally, a pop-out version of the label is shown when hovering over the node.

The use of outline for PaisleyTree is optional (See Figure 5). Using the outline makes a stronger resemblance to the traditional pattern, but with the price of possible cluttering with labels. The gray cones are removed when the outline is drawn. The gray cones have the advantage of explicitly indicating the node's location in its ancestor.

Each of the intermediary nodes follows the same internal layout although slightly smaller. Each node is an expansion of a child within its parent node. To make this relationship more obvious, each node is connected to its own representation in its parent's node by a slightly transparent grey cone (Figure 1 and 4).

One of PaisleyTree's design goals was to support trees with arbitrary depth and breadth while preserving size and readability. The layout described above, however, breaks for certain cases. If the tree has more levels than there are spaces available for node circles on the Paisley curve, then not all levels can be displayed. Similarly, when the number of children or grandchildren grows large, they are placed very close to one another, eventually becoming unreadable and impossible to interact with. For these reasons, we provide *ellipses*, to act as placeholders for parts of the tree. These ellipses exist in two forms: *ancestral ellipses* (see Figure 6, right) that contain levels of the tree that are not shown as node circles; and two *descendant ellipses* (see Figure 6, left) that contain excess leaf- and non-leaf children.

Ancestral ellipses lie on the Paisley spiral curve, taking the place of one or more node circles, and are displayed as grey shapes pointing towards the origin of the first node circle they contain. All contained nodes are displayed as small circles arranged in a spiral shape. The first contained circle (i.e., the closest to the root) is placed at the right end of the spiral (see Figure 6, right). Descendant ellipses hide excess child nodes within a node circle. They are displayed as grey shapes at the ends of both the leaf- and non-leaf regions of the spiral. Numeric labels indicate the number of contained nodes. Interaction provides access to nodes in either type of ellipses.

3.3. Interaction

In this section we describe interaction with PaisleyTrees in desktop environments. Adaptations for touch-based devices are described in section 5.3. Interaction in PaisleyTrees falls into the two main categories: tree navigation and adjusting ellipses.

A PaisleyTree is initialized as a single node circle, representing the root node (see Figure 7, left). By clicking one of the non-leaf nodes, that node is expanded and made the current NOI (see Figure 7, left to middle). To make the transition comprehensible, the node and its connecting cone grow out of the position from which it was selected using an animated transition (cf. [29]) while the root node itself is shifted up one space. Thus, clicking on nodes allows for quick navigation through the tree. Choosing a child that is a leaf node results in the new NOI having no further descendants. All displayed nodes (including the root node) are interactive and can be selected as the NOI (See Figure 8).

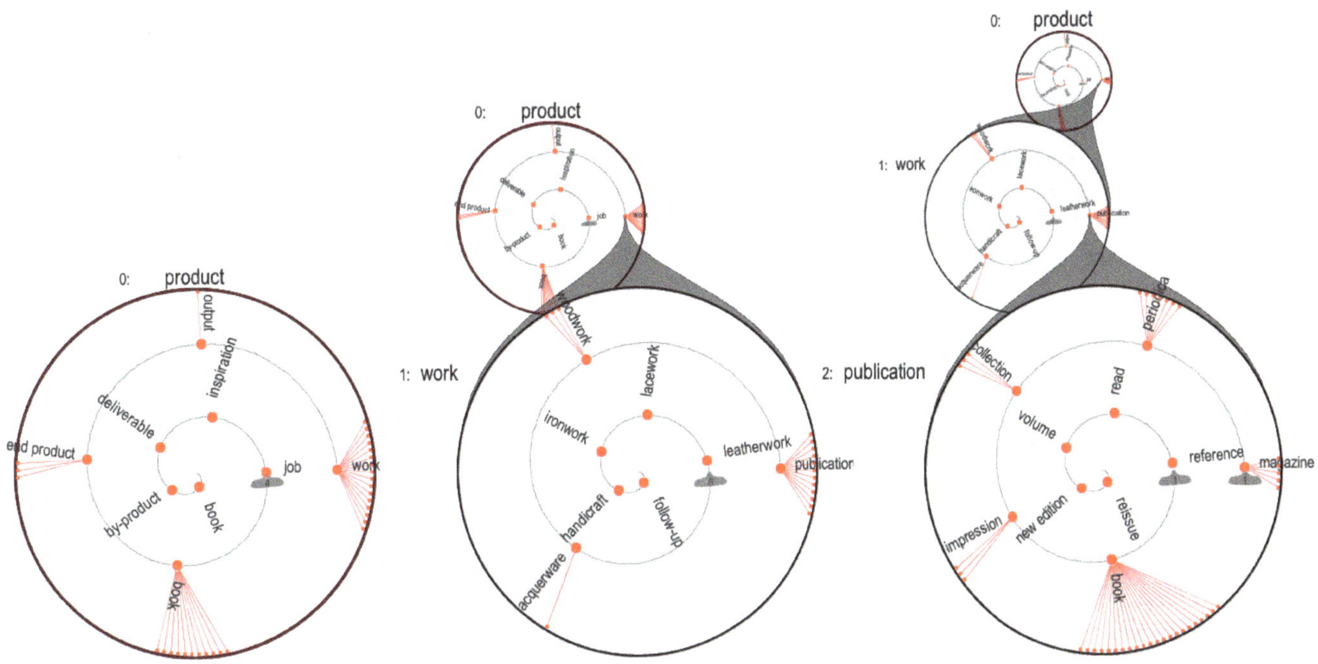

Figure 7. Drilling into a PaisleyTree: Each node circle in a PaisleyTree represents one node, with the bottom-most node representing the current node-of-interest. The node's children are shown along a spiral in the node circle, while grandchildren are shown along the outer perimeter. By consecutively clicking on a sequence of children nodes the most recently selected child node becomes the current node-of-interest and expands into a new node circle, supporting drill down in the tree.

Changing the NOI can also cause the creation or dissolution of an ancestral ellipsis. A limited number of node circles can be displayed within a PaisleyTree at any one time, depending on the on-screen resolution (we commonly use three to seven levels). Every time a node within a node circle becomes the new NOI, all visible node circles shift up one space. Once the limit of visible node circles is reached, the system automatically creates an ancestral ellipsis below the root node and adds the node circle directly below the root to it.

We additionally enforce a limit on the number of children nodes displayed along the spiral within a node circle. When the actual number of children exceeds the limit, the necessary ellipses automatically appear. This prevents overlap and maintains interactivity and legibility. On desktops this fixed maximum can be adjusted, so customization based on screen resolution and preferences is possible.

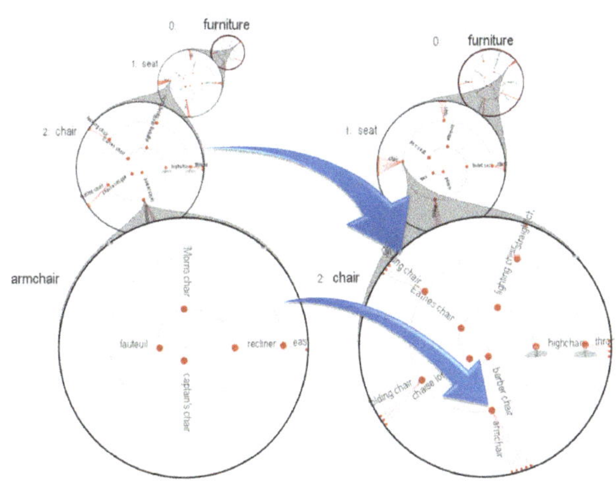

Figure 8. *Clicking on the 'chair' node circle (left) makes it the node of interest (NOI) (right). The current NOI, 'armchair', (left) is reduced to its child position in the 'chair' circle. Ancestral elision is not required here.*

3.4. Implementation

We implemented the desktop version of PaisleyTrees in Java, relying on Processing as the drawing backend. We also ported a version of PaisleyTrees to the mobile Android operating system with suitable adaptations for touch input, rather than requiring a direct hit from a cursor, it returns the closest node within a radius of the touch area.

4. Design Rationale

The strict goals of producing a size-invariant visualization technique while keeping visual elements readable required us to take the uncommon approach of combining tree-cuts with a hybrid layout. In this section we provide the background for our design influences and decisions.

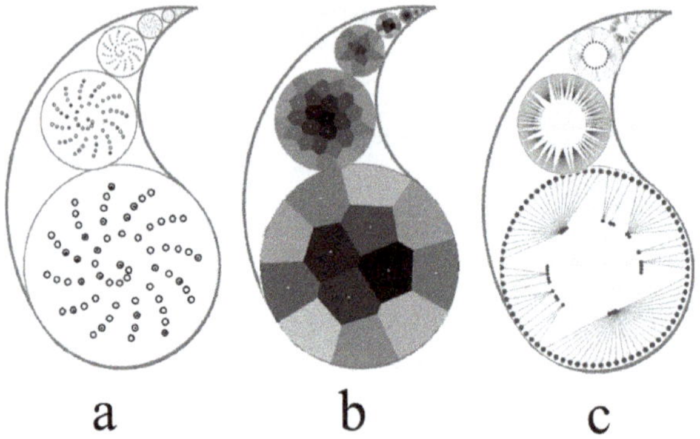

Figure 9. *Different design approaches for layouts of descendant nodes: (a) Phyllotactic, (b) Voronoi and (c) Circle.*

4.1. Hybrid Layout

Combining different layout methods into a hybrid layout incorporates the beneficial aspects of different approaches into one coherent visualization. Existing hybrid approaches (e.g. Elastic Hierarchies [21]) are good examples of this. We arrived at PaisleyTree's hybrid design by considering the advantages and disadvantages of the three main layout approaches and combining them accordingly. For example, nesting methods work well for emphasizing parent-children relationships. We therefore used nesting to represent children and grandchildren inside each node circle. On the other hand, adjacency methods are especially effective at presenting hierarchical structure, so we employed adjacency for the ancestral hierarchy for each selected node. Finally, we use the idea behind node-link layouts to explicitly connect non-leaf children in a node circle with their respective children along the perimeter.

For the NOI, we present its ancestors as a sequence of shrinking circles positioned on the Paisley curve which is part of a spiral (see Section 3.1 for its mathematical derivation). A circle's position on the curve expresses its level in the hierarchy and its distance from the NOI. Therefore, the farther the ancestors are from the NOI, the closer to the apex of the curve they are positioned. The NOI is, by definition, the most interesting node at a given moment, so we mapped the size of ancestor node circles to its distance from the NOI.

Children within node circles are arranged along a spiral to make effective use of the circle's area. As shown in Figure 9 a and b, we considered other arrangements, such as Phyllotactic and Voronoi diagrams, using a recursive nested layout. In these layouts the space assigned to grandchildren is very small. Therefore, we use a node-link layout for children/grandchildren connections. Figure 9 c, shows an early variation of this idea. However, the use of spirals makes the arrangement of nodes more organized and fits better with the traditional form of the Paisley pattern. Spirals exhibit a type of symmetry called *dilational symmetry*, that refers to the symmetry of self-similar elements at different scales. Usually, symmetry creates visual appeal and is known to positively affect information comprehension [30] and can engage and motivate the viewer to devote more time and attention to their information exploration tasks [31]. In addition, the use of spirals provides an effective means for handling elision for descendant nodes. Finally, the ancestral spiral curve and its shrinking circles produce a slight illusion of perspective, as if the circles were shrinking into the distance.

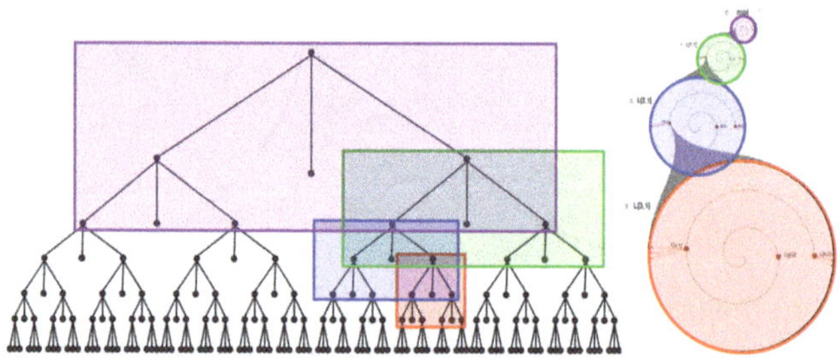

Figure 10. *A tree with 190 nodes and 7 levels in both node-link and PaisleyTree layout. In PaisleyTrees, each node circle shows three levels: a node and all of its children and grandchildren.*

As noted, nodes of the tree are represented redundantly in PaisleyTrees. As visible in Figure 10, nodes can appear repeatedly, either as node circles or as nodes within a node circle. The same node can have up to three concurrent representations in the visualization: as a grandchild, as a child node within the next node circle and as its own node circle. While this concurrent node representation might seem unnecessary, we made the local tree structure explicit with the intention of aiding navigation interaction.

4.2. Use of elision

One of PaisleyTree's goals is to to provide a new alternative for the presentation of trees of arbitrary size. Problems with visualizing large trees arise from two sources: too many children on one level (large breadth) and too many levels (large depth). To keep visual elements at a reasonable size, our strategy was to elide some of the nodes when there are too many in either dimension. Using elision for both the ancestral and descendant dimensions, an arbitrary number of nodes and levels can be implied in the representation - at the cost of increased interaction. Ancestral and descendant ellipses behave differently, in that ancestral ellipses cannot be freely configured. Interaction only allows stepwise upwards or downwards movement, and in the case of too many circles within the elision (see Section 3.3) not every contained circle can be extracted.

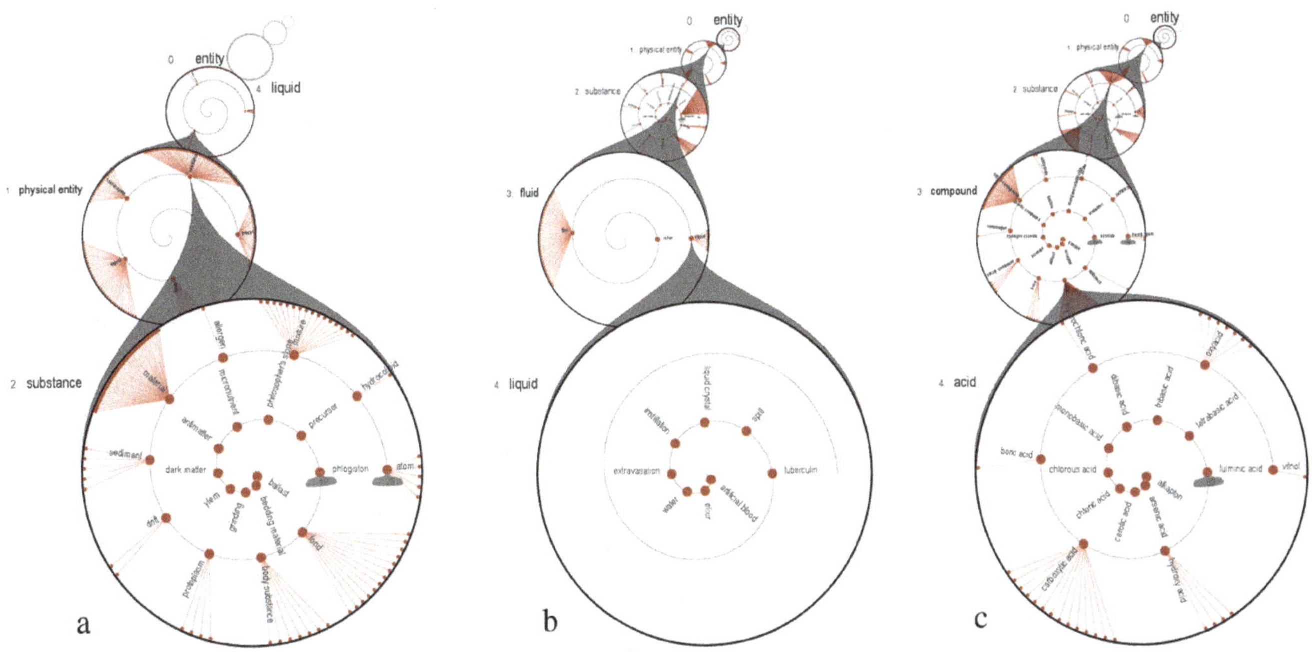

Figure 11. *Select interaction steps in finding the* **hydrochloric acid** *node in the* **entity** *subtree of the WordNet [32] tree structure.*

However, as all node circles within an ellipsis remain reachable, it is still possible to click on one and make it the NOI. The situation is different for descendant ellipses, where the only way to reach a hidden node is through bringing other sibling nodes into the ellipsis. The circular dragging gesture that is required can be repeated indefinitely which allows relatively fast access and also works well on touch-based devices.

5. Examples

In this section, we describe three applications that demonstrate the utility of PaisleyTrees: an exploration of Word Net, a file tree browser for mobile devices, and location selection in mapping and GPS applications.

5.1. Exploration of Word Net

The following provides a brief example of how the PaisleyTree node-focused approach to tree traversals functions when traversing the *entity* (root) subtree of the WordNet [32] tree structure.

Consider Bob, who would like to investigate the *hydrochloric acid (HCl)* node, but is unsure where this node is located. He begins at the root *entity* and traverses down through the nodes, shifting the NOI to *physical entity* and then to *substance* (Figure 11a). Children of *substance* include *chemical irritant, fluid,* and *medium,* all of which Bob suspects may lead to *HCl.* He first selects *fluid* which, leads to *liquid,* whose twelve children do not include *HCl* (Figure 11b). Realizing that he has chosen an incorrect *substance* he would like to return to the *substance,* node. To do this, Bob clicks on the *substance* ancestor (two nodes above the current NOI in Figure 11b), returning to this node as the NOI (Figure 11a), allows him to choose a different child. This time Bob selects *compound,* leading him to *acid,* and then to his target, *hydrochloric acid* (Figure 11c).

When searching for nodes that have not been visited previously, a return to the upper nodes of the PaisleyTree is helpful after traversing an incorrect branch. Also, alphabetical sorting of child nodes may ease searching for specific nodes that may or may not be present.

5.2. Location Selection

A task commonly faced on mobile devices such as GPS systems and smartphones is entering a target location. The common approach is to use an on-screen, touch-based keyboard which can be very small, depending on the size of the device. Also, GPS devices often rely on long, paginated lists of potential target locations, adding further interaction overhead. A different approach to target selection is to present targets in a large tree based on location hierarchy: from countries through to states/provinces to cities, streets, houses, etc. The resulting tree for this approach is naturally quite broad, which makes it problematic for common tree visualizations.

To test the capabilities of PaisleyTrees we built such a tree from countries to cities via intermediate states/provinces. The resulting tree has a total of 334,681 city nodes that are either directly below a country node (independent cities/regions) or children of states/provinces. We added two more layers

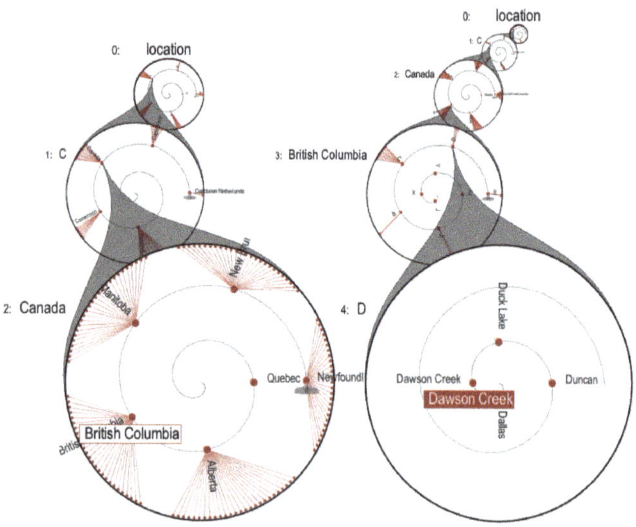

Figure 12. Selecting a location by navigating through a country-state-city tree.

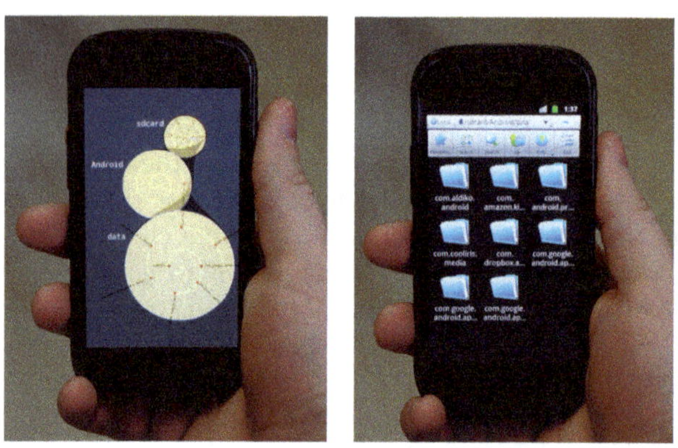

Figure 13. Comparison between a file system as PaisleyTree (left) and a popular mobile file tree browser (right).

containing the 26 letters of the Latin alphabet to make navigation in longer lists easier (regions such as 'England' in Great Britain have more than 1700 cities and towns). One alphabet layer was added above the country level and another one above the city level. Figure 12 shows an illustrative navigation from Canada, through British Columbia, to the town of Dawson Creek. In this scenario, the leaf node for Dawson Creek is the navigation target and would be selected.

5.3. Mobile File Browser

PaisleyTrees have several useful features that are suited for file browsing: fixed screen size prevents the need for zooming and panning; there exists a clear division between leaf-nodes (files) and non-leaf nodes (subdirectories); easy access to ancestor nodes and the root supports easy navigation to higher level directories; and the presentation of children and grandchildren for each node assists in the search for useful subdirectories.

The ability of PaisleyTrees to work entirely within strict size limitations makes PaisleyTree file browsing ideally suited for phones with touch displays. Figure 13 provides a comparison between the popular Android-based file browser ES File Explorer (http://www.estrongs.com/en/products/es-file-explorer.html) with our Paisley mobile file browser. The Paisley file browser is able to show multiple levels of the ancestral directory structure.

Selection and elision panning operations within the NOI are within reach of a thumb at the bottom of the display, which makes it ideally suited for one-handed interaction.

A fully-functional Paisley file browser requires additional interface elements to enable the full range of file system operations, such as sorting and file manipulation operations (i.e., creating, moving, deleting, executing). Additionally, some means of visually distinguishing empty subdirectory leaf nodes from file nodes would be beneficial (e.g., using icons instead of plain circles).

6. Discussion

PaisleyTrees represent an uncommon approach to the task of tree visualization. While they incorporate ideas from hybrid layout systems (e.g., [21]) and tree-cut based visualizations (e.g., [17]), the overall result is decidedly different. In this section we discuss and compare PaisleyTrees to other, more traditional tree visualizations.

In order to emphasize the differences between PaisleyTrees and existing visualization techniques we compared it to two other representative tree layouts. As an example of an adjacency-based layout, we used a radial space-filling layout, similar to the Sunburst technique [15] as implemented in the Lark system [33]. This radial space-filling method, shrinks visual elements to cope with increased size, similar to other implicit visualization techniques such as Treemaps. Growth happens outwards for each new level of the tree. A second comparison technique is SpaceTree [17]. SpaceTrees, which also work with tree-cuts, rely on a more conventional node-link-based layout. While they save space by only displaying parts of the tree, once a sufficient number of branches has been expanded, space requirements are comparable to a regular node-link diagram. They share, however, one of the advantages of PaisleyTrees in that they allow rapid navigation of a path through the tree.

Figure 14 shows all three systems, each displaying two different trees. The top row displays the *product* tree, which contains 197 nodes and a total of nine levels, while the bottom row shows the *entity* tree, which

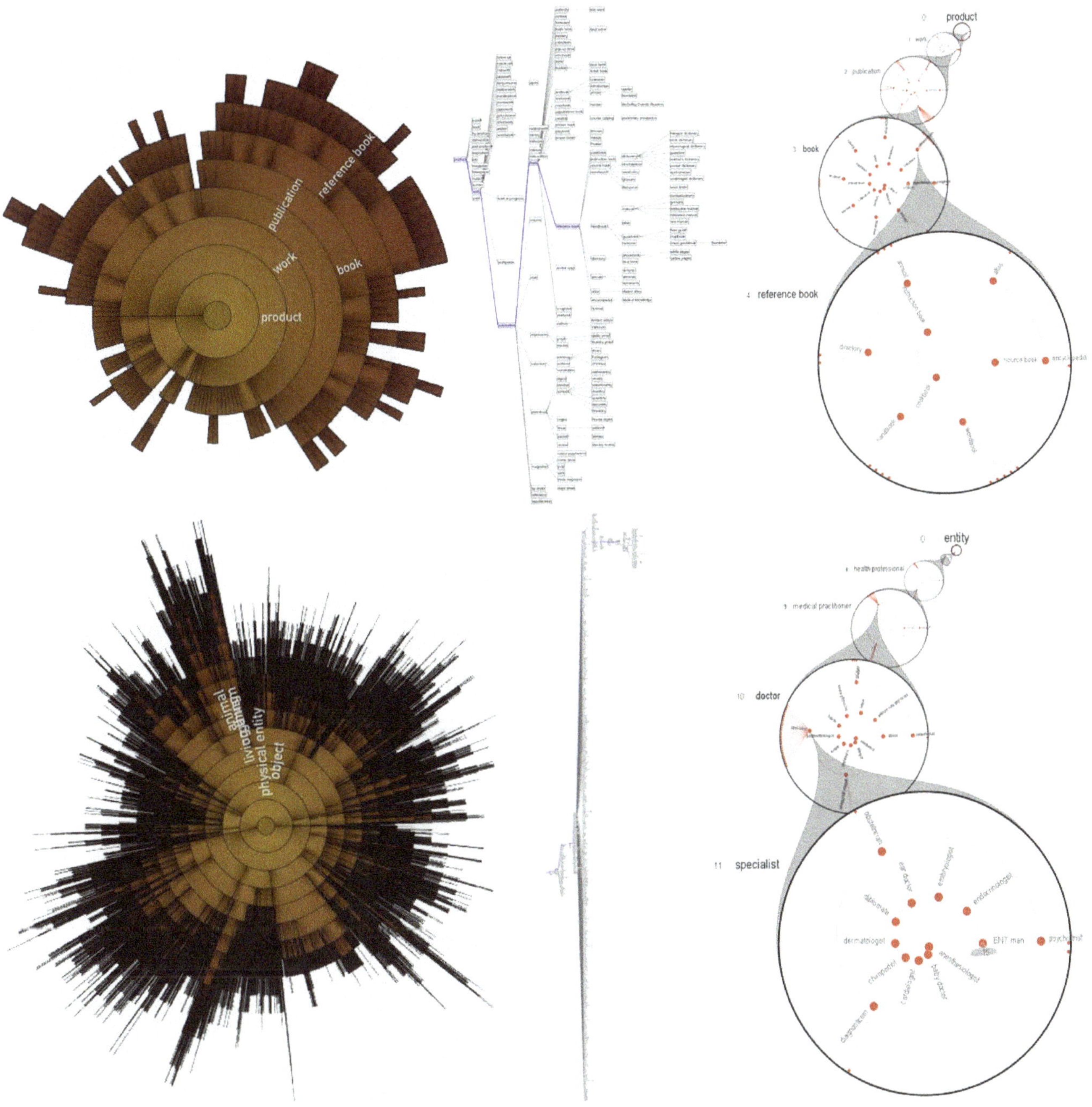

Figure 14. Comparison of three tree visualization techniques for a small tree (top: 197 nodes, navigated to level 4) and a large tree (botton: 75,000 nodes, navigated to level 11). Left: Sunburst/Lark [15, 33]. Middle: SpaceTree [17]. Right: PaisleyTrees.

is considerable larger holding 75,000 nodes with 18 levels of depth. Visually, comparing the results for these three tree types clearly demonstrates the differences in approach between these three tree visualization techniques. Notice that SpaceTree examples have been created through its default setup. The arrangement can be improved interactively using zoom/pan and wrap layout tool. The biggest difference between the radial layout, and the other two techniques is its goal of displaying the complete tree instead of a tree-cut. This allows the technique to answer structural queries and provide an idea of the overall structure of the data without interaction. The approach, however, breaks down once the tree reaches a certain size: while all 197 nodes of the smaller *product* tree are somewhat visible and shown surrounded by their neighbors, in the visualization of the much larger *entity* tree many nodes become so small as to be virtually unreadable. The Lark implementation only displays labels once a tree element has been clicked, which becomes impossible along the

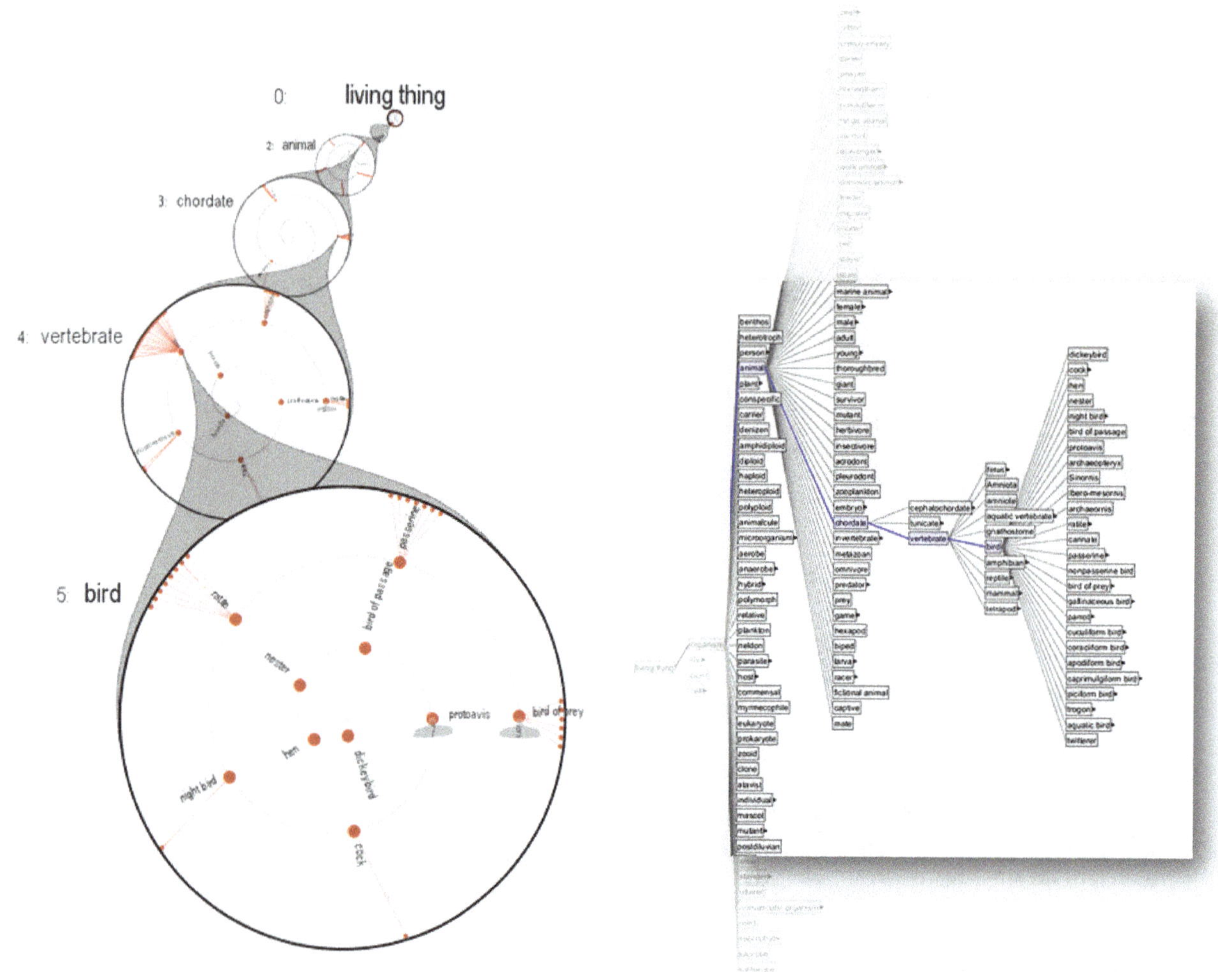

Figure 15. Comparing PasileyTree and SpaceTree: Left: in PaisleyTree the overall size is fixed, so the full structure stays on screen. Right: SpaceTree requires interactive adjustments to view the full structure (in this image the part displayed on screen has been highlighted).

outer rings of large trees where nodes have become very small. The display size is kept relatively compact but visual elements are shrunk in the process, leading to a reading that is mostly a structural impression, rather than providing access to individual nodes. Similar to other techniques that aim to visualize the complete tree, the radial space filling layout is able to convey a picture of a tree's structure, but struggles with larger trees. More importantly, regardless of the size of the tree, there is no mechanism for visually re-focusing the tree based on an NOI. For example, in Figure 14 top-left, the selection of "reference book" as the NOI in Lark only indicates labels of ancestral path, while in PaisleyTree the entire visualization refocuses (and reorganizes) based on the NOI and its proximity.

By taking into account only parts of the tree, SpaceTrees cope much better with larger trees. Due to being based on node-link diagrams, they still grow in size while exploring the tree, though typically gradually. Additionally, on-screen elements remain legible and may be interacted with the whole time. Zooming and panning allows all parts of the tree to be reached. SpaceTree also features several improvements to handle larger trees: the current level is displayed in two columns to use space more efficiently and structural previews are included, such as small triangles that depict the size and depth/breadth of the subtree.

6.1. Comparison between SpaceTree and PaisleyTree

The main conceptual difference between PaisleyTree and SpaceTree is the focusing strategy of these two visualizations. The zoom/pan method in SpaceTree is a linear approach for presenting the tree-cut containing the NOI, while in PaisleyTree, NOI has pivotal role for defining the tree-cut. In SpaceTree, the interaction for

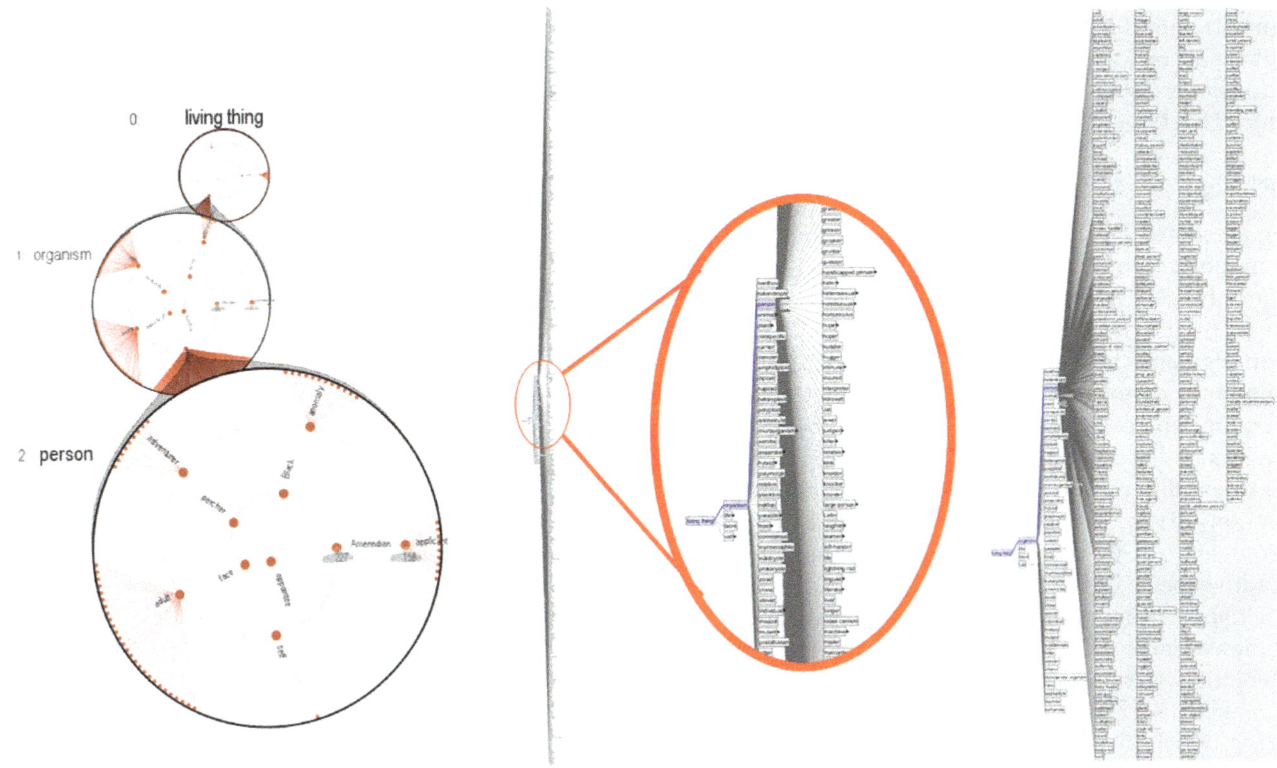

Figure 16. This Figure shows comparison of the descendant tree-cut of PaisleyTree and similar tree-cut of SpaceTree. The node *person* has 163 non-leaf and 232 leaf children.

selecting NOI requires the user to interactively zoom and pan to create a good view of the tree-cut. For larger trees, it is also necessary to interactively organize the vertices, as the default setup does not always provide a suitable arrangement (see Figure 15). In PaisleyTree the only required interaction is to select NOI and the tree-cut is organized automatically based on this NOI.

Keeping overall structure (ancestral tree-cut). In PaisleyTree, we intentionally create two tree-cuts of the NOI: ancestral tree-cut and descendant tree-cut of the NOI. The layout and strategy used for these tree-cuts are different. For the ancestral tree-cut, we present the context and overall structure from NOI to the root. This has been a positive aspect of radial space filling layouts. In essence, the ancestral tree-cut is constructed in proximity to the ancestral path from NOI to the root. A closer zoom in SpaceTree loses the context as shown in Figure 16.

Descendant Tree-cut. For visualizing descendant tree-cut, we use a hybrid method to reduce clutter and make a better use of space (see Figure 16). Using nested layout for the children removes the need for drawing a large set of edges. Furthermore, using circular space and the radial arrangement of the children, provides a compact and well organized structure. This arrangement is generated algorithmically. Our current implementation of PaisleyTree can only visualize up to two levels of descendant subtree. Although, other nested layouts such as ShamsehTree [12] can be used for more than two levels, they may not be useful for larger trees.

Aesthetic factor. An exploration of aesthetics has been one of the goals for designing the PaisleyTree layout. In addition to the use of Paisley pattern, circles and spirals are important elements of the PaisleyTree layout. All these elements influence the aesthetics of the visualization [31, 34–36]. Perhaps most importantly, we have used the exploration of a different aesthetic, in this case traditional Paisley patterns, to help us think out of the box, to help us create an alternate layout with different presentation and interaction possibilities that expands the options for large tree layout.

Elision. The two types of elisions used in PaisleyTree are unique. While SpaceTree does provide elision for descendant nodes, it does not have elision for ancestral nodes. PaisleyTree's use of ellipses helps in handling an NOI with extreme degree in either number of children or a very long ancestral path. In this way PaisleyTrees can display long ancestral paths as demonstrated in Figure 15. In SpaceTree, the elision is applied to parts of the descendant subtree of NOI as shown in Figure 16 right. The small gray triangles in front of the nodes are ellipses that show the existence of descendant subtrees. Since SpaceTree displays all the children of the NOI, it has difficulty with nodes of high degree (see Figure 16 middle). While PaisleyTree elision for the descendant subtree can handle high degree nodes, more interactions may be needed to extract the hidden nodes (see Figure 16 left).

7. Conclusion

We have presented PaisleyTrees, a novel tree visualization technique intended to offer an alternate solution to the problem of tree visualization techniques that grow in size with the underlying data or shrink their on-screen elements indefinitely. PaisleyTrees combine three different methods to obtain a solution that establishes size-invariance (1) we use a hybrid combination of node-link, adjacency and nested layouts, to efficiently present different parts of the tree in one uniform visualization, (2) we display a tree-cut instead of the whole tree, and (3) we introduce elision on the depth and breadth aspects of the tree to indicate access to the nodes of the tree that would expand the visualization beyond the available screen space. Animated transitions and interaction techniques that also work on mobile devices make PaisleyTrees applicable to both small and large trees. PaisleyTrees are promising in that they directly tackle the problem of size-invariance for tree visualizations. Future work could extend the principles we described here and apply them to other types of data. The idea of hybrid layouts could be applied to other types of data-sets with an existing library of visualization techniques, such as graphs. Similarly, PaisleyTrees focus on a single path through the tree, but a variation that also supports access to different branches and allows direct comparison, could be a useful exploration for future.

Acknowledgements

This research was supported in part by NSERC, SMART Technologies, AITF, NSERC Surfnet, and GRAND NCE.

References

[1] Di Battista, G., Eades, P., Tamassia, R. and Tollis, I. (1999) *Graph Drawing: Algorithms for the Visualization of Graphs.* (Prentice-Hall).

[2] Johnson, B. and Shneiderman, B. (1991) Tree-maps: a space-filling approach to the visualization of hierarchical information structures. In *Proceedings of the 2nd Conference on Visualization, (VIS'91)* (Los Alamitos, CA, USA: IEEE Computer Society Press): 284–291.

[3] Etemad, K. and Carpendale, S. (2010) Symmetry and node focused visualization of large trees. In *Poster: at GRAND NCE Annual Conference, GRAND.*

[4] Schulz, H.J., Hadlak, S. and Schumann, H. (2010) The Design Space of Implicit Hierarchy Visualization: A Survey. *IEEE transactions on visualization and computer graphics* **17**(4): 393–411. doi:10.1109/TVCG.2010.79, URL http://www.ncbi.nlm.nih.gov/pubmed/20498508.

[5] Reingold, E. and Tilford, J.S. (1981) Tidier drawings of trees. *Software Engineering, IEEE* **SE-7**(2): 223–228. doi:10.1109/TSE.1981.234519, URL http://ieeexplore.ieee.org/lpdocs/epic03/wrapper.htm?arnumber=1702828http://ieeexplore.ieee.org/xpls/abs_all.jsp?arnumber=1702828.

[6] Wetherell, C. and Shannon, A. (1979) Tidy drawing of trees. *IEEE Transactions on Software Engineering* **5**(5): 514–520.

[7] Beaudoin, L., Parent, M.A. and Vroomen, L.C. (1996) Cheops: A compact explorer for complex hierarchies. In *Proceedings VIS '96*: 1–14. URL http://ieeexplore.ieee.org/xpls/abs_all.jsp?arnumber=567745.

[8] Robertson, G., Mackinlay, J. and Card, S. (1991) Cone Trees: Animated 3D Visualizations of Hierarchical Information. In *Proceedings CHI '91*: 189–194. URL http://dl.acm.org/citation.cfm?id=108883.

[9] Neumann, P., Carpendale, S. and Agarawala, A. (2006) Phyllotrees: Phyllotactic patterns for tree layout. In *Proceedings of Eurographics / IEEE VGTC Symposium on Visualization, (EUROVIS'06)* (Eurographics): 59–66. doi:http://dx.doi.org/10.2312/VisSym/EuroVis06/059-066.

[10] Hao, L. and Fogarty, J. (2008) Cascaded treemaps: examining the visibility and stability of structure in treemaps. In *Proceedings of Graphics Interface, (GI'08)* (Toronto, Ontario, Canada): 259–266.

[11] Boardman, R. (2000) Bubble trees, the visualization of hierarchical information structures. In *Extended Abstracts on Human Factors in Computing Systems, (CHI'00)* (New York, NY, USA: ACM Press): 315–316. doi:http://doi.acm.org/10.1145/633292.633484.

[12] Etemad, K. and Carpendale, S. (2009) Shamsehtrees: Providing hierarchical context for nodes of interest. In *Proceedings of Bridges 2009: Mathematics, Music, Art, Architecture, Culture* (Tarquin Books): 293–300.

[13] Collins, C., Carpendale, S. and Penn, G. (2009) Docuburst: Visualizing document content using language structure. *Computer Graphics Forum* **28**(3): 1039–1046.

[14] Chevalier, F., Auber, D. and Telea, A. (2007) Structural analysis and visualization of c++ code evolution using syntax trees. In *Proceedings of the 9th International Workshop on Principles of Software Evolution, (IWPSE'07)* (New York, NY, USA: ACM Press): 90–97. doi:http://doi.acm.org/10.1145/1294948.1294971.

[15] Stasko, J., Catrambone, R., Guzdial, M. and Mcdonald, K. (2000) An evaluation of space-filling information visualizations for depicting hierarchical structures. *International Journal of Human-Computer Studies* **53**(5): 663–694. doi:10.1006/ijhc.2000.0420, URL http://linkinghub.elsevier.com/retrieve/pii/S1071581900904208.

[16] Andrews, K. and Heidegger, H. (1998) Information slices: Visualising and exploring large hierarchies using cascading, semi-circular discs. In *Proc of IEEE Infovisï£¡ 98 late breaking Hot Topics*: 9–11.

[17] Plaisant, C., Grosjean, J. and Bederson, B.B. (2002) Spacetree: Supporting exploration in large node link tree, design evolution and empirical evaluation. In *Proceedings INFOVIS '02*: 57–64.

[18] Card, S. and Nation, D. (2002) Degree-of-interest trees: A component of an attention-reactive user interface. In *Proceedings of the Working Conference on Advanced Visual Interfaces* (ACM): 231–245. URL http://dl.acm.org/citation.cfm?id=1556300.

[19] Heer, J. and Card, S. (2004) DOITrees revisited: scalable, space-constrained visualization of hierarchical data. In *Proceedings of the working conference on Advanced visual interfaces* (ACM): 421–424. URL http://dl.acm.org/citation.cfm?id=989941.

[20] Li, H. and Abe, N. (1998) Generalizing case frames using a thesaurus and the MDL principle. *Computational linguistics* **24**(2): 217–244. URL http://dl.acm.org/citation.cfm?id=972734.

[21] Zhao, S., McGuffin, M. and Chignell, M. (2005) Elastic hierarchies: Combining treemaps and node-link diagrams. In *Proceedings INFOVIS '05*: 57–64. URL http://ieeexplore.ieee.org/xpls/abs_all.jsp?arnumber=1532129.

[22] Linsen, L. and Behrendt, S. (2011) Linked treemap: a 3D treemap-nodelink layout for visualizing hierarchical structures. *Computational Statistics* : 679–697doi:10.1007/s00180-011-0272-2, URL http://www.springerlink.com/index/L1327352V579R838.pdf.

[23] Shin, H., Park, G. and Han, J. (2011) Tablorer - An Interactive Tree Visualization System for Tablet PCs. *EuroVis 2011* **30**(3): 1131–1140. doi:10.1111/j.1467-8659.2011.01962.x, URL http://onlinelibrary.wiley.com/doi/10.1111/j.1467-8659.2011.01962.x/full.

[24] Karstens, B., Kreuseler, M. and Schumann, H. (2003) Visualization of complex structures on mobile handhelds. In *Proceedings Workshop on Mobile Computing*. URL http://citeseerx.ist.psu.edu/viewdoc/download?doi=10.1.1.112.1960&rep=rep1&type=pdf.

[25] Hao, J. and Zhang, K. (2007) A mobile interface for hierarchical information visualization and navigation. In *Proceedings ISCE '07* (IEEE): 1–7. doi:10.1109/ISCE.2007.4382214, URL http://ieeexplore.ieee.org/xpls/abs_all.jsp?arnumber=4382214.

[26] Scruton, R. (2009) *Why Beauty Matters* (BBC2).

[27] Skog, T., Ljungblad, S. and Holmquist, L.E. (2003) Between aesthetics and utility: designing ambient information visualizations. In *Information Visualization, 2003. INFOVIS 2003. IEEE Symposium on* (IEEE): 233–240.

[28] Eric, W. (2009), Fermat's spiral. mathworld–a wolfram web resource., http://mathworld.wolfram.com/FermatsSpiral.html.

[29] Elmqvist, N., Moere, A.V., Jetter, H.C., Cernea, D., Reiterer, H. and Jankun-Kelly, T. (2011) Fluid interaction for information visualization. *Information Visualization* **10**(4): 327–340. doi:10.1177/1473871611413180, URL http://ivi.sagepub.com/lookup/doi/10.1177/1473871611413180.

[30] Attneave, F. (1955) Symmetry, information, and memory for patterns. *American Journal of Psychology* **68**(2): 209–222. doi:http://www.jstor.org/stable/1418892.

[31] Zhang, K. (2007) From abstract painting to information visualization. *Computer Graphics and Applications* **27**(3): 12–16. doi:10.1109/MCG.2007.58.

[32] Fellbaum, C. (1998) *WordNet: An Electronic Lexical Database.* (Cambridge, MA, USA: MIT Press).

[33] Tobiasz, M., Isenberg, P. and Carpendale, S. (2009) Lark: coordinating co-located collaboration with information visualization. *IEEE transactions on visualization and computer graphics* **15**(6): 1065–72. doi:10.1109/TVCG.2009.162, URL http://www.ncbi.nlm.nih.gov/pubmed/19834173.

[34] Etemad, K., Carpendale, S. and Samavati, F. (2014) Spirograph inspired visualization of ecological networks. In *Proceedings of Computational Aesthetics in Graphics, Visualization, and Imaging* (ACM): 81–91.

[35] C. Bennett, C., Ryall, J., Spalteholz, L. and Gooch, A. (2007) The aesthetics of graph visualization. *Computational Aesthetics in Graphics, Visualization, and Imaging* 27: 1–8.

[36] Coleman, M. and Parker, D. (1996) Aesthetics-based graph layout for human consumption. *Softw. Pract. Exper.* **26**(12): 1415–1438. doi:http://dx.doi.org/10.1002/(SICI)1097-024X(199612)26:12<1415::AID-SPE69>3.3.CO;2-G.

4

Social retrieval of music content in multi-user performance

Maurizio Mancini[*,1], Gualtiero Volpe[1], Giovanna Varni[1], Antonio Camurri[1]
[1]Casa Paganini - InfoMus, DIBRIS, University of Genoa, Italy

Abstract

An emerging trend in interactive music performance consists of the audience directly participating in the performance by means of mobile devices. This is a step forward with respect to concepts like active listening and collaborative music making: non-expert members of an audience are enabled to directly participate in a creative activity such as the performance. This requires the availability of technologies for capturing and analysing in real-time the natural behaviour of the users/performers, with particular reference to non-verbal expressive and social behaviour. This paper presents a prototype of a non-verbal expressive and social search engine and active listening system, enabling two teams of non-expert users to act as performers. The performance consists of real-time sonic manipulation and mixing of music pieces selected according to features characterising performers' movements captured by mobile devices. The system is described with specific reference to the SIEMPRE Podium Performance, a non-verbal socio-mobile music performance presented at the Art & ICT Exhibition that took place in Vilnius (LI) in November 2013.

Keywords: personalised social media experience in mobile devices, embodied cooperation, expressive and social features, music retrieval

1. Introduction

Along with the widespread dissemination of smartphones or, in general, of mobile devices endowed with increasing computational power, users are enabled to participate more actively in creative experiences by means of their natural (expressive and social) behaviour measured from data mobile devices capture. In interactive music performance a peculiar expression of such an emerging trend consists of the audience directly participating in the performance by means of mobile devices. This is obtained, for example, by distributing specific apps during the performance (e.g., see [1] for a technique for optimised distributions of apps in an audience) or by exploiting existing apps or social media (e.g., see [2] for an example of real-time sonification of tweets).

This is a step forward with respect to concepts like active listening and collaborative music making. In active listening, users (e.g., performers and/or customers) are enabled to adapt/modify the music content they are listening to. In collaborative music making, a group of users is enabled to create and play together a (new) music piece. Starting from such concepts and somewhat combining them, non-expert persons (e.g., members of an audience) are enabled to directly participate in a creative activity such as performing music: they can either become performers or collaborate with professional performers in a creative joint action.

Developing such novel forms of music performances requires the availability of technologies for capturing and analysing in real-time the natural behaviour of the users/performers, with particular reference to non-verbal expressive and social behaviour. Indeed, music is a media for expressing emotional states, moods, personality traits and stances [3], and music performance is an inherently expressive and social activity.

This paper presents a prototype of a non-verbal expressive and social search engine and active listening system, enabling two teams of non-expert users to act as performers, where the performance consists of real-time sonic manipulation and mixing of music pieces selected according to features characterising performers' movements captured by mobile devices.

The system is described with specific reference to how it was used for the *SIEMPRE Podium Performance*, a non-verbal socio-mobile music performance Casa Paganini - InfoMus presented at the Art & ICT Exhibition[1] that took place in Vilnius (LI) in November 2013. The exhibition was included in the program of the EU ICT 2013 event, organised by the European

*Corresponding author. Email:maurizio.mancini@unige.it

[1]http://ec.europa.eu/digital-agenda/en/ict-2013-exhibition

Commission in partnership with the Lithuanian Presidency of the Council of the EU, the Europe's biggest digital technology event where the most recent research and developments on information and communication technologies are presented. The performance is a joint action involving two teams of three performers using smartphones. Performers are not expert musicians and may represent members of a generic audience. They have to compete by shaking rhythmically their smartphones or by moving/dancing while wearing tightly their smartphones to their body, to choose the music to be played actively in the common environment. Performers' movement information is acquired by the gyro and 3-axes accelerometers embedded in the smartphones. The team who is expressing the most coherent expressivity in terms of movement tempo, fluentness and motion energy, is the one that makes music emerge. That is, we aim to model the following analogy: the more a group of music players moves in a coherent way during performance, the more enjoyable (for the both audience and players) a music performance is.

After discussing some related work in Section 2, the concept of the performance is introduced in Section 3. Section 4 presents the architecture of the prototype followed in Sections 5 and 6 by the expressive and social feature extraction components as they were used in the SIEMPRE Podium Performance. Finally, Section 7 presents the music database for the performance and the components for song selection and interaction management.

2. Related work

This paper focuses on how the non-verbal expressive joint action of a group of people extends the action affordances that are possible for a single individual [4]. The trend of game industry (e.g., Nintendo and Microsoft) shows that even a simplified non-verbal expressive joint action, i.e., one that does not take into account measures of emotional and social content in a constrained ecological setting, can be successful in engaging users in video games (e.g., see [5][6][7]). However, merging non-verbal expressive joint action and mobile technologies is still under-investigated and opens new perspectives for future *User Centric Media* [8].

Systems that explicitly measure social signals and perform analysis of social interaction are more and more emerging in the HCI community (e.g., see [9][10][11] and [12] for a survey on social signal processing), but are still missing embodied cooperation and collaborative features, e.g., they do not allow multiple users to collaborate in performing common tasks or reach common goals. Conversely, existing music making and information retrieval systems do not yet take into account features from social signal analysis, e.g., entrainment, empathy, and so on.

2.1. Socio-mobile signal analysis

Several works used mobile devices like smartphones and tablets to analyse social relationships between users. Lukowicz et al. [13] demonstrated that social information they define as "behavior demographics", that is, preferences and risks of segments of a population, could be computed by looking at people's mobile phone traces. In another work (see [14]) they investigated the evolution of social relationships in a student dormitory by monitoring the position of 70 users every 6 minutes over nine months. They collected data from mobile phones such as phone calls and text messages between people living in the dormitory, physical proximity (detected by Bluetooth signals) and position in the dormitory (detected by WiFi signals). This data was used to estimate friendships and their evolution over time (e.g., persons are more likely to become friends if they frequently share the same space) and to monitor social activities. In [15], different types of conversations between users, mainly brainstorming and decision making, were distinguished using mobile wearable devices embedding a microphone and a radio transmitter that communicated the recorded data to radio hotspots. When concerning a better understanding of interpersonal creative interaction (such as playing music), authors in [16] state that detecting finer-grained physiological process is determinant. Their MobileMuse, an unobtrusive sensor package for mobile physiological signal acquisition, aims at collecting seamlessly temperature, GSR, pulse oximetry, and triaxial accelerometer to characterise critical component of social interaction such as emotional contagion.

2.2. Collaborative music systems

Collaborative music systems have a twofold focus: music making and music listening. Traditionally conceived for laptops and workstations, mobile versions of these systems recently appeared.

Concerning music making, Blaine and Fels [17] claim that such systems commonly restrict musical control, facilitating novices' participation in the musical experience. Further, as Jordà argues [18], multi-user instruments facilitate responsiveness and interaction between each performer and the instrument, and also between performers. Among music making systems, the ReacTable [19] allows a group of people to share control of a modular synthesiser by manipulating physical objects on a round table; Audiopad is a composition and performance instrument for collaborative electronic music which tracks the positions of objects on a tabletop surface and converts their motion into music [20]; TinyTune is a collaborative musical instrument using sensor motes [21]; JamMo is a mobile technological tool for music making for young children [22]; Malleable Music Making [23] is an interactive mobile system allowing multiple users to collaboratively create music over wireless networks; Beatbugs [24] are mobile instruments that can be used in collaborative networks to form large-scale collaborative compositions.

As for music listening, Camurri [25] proposed an early pioneering system where the user full-body rhythmic movements were analysed in real-time and compared with the beat of a song (extracted from the MIDI music signal). In the game Ghost In The Cave [26], two groups of users compete to collaboratively communicate emotions by full-body movement and singing voice, in order to achieve the goals of the game. Leman et al. [27] reworked the concept of social music game: the movement beat of multiple users was extracted and compared with the beat of the music the users were listening to. Users could compete among them or collaborate to win the game. Stokholm and Pasquier [28] implemented a system mixing audio representations of the mood of several users to increase collaboration and empathy. Vinyes and colleagues developed the Audio Explorer system, enabling users to concurrently modify the audio mixing of a piece of music downloaded from the Web and to share the resulting content [29]. MoodifierLive [30] is a mobile phone application that allows up to 4 users to collaboratively influence the expressive play back of MIDI content. Users move their mobile phones by taking control of one of the four music parameters: Overall Tempo, Overall Sound level, Overall Articulation, and Phrasing.

3. Concept

The SIEMPRE Podium Performance is a music performance involving two teams of three performers. These are non-expert persons rather than, e.g., experienced musicians. The performance is a kind of competitive jukebox where each team competes with the other one with the goal of making themselves and the whole audience listening to the music piece the team likes. Performers express their choices by shaking rhythmically their smartphones.

The performance starts with the smartphones laid down on a table. No music is played. The six performers are on the stage, each team occupying one side of it (see Figure 1). One performer takes one smartphone and starts shaking it. A percussive sound is reproduced, following the tempo of the movement. After a while, the other performers take their smartphone and start shaking them. Further percussive sounds are played back following the tempo of each performers. Percussive sound are different for each performer,

Figure 1. The SIEMPRE Podium Performance presented at the Art & ICT Exhibition that took place in Vilnius (LI) in November 2013. Two teams of three performers are interacting by shaking their mobile phones: on the screen some technical data about participants is shown; team 2 is exhibiting higher cohesion and synchronisation, thus the music piece retrieved by team 2 will emerge over the one selected by team 1.

enabling performers and the audience to recognise who is producing each sound.

As soon as two performers in one team synchronise with each other, i.e., they shake their smartphone at almost the same tempo, a song is selected. The selected song depends on the expressive qualities of the movements of the pair, namely: (i) the tempo of their movements, and (ii) the current position of their movement in an expressive space, the energy-fluentness space. The song exhibiting the most similar characteristics is selected from a database. However, since the team is not fully collaborating (one performer is not synchronised with the other two), they listen just to the rhythmic accompaniment of the song, rather than to the whole piece. When the pair or one of its component exhibits enough leadership to involve the third team member so that all the three performers synchronise to the same tempo, the song emerges with all its voices. The other team has now to take the leadership of the performance to make its preferred music emerge. To do so, team members need to synchronise at a given tempo and they have to do it better than the other team. When the winning team changes its tempo, energy, and/or fluentness, or the other team takes the leadership of the performance, a new song emerges and is, again, modulated by the winner team.

The performance music database consists of a collection of European traditional folk music, from Portuguese fado to Lithuanian sutartine (see Section 7).

3.1. Collaboration paradigm evaluation

In *Sync4All*, a reduced version of the performance described in [31], a single team of four people has to try to completely synchronise to explore a music piece. In this version of the performance, a single music instrument is associated to each pair of performers in the team: the more they synchronise their movements, the more the listening experience is enriched by adding different instruments and voices.

Sync4All was presented and evaluated during the public exhibition "Festival della Scienza" hosted in Genoa (Italy) on November 2010. The evaluation was carried out via questionnaires. The analysis of the colected data was aimed at finding out which is the overall attitude of the users to the application, and how the participants perceived the level of collaboration and music embodiment.

Seventy-two individuals tested the application as volunteers. The questionnaire was filled up by 70 participants (38 male and 32 female) coming both from European (60 people from Italy, 4 from France, 1 from Serbia) and extra-European countries (1 from Algeria, 1 from Brazil, and 1 from Ecuador). Mean age was 22.9y (std=14.1y). Participants were clustered in basic and advanced users, depending on their skills in using mobile phones daily.

Results showed that both basic and advanced users (i) have a very positive attitude to Sync4All, (ii) perceived the level collaboration in the group and (iii) were aware that music changed according to it. Free comments and suggestions further confirmed that the participants' opinion on Sync4All was very satisfactory. One user suggested to implement a version of Sync4All in which the music tempo is aligned with the rhythm of the team: this feature is part of the current performance set-up. Another user argued that she felt constrained by the gesture to be performed in the application.

4. System set-up

Figure 2 sketches the system set-up. It encompasses several components: 6 *mobile devices* acting as clients, a laptop running EyesWeb XMI[2] acting as a server, and a *music database*.

Mobile devices capture acceleration, perform pre-processing, and send it via network to the EyesWeb server. The server carries out expressive and social features extraction from such data. Features are used to select a song for each team and to determine the winning team. Both expressive and social features are used to perform audio mixing between the two songs.

Details are provided in the following sections. To make the description concrete, components are discussed with reference to the techniques used to carry out the SIEMPRE Podium Performance presented in Section 3. However, the system can be generalised to implement further instances of expressive and social retrieval and active listening applications. For example, the system can be extended to support active experience of audiovisual cultural content, e.g., for museums or consumer cultural applications.

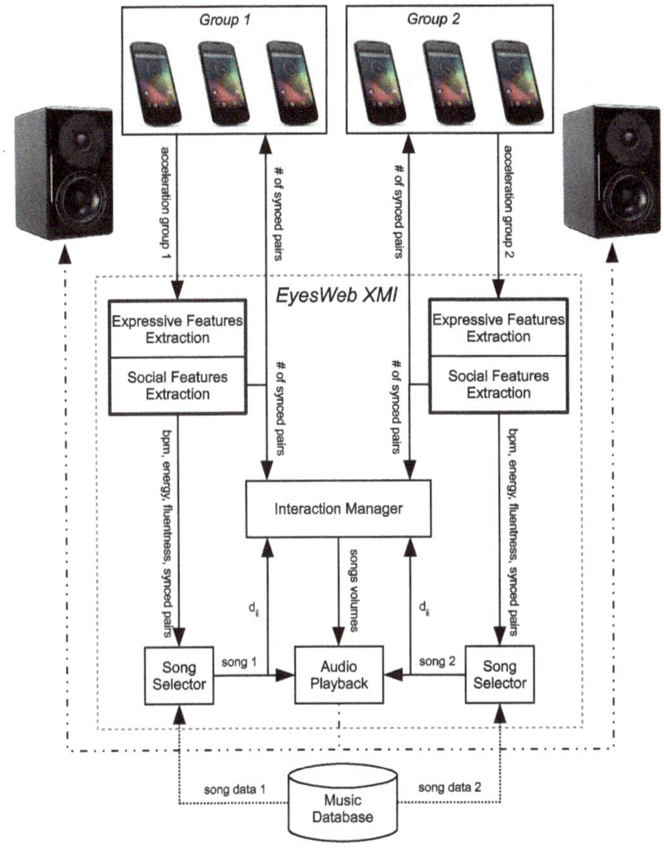

Figure 2. System set-up.

5. Mobile devices

Mobile devices are end-user devices (e.g., Android phones) equipped with sensors including accelerometers, gyroscopes, cameras, GPS, and so on. They are connected to a wireless local area network through UDP. Mobiles run a dedicated Android application to filter out gravity acceleration and to compute the magnitude of the 3D acceleration:

$$A = \frac{\sqrt{A_x^2 + A_y^2 + A_z^2} - g}{A_{MAX}}; \qquad (1)$$

where: A_x, A_y, A_z are the components of the detected acceleration along the three axes, and g is the magnitude of gravity acceleration. The acceleration A is sent to EyesWeb XMI through UDP at 50 fps.

[2] http://www.infomus.org

As depicted in 2, the *Social Features Extraction* module transmits the number of synchronised pairs to the mobile devices. Even if the concept of synchronisation is widely debated in the literature and several kinds of synchronisation are defined, in the framework of this performance synchronisation is simply intended with respect to tempo, i.e., two performers are synchronised if they shake their mobile devices at the same tempo. If the number of synchronised pairs is zero, that is, the three mobile devices of the team are moving with different tempi, then each mobile device plays a short percussive beat sound synchronised with the acceleration peaks corresponding to the user's movement strokes.

6. Expressive and social feature extraction

Expressive and social feature extraction has been implemented in the EyesWeb XMI platform and in particular in the EyesWeb Expressive Gesture Processing Library [32]. EyesWeb XMI is a software platform that allows developers to implement software modules for automatic analysis of user's expressive movement in an intuitive, visual way.

In the performance, EyesWeb XMI receives acceleration data from the mobile devices and manages the extraction of expressive and social features and the formulation of queries for search and retrieval of songs. Further, it controls the active experience of the music content, i.e., it modulates the audio mixing. The following describes the core feature extraction modules: the *Expressive Features Extraction* and the *Social Features Extraction* modules.

6.1. Expressive Features Extraction

When considering creative activities such as music playing, expressivity of human behaviour is a key component to analyse. Research works such as [33], [34] and [35] have shown that detecting and quantifying speed, acceleration, energy, smoothness, or impulsivity of one's movement may inform about emotional states.

In the SIEMPRE Podium Performance expressive feature extraction follows the multi-layered framework for analysis of expressive gesture developed by Camurri and colleagues [32]. In particular, the performance exploits the following features which are extracted for each team: (i) the tempo of the movements, (ii) the position in the energy-fluentness expressive space. Tempo was selected in order to make the performers and the audience immediately and macroscopically perceive that movement controls music selection, slow pieces being associated with low movement tempo, and quick pieces being associated with high movement tempo. Indeed, the relationship between music and movement tempo is one of the most direct and intuitive even for non-expert participants (consider for example foot tapping in time with music). The position in the energy-fluentness space is used to distinguish subtler expressive nuances between movements sharing the same tempo. Such nuances are reflected in the expressive content of the selected song. The energy-fluentness space was selected since it enables cross-modal mappings of movement features into music features. It is related to kinaesthetic, or energy-velocity spaces successfully used as control spaces to analyse and synthesise music performances, and in particular to the energy-articulation space [36]. The energy-fluentness space was already used for *ExpressiveHiFi* [37], a system where, besides the usual knobs for volume, balance, and so forth, users are also provided with additional knobs to control by their movement the expressive content of the music they are listening to.

Expressive features are extracted starting from the acceleration data mobile devices capture. The algorithms for automatic extraction of tempo, energy, and fluentness have been implemented using the EyesWeb Expressive Gesture Library [38]:

- *Tempo* (*bpm*): movement *beats per minute* (*bpm*) is computed by the FFT of the 3D acceleration magnitude A. The FFT amplitude vs. samples spectrum is analysed to find the harmonics exhibiting the maximum (A_1) and second highest (A_2) amplitudes. The first one's period (in samples) is converted to *bpm* as it follows:

$$bpm = samples * (fps/L) * 60; \quad (2)$$

  where L is the length of the buffer on which the FFT is computed. Then, *bpm* is valid if and only if:

  - the mobile device has been moved in the last 2 seconds, that is, movement energy ME is higher than a threshold;
  - $bpm > 40$;
  - the current *bpm* value has been maintained constant for the last 2 seconds;
  - the *bpm* confidence level C is higher than a threshold, where:

$$C = |A_1 - A_2|/A_1; \quad (3)$$

    That is, the difference between the highest peak amplitude (A_1) and the second highest peak (A_2) amplitude normalised by the highest peak amplitude (A_1). Figure 3 shows an example of high and low confidence levels.

- *Movement Energy* (ME): acceleration data captured by the mobile device's accelerometer as

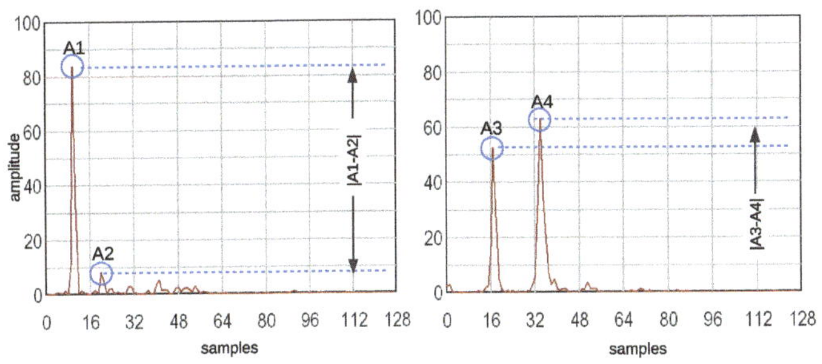

Figure 3. Example of bpm extraction and confidence level computation using FFT: (graph on the left) the difference between the highest peak amplitude (A_1) and the second highest peak amplitude (A_2) normalised by the highest peak amplitude (A_1) tends to 1; (graph on the right) the amplitudes are close, that is, the normalised difference tends to zero.

described in Section 5 is converted to Movement Energy by integrating acceleration over a time window $[t_1, t_2]$:

$$ME = \int_{t_1}^{t_2} Adt; \tag{4}$$

- *Fluentness* (*Fl*): the multiplicative inverse of acceleration variance over a time window $[t_1, t_2]$ is computed:

$$Fl = 1/\sigma_A; A \in [t_1, t_2]; \tag{5}$$

6.2. Social Features Extraction

This module checks whether a pair of performers in a team is tuned to the same tempo, i.e., whether they shake their mobile devices approximately with the same number of bpm. This feature is regarded as an indicator of the degree of synchronisation between a pair of performers in a team. Especially in the field of music and dance, synchronisation is used to measure the level of collaboration and entitativity of a group of people (that is, their tendency to act as a single entity) performing a joint activity [39][31].

In the proposed system, the synchronisation between a pair of performers is computed starting from the bpm they are expressing by moving their mobile devices. All the possible pairs of performers in a team are considered. In the SIEMPRE Podium Performance teams are composed by three performers, that is, there exist three pairs of performers. Synchronisation between two performers is defined as:

$$T_{u1,u2} = \begin{cases} 1, & \text{if } |bpm_{u1} - bpm_{u2}| < 15 \\ 0, & \text{otherwise} \end{cases} \tag{6}$$

The tolerance of 15 bpm has been chosen according to the music database content described in 7: it corresponds to the minimum step between songs in the database. If two performers exhibit periodic movements with the same tempo (with a tolerance of 15 bpm) but shifted in phase (e.g., opposite phases, that is, 180 degrees) they will considered as synchronised. If, instead, two performers exhibit a bpm scaled by an a factor (e.g., 40 vs 80 bpm) they will be considered not in sync. The total number of synchronised pairs for each team is then sent to the *Interaction Manager* module and to the mobile devices. Based on that, audio effects (beat sounds) or songs are played back to the performers and the audience.

7. Music database, song selection and playback

The music database used in the performance consists of 46 pieces of European traditional folk music and related annotations. Songs are organised according to their tempo. This ranges from about 40 bpm, corresponding to slow pieces (*lento*), to about 200 bpm, corresponding to extremely fast pieces (*presto*). The minimum step between songs is 15 bpm. Songs are chosen so that they are clearly characterised by their tempo, i.e., tempo is quite steady and pieces do not encompass significant increase (*accelerando*) or decrease (*ritardando*) of tempo. Slow tempo pieces include for example some Polish folk music and some Romanian dances, extremely fast pieces include for example Scottish and Spanish dances. All songs are polyphonic, i.e., they include multiple musical voices (musical instruments).

Each song in the database is further annotated with respect to energy and fluentness. Both of them are quantised on three levels: *Low*, *Medium*, and *High*. Annotations for tempo, energy, and fluentness were performed by three music experts. Annotations are stored in metadata according to a tree-based hierarchical structure organised on three layers (see Figure 4):

- at the *top layer (root)*, one text file in the root folder of the database contains in each row a pair where the first item is a tempo in bpm and the second item is the path of a text file at the intermediate layer.
- at the *intermediate layer*, one text file for each tempo contains in each row a record of three items, where the first and second items are annotated values of energy and fluentness respectively (0 for *Low*, 0.5 for *Medium*, and 1 for *High*) and the third item is the path of a text file at the bottom layer.
- at the *bottom layer (leaves)*, one text file for any available combination of tempo, energy, and

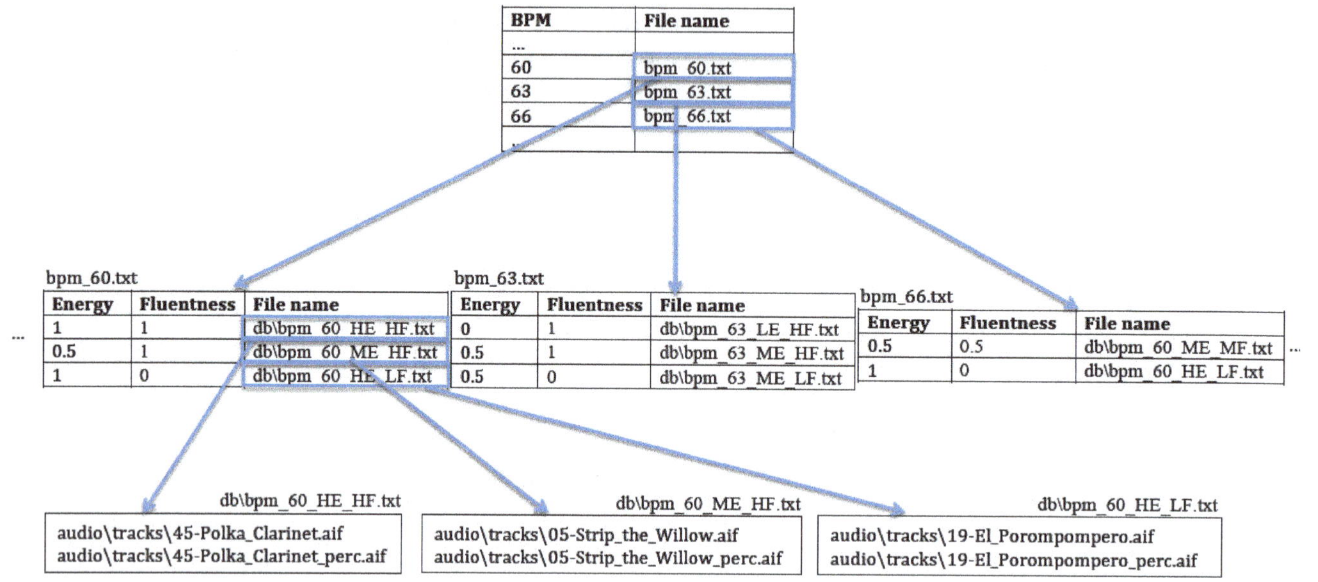

Figure 4. The tree structure for the metadata of the music database. At the root level, each tempo (measured in bpm) is associated with a file including annotations for energy and fluentness. Each combination of those is linked to a leaf including the list of music pieces exhibiting such combination of characteristics. This structure enables retrieving the result of a query by performing a visit of the tree in constant time, i.e., always requiring three steps. The files do not contain all the possible bpm values and all the possible combinations of energy and fluentness levels, but only those for which a song is available in the database.

fluentness contains the paths of the audio files of the songs exhibiting such combination of characteristics.

If on the one hand, such a multi-layered structure for metadata may require to store a possibly high number of text files, on the other hand it is optimised for enabling a fast retrieval of the selected song. Indeed, the result of a query by the *Song Selector* is retrieved by performing a visit of the metadata tree from the root to the leaf containing the songs with the desired characteristics. The visit is performed in constant time, i.e., it always needs three steps: given the desired tempo, the text file at the root layer is accessed and the path of the file including the annotations for energy and fluentness is obtained. From such a file, depending on the required level of energy and fluentness, the path of the text file including the list of available songs with those characteristics is retrieved. Finally, one song in the list is randomly selected for the performance.

It should be noted (see Figure 4) that the root file does not include all the bpm values in the range from 40 to 200, but only the tempos for which songs are available in the database. Accordingly, the intermediate layer files do not necessarily include all the eight possible combinations of *Low*, *Medium*, and *High* energy and fluentness, but only those for which songs are available in the database. In case the required tempo is not available in the database, the nearest available bmp value is selected. Similarly, the piece at the shortest distance with respect to the extracted values of energy and fluentness is retrieved (see Section 7.1).

As the file names in the leaves of the tree shows, two audio files are associated to each song. The first one is the polyphonic song itself. The second one is a rhythmic accompaniment composed for the purpose. The rhythmic accompaniment helps performers to synchronise with the rhythm of the song and can be seamlessly integrated in the song when the two components are mixed together.

7.1. Song Selector

The system set-up includes two *Song Selectors*, one for each team. Each of them receives the movement features extracted from each performer in one team and the number of synchronised pairs in the team, and selects one song for that team. The selected song is the song retrieved using the movement features of the winner performer in the team. The winner performer is identified as follows:

- for each performer $k = 1, 2, 3$, the minimum distance d_k between the tempo of the movement of the kth performer bpm_k and the tempo of each song bpm_i available in the music database is computed as:

$$d_k = \min_{i=1...N} d(bpm_k, bpm_i)\ k = 1, 2, 3; \tag{7}$$

where N is the number of songs in the database and d is the Euclidean distance:

$$d(bpm_k, bpm_i) = \sqrt{(bpm_k - bpm_i)^2}. \tag{8}$$

- the winner performer $\hat{k}$ is identified as the performer belonging to a synchronised pair having the lowest minimum distance $d_{\hat{k}}$:

$$\hat{k} = \arg\min\{d_1, d_2, d_3\}; \quad \text{s.t. } \exists p_j \in P : \hat{k} \in p_j \tag{9}$$

  where $p_j = (a, b)$ is the pair of performers a and b, $a, b = 1, 2, 3$, and P is the set of currently synchronised pairs. This means that: if performers are not synchronised, no winner is identified and no song selected; if two performers out of three are synchronised, the winner is the performer exhibiting the lowest minimum distance between the two synchronised performers; if all three performers are synchronised, the performers exhibiting the lowest minimum distance is the winner.

Once the winner performer of the team is identified, the selected song for her team is retrieved as follows:

- the set S of songs in the database whose tempo is at the minimum distance from the tempo of the winner $\hat{k}$ is computed (more than one song could have the same minimum distance):

$$S = \{\hat{i} = \arg\min_{i=1...N} d(bpm_{\hat{k}}, bpm_i)\} \tag{10}$$

- the selected song $\hat{s}$ is the song in S whose annotated energy and fluentness are at minimum distance with respect to the energy and fluentness of the movement of the winner $\hat{k}$, that is:

$$\hat{s} = \arg\min_{s \in S} d(x_{\hat{k}}, x_s)\} \tag{11}$$

  where $x_{\hat{k}} = [ME_k, Fl_k]$ is the vector whose components are the energy and the fluentness of the motion of the winner performer $\hat{k}$, $x_s = [ME_s, Fl_s]$ is the annotated energy and fluentness of the song $s \in S$, and d is the Euclidean distance.

The selected song $\hat{s}$ and the distance $d_{\hat{k}}$ are sent to the *Interaction Manager* described in Section 7.2

7.2. Interaction Manager and Audio Playback

The input of the *Interaction Manager* module consists of: each team's number of synchronised pairs and distance $d_{\hat{k}}$. For the sake of clarity, $d_{1\hat{k}}$ is the winner's distance for team 1 and $d_{2\hat{k}}$ is the same distance for team 2, see Section 7.1 for details about distance computation. Its output consists of a vector:

$$V = [V_{p1}, V_{r1}, V_{p2}, V_{r2}] \tag{12}$$

where: (i) V_{pi} is the volume at which the **polyphonic** version of the winner song for team i has to be played at; (ii) V_{ri} is the volume at which the **rhythmic** version of the winner song for team i has to be played at. This vector is sent to the *Audio Playback* module, together with the songs associated to each team. The audio volumes are adjusted depending on the values stored in V and the final audio output is played back to the users through speakers. The content of V depends on the amount and type of intra- and inter-teams interaction as it follows:

1. *intra-team* (for each team i):
   (a) *0 synchronised pairs*: $V_{pi} = V_{ri} = 0$; i.e., in this case the mobile devices will produce beat sounds, see Section 5);
   (b) *1 or 2 synchronised pairs*: $V_{pi} = 0, V_{ri} = 1$;
   (c) *3 synchronised pairs*: $V_{pi} = 1, V_{ri} = 0$;
2. *inter-team*:
   (a) *3 synchronised pairs for both teams*: if $(d_{1\hat{k}} < d_{2\hat{k}})$ then $V_{p2} = 0$ else $V_{p1} = 0$;
   (b) *3 synchronised pairs for just one team j*: then $V_{pi} = 0, i \neq j$;

The goal of the SIEMPRE Podium Performance is to engage performers in joint music content exploration and retrieval. The music retrieved by the team exhibiting higher participants' cohesion and synchronisation will emerge: (steps 1a-1c) a higher number of synchronised pairs inside the team (higher cohesion) will increase the audio polyphony; (steps 2a and 2b) a higher similarity between the team participants and audio rhythm (synchronisation) will make the corresponding team's song emerge over the other team's one.

8. Conclusion

This paper presented the SIEMPRE Podium Performance and the system developed for it. The system includes algorithms for extraction of social and expressive features characterising natural behaviour of performers and techniques for automated retrieval of and active listening to music. The system enables non-expert members of an audience to be involved in the show and allow them to interpret the music they like. The performance was repeated two times at the Art & ICT Exhibition 2013. It may contribute to open novel directions in both research and market niches for future User Centric Media. Next developments will focus on extraction of further social features, on query formulation, and on other kinds of capture devices (e.g., LeapMotion). For example, our performance is percussive feedbacks inducing users to perform rhythmic gestures to interact between them: other types of feedback, e.g., a violin melody, could be introduced to trigger fluent movements in the performers requiring the extraction of different expressive and social features. By adding new gestures types, users could implicitly choose their audio instrument and interpret the output melody, like in improvisational music. Results from the research described in this paper include new developments to the EyesWeb software platform, to the library for real-time analysis of human behaviour and to libraries for real-time processing of movement and audio content on mobiles. These improvements are currently exploited in scientific experiments and artistic projects in the framework of the EU Culture Project MetaBody[3].

Acknowledgements

This research originated from the results obtained in the EU FP7 ICT FET Project SIEMPRE, and is currently partially funded by the EU Culture Project MetaBody (Agreement Number 2013-1572/001-001). We thank our colleagues Paolo Coletta, Alberto Massari, and Simone Ghisio for their precious contributes to software development. We thank Corrado Canepa and Giacomo Lepri for their contribute to the design of the music content. Finally, we thank Teresa De Martino and Beatrice Marquez-Garrido for their kind support and for the permission to use in this paper their ICT2013 Vilnius photos of our demo.

References

[1] Lee, S.W., Essl, G. and Mao, Z.M. (2014) Distributing mobile music applications for audience participation using mobile ad-hoc network (manet). In *NIME'14 (New Interfaces for Musical Expression)*.

[2] Dahl, L., Herrera, J. and C., W. (2011) Tweetdreams: Making music with the audience and the world using real-time twitter data. In *NIME'11 (New Interfaces for Musical Expression)*.

[3] Camurri, A., Volpe, G., Vinet, H., Bresin, R., Fabiani, M., Dubus, G., Maestre, E. *et al.* (2010) User-centric context-aware mobile applications for embodied music listening. In *User Centric Media* (Springer), 21–30.

[4] Marsh, K., Richardson, M. and Schmidt, R. (2009) Social connection through joint action and interpersonal coordination. *Topics in Cognitive Science* **1**: 320–339.

[5] Bianchi-Berthouze, N., Kim, W.W. and Patel, D. (2007) Does body movement engage you more in digital game play? and why? In *Proceedings of Affective Computing and Intelligent Interaction*: 102–113.

[6] Lindley, S.E., Couteur, J.L. and Berthouze, N. (2008) Stirring up experience through movement in game play: effects on engagement and social behaviour. In *Proceedings of ACM Conference on Human Factors in Computing Systems*: 511–514.

[7] Mancas, M., Madhkour, R.B., De Beul, D., Leroy, J., Riche, N., Rybarczyk, Y.P. and Zajéga, F. (2011) Kinact: a saliency-based social game. In *Proceedings of the 7th International Summer Workshop on Multimodal Interfaces eNTERFACE11*, **8**.

[8] I. Laso-Ballesteros and P. Daras (Eds.), User Centric Future Media Internet, EU Commission, September 2008.

[9] Pentland, A. (2007) Social signal processing. *IEEE Signal Processing Magazine* **2**(4): 108–111.

[10] Hung, H., Huang, Y., Friedland, G. and Gatica-Perez, D. (2011) Estimating dominance in multi-party meetings using speaker diarization. *IEEE Transactions on Audio, Speech & Language Processing* **19**(4): 847–860.

[11] Hung, H. and Gatica-Perez, D. (2010) Estimating cohesion in small groups using audio-visual nonverbal behavior. *IEEE Transactions on Multimedia* **12**(6): 563–575.

[12] Vinciarelli, A., Pantic, M., Bourland, H. and Pentland, A. (2008) Social signals, their function, and automatic analysis: a survey. In *Proceedings of the 10th International Conference on Multimodal Interfaces, ICMI08*: 61–68.

[13] Lukowicz, P., Pentland, S. and Ferscha, A. (2012) From context awareness to socially aware computing. *Pervasive Computing, IEEE* **11**(1): 32–41.

[14] Dong, W., Lepri, B. and Pentland, A. (2011) Modeling the co-evolution of behaviors and social relationships using mobile phone data. In *Proceedings of the 10th International Conference on Mobile and Ubiquitous Multimedia* (ACM): 134–143.

[15] Jayagopi, D., Kim, T., Pentland, A. and Gatica-Perez, D. (2012) Privacy-sensitive recognition of group conversational context with sociometers. *Multimedia systems* **18**(1): 3–14.

[16] Bortz, B., Salazar, S., Jaimovich, J., Knapp, R. and Wang, G. (2012) Shemp: A mobile framework for shared emotion, music, and physiology. In *Proceedings of the Third International Workshop on Social Behaviour in Music, 14th ACM International Conference on Multimodal Interaction*.

[17] Blaine, T. and Fels, S. (2003) Collaborative musical experiences for novices. *Journal of New Music Research* **32**: 411–428.

[18] Jordà, S. (2005) Multi-user instruments: models, examples and promises. In *Proceedings of the 2005 conference on New interfaces for musical expression*, NIME '05 (Singapore, Singapore: National University of Singapore): 23–26.

[3]www.metabody.eu

[19] Jordà, S., Geiger, G., Alonso, M. and Kaltenbrunner, M. (2007) The reactable: exploring the synergy between live music performance and tabletop tangible interfaces. In *Proceedings of the 1st international conference on Tangible and embedded interaction*, TEI '07 (New York, NY, USA: ACM): 139–146.

[20] Patten, J., Recht, B. and Ishii, H. (2002) Audiopad: a tag-based interface for musical performance. In *Proceedings of the 2002 conference on New interfaces for musical expression*, NIME '02 (Singapore, Singapore: National University of Singapore): 1–6.

[21] Newman, B., Sanders, J., Hughes, R. and Jurdak, R. (2010) Tinytune, a collaborative sensor network musical instrument. In *Proceedings of the 8th ACM Conference on Embedded Networked Sensor Systems*, SenSys '10 (New York, NY, USA: ACM): 375–376.

[22] Myllykoski, M. and Paananen, P. (2009) Towards new social dimensions for children's music making - jammo as a collaborative and communal m-learning environment. In *Proceedings of the 7th Triennial Conference of European Society for the Cognitive Sciences of Music*, ESCOM 2009: 366–371.

[23] Tanaka, A. (2004) Malleable mobile music. In *Adjunct Proceedings of the 6th International Conference on Ubiquitous Computing (UBICOMP)*.

[24] Weinberg, G., Aimi, R. and Jennings, K. (2002) The beatbug network: a rhythmic system for interdependent group collaboration. In *Proceedings of the 2002 conference on New interfaces for musical expression* (National University of Singapore): 1–6.

[25] Camurri, A. and Ferrentino, P. (1999) Interactive environments for music and multimedia. *Multimedia Systems* **7**(1): 32–47.

[26] Lindström, E., Camurri, A., Friberg, A., Volpe, G. and Rinman, M.L. (2005) Affect, attitude and evaluation of multisensory performances. *Journal of New Music Research* **34**(1): 69–86.

[27] Leman, M., Demey, M., Lesaffre, M., van Noorden, L. and Moelants, D. (2009) Concepts, technology and assessment of the social music game 'sync-in team'. In *Proceedings of the 12th IEEE International Conference on Computational Science and Engineering*.

[28] Stockholm, J. and Pasquier, P. (2009) Reinforcement learning of listener response for mood classification of audio. In *Proceedings of the 12th IEEE International Conference on Computational Science and Engineering*.

[29] Vinyes, M., Bonada, J. and Loscos, A. (2006) Demixing commercial music productions via human-assisted time-frequency masking. In *Proceedings of the 120th AES Convention*.

[30] Fabiani, M., Dubus, G. and Bresin, R. (2011) MoodifierLive: Interactive and collaborative expressive music performance on mobile devices. *Proceedings of NIME 2011* : 116–119.

[31] Varni, G., Mancini, M. and Volpe, G. (2012) Embodied cooperation using mobile devices: presenting and evaluating the sync4all application. In *AVI*: 312–319.

[32] Camurri, A., Mazzarino, B. and Volpe, G. (2004) Expressive interfaces. *Cognition, Technology and Work* **6**: 15–22.

[33] Wallbott, H.G. and Scherer, K.R. (1986) Cues and channels in emotion recognition. *Journal of Personality and Social Psychology* **51**(4): 690–699.

[34] Wallbott, H.G. (1998) Bodily expression of emotion. *European Journal of Social Psychology* **28**: 879–896.

[35] Gallaher, P.E. (1992) Individual differences in nonverbal behavior: Dimensions of style. *Journal of Personality and Social Psychology* **63**(1): 133–145.

[36] Canazza, S., Poli, G., RodÃă, A. and Vidolin, A. (2003) An abstract control space for communication of sensory expressive intentions in music performance. *Journal of New Music Research* **32**(3): 281–294.

[37] Camurri, A., Volpe, G., De Poli, G. and Leman, M. (2005) Communicating expressiveness and affect in multimodal interactive systems. *Multimedia, IEEE* **12**(1): 43 – 53.

[38] Camurri, A., Mazzarino, B. and Volpe, G. (2004) Analysis of expressive gesture: The eyesweb expressive gesture processing library. *Lecture notes in computer science* .

[39] Glowinski, D., Mancini, M., Cowie, R., Camurri, A., Chiorri, C. and Doherty, C. (2013) The movements made by performers in a skilled quartet: a distinctive pattern, and the function that it serves. *Frontiers in Psychology* **4**(841).

5

Characterisation of gestural units in light of human-avatar interaction

I. Renna[1,*], S. Delacroix[2], F. Catteau[1], C. Vincent[1], D. Boutet[1]

[1]Structures Formelles du Langage, UMR 7023 (CNRS / Université Paris 8)

[2]Laboratoire d'Analyse du Mouvement, Institut National de Podologie, Paris

Abstract

We present a method for characterizing coverbal gestural units intended for human-avatar interaction. We recorded 12 gesture types, using a motion-capture system. We used the markers positions thus obtained to determine the gestural units after stroke segmentation. We complement our linguistic analysis of gestures with an elaboration of our biomechanical hypotheses, our method of segmentation, our characterization hypotheses and the results obtained.

Keywords: Gesture units, stroke segmentation, gesture characterization.

1. Introduction

Characterization of the meaning of gestures is traditionally based on body-oriented descriptions [36] capturing the gestural elements according to a global description of bodily reference points. We aim to show that the meaning of different Gestural Units (GUs) can be defined on the basis of forms along the upper limb, using multiple reference points that are not limited to body-orientation, but oriented via each of the segments (hand, forearm, arm), thus facilitating an automatic characterization of gestures to exploit in a human-avatar context.

This work takes place in the CIGALE project, whose final goal is to create novel human-avatar interactions in the context of theatrical performances starting from the off-line analysis, characterization and classification of 4 different datasets of gestures we recorded using a motion-capture system; one of these datasets is presented (Sec. 3) and exploited in this work. Once the off-line study is completed, the project will move to the on-line analyses in order to evaluate the actual possible interaction between humans and the avatar, whose behavior has been built exploiting the off-line gesture analysis.

After presenting the state of the art from a linguistic and engineering point of view (Sec. 2), we describe the database and the adopted biomechanical model (Sec. 3). Section 4 provides an overview of the linguistic framework within which our semantic characterization of co-verbal gestures is presented. The gesture significant part (*stroke* [31], Sec. 5) segmentation represents a necessary preliminary to such characterization, since it is impossible to characterize the meaning of a gesture without knowing when it occurs. For this purpose, an automatic segmentation is presented and tested against the segmentation of two coders, which serves as ground truth, in line with the highest standard methodes of both domains: robotics and linguistics (Sec. 5). Once this operation is validated, automatic characterization relies on centering with respect to the motion variation of the degree of freedom (DOF) – the prono-supination (Sec. 6). We conclude summing up our results and presenting future work (Sec. 7).

*Corresponding author. Email: ilaria.renna@gmail.com

2. State of the art

Gesture segmentation in concatenated units inspired by Kendon's work [25, 26, 30] is widely used: a Gesture Unit consists of a series of Gesture Phrases which are themselves composed by Gesture Phases. Each latter unit includes the core-meaning of the gesture, named *stroke* (see Sec. 5). Other studies try to set up syntagmatic rules system for movements phases available for both gesture and sign of sign languages [31]. Minor differences exist between these approaches. In the linguistics part of the present study, we adopt Kendon's terminology.

According to their meaning or function, gestures are classified in several categories. McNeill [37], following overall Kendon's classification [27], differentiates gestures in beats, which punctuate the discourse, deictics which includes pointing gestures, iconics which are "images of concrete entities and/or actions" and metaphorics which show "images of the abstract".

Our study concerns gestures which belong to both iconic and metaphoric categories. We adopt Kendon's type gesture of quotable gestures defined as "those standardized gestures which have fairly stable meanings within a given community and which, on the whole, appear to serve in the place of a complete speech act of same sort" [28].

Gestures/actions segmentation is necessary to cut streams of motions into single instances that are consistent to the set of initial model hypotheses and that can be used as training sequences for recognition. In computer vision, different techniques are used to prepare data for gesture recognition and the segmentation concerns image processing methods to extract features [12, 13, 33, 48] to represent the spatial structure of gestures. Such features are then exploited to learn the temporal structure of gestures with different methods, e.g. Hidden Markov Models (HMMs) [7, 20, 50, 54], Baum-Welch [3], parametric models [21, 56] and others.

When temporal segmentation is needed, different kinds of methods can be used to investigate motion profiles and trajectories to recognize human gestures. A general approach for segmenting actions is based on *concatenating action grammars* to model transitions in a gesture or between consecutive gestures [53]. Concatenative grammars can be built, for instance, by joining all models in a common start and end node and by adding a loop-back transition between these two nodes; segmentation and labeling of a complex action sequence is then computed as a minimum-cost path trough the network using dynamic programming techniques. Some works [34, 43] use such networks for action recognition based on HMMs, others [38, 48] on Conditional Random Fields (CRFs) or on semi-Markov models [46].

Another strategy for recognizing gestures consists in dividing video sequences into multiple, overlapping segments, using a *sliding window*; classification is then performed sequentially on all the candidate segments, and peaks in the resulting classification scores are interpreted as gesture locations. Sliding window are used in many template-based representations [17, 59], in combination with dynamic time warping (DTW) [13, 39] and even grammars [4]. For example, Abdelkader et al. [1] propose a template-based approach using DTW to align the different trajectories using elastic geodesic distances on the shape space; the gesture templates are then calculated by averaging the aligned trajectories.

A common strategy is to use a generic segmentation method based on detecting *motion boundaries*, then separately classifying the resulting segments. Such motion boundaries are typically defined as discontinuities and extrema in acceleration, velocity, or curvature of the observed motions. For example, Ogale et al. [40] segment action sequences by detecting minima and maxima of optical flow inside body silhouettes; Zhao et al. [58] calculate velocity and treat local minima in the velocity as gesture boundaries; Wang et al. [51] treat local minima in acceleration as a gesture boundary, allowing them to construct a motion alphabet whose "characters" of this motion are then combined using a HMM; Kahol et.al. [24] tested a user centric gesture segmentation algorithm and developed observer profiles based on how individual users segment motion sequences, encoding gesture boundaries as a binary vector of hierarchically connected body segment activities. Boundary detection methods are attractive because they provide a generic segmentation of the video, which is not dependent on the gestures classes; some precautions are needed because they are not stable across viewpoints and they are easily confused by the presence of multiple, simultaneous movements.

Movements primitives can also be extracted as joint trajectories using Principal Component Analysis (PCA) [18, 23]. In Lim et al. [32] each movement primitive is represented and stored as a set of joint trajectory basis functions that are then extracted via a PCA of human motion capture data. In [2], gestures computed from inertial sensors are defined by hand paths as a discrete time sequence in the Cartesian space; these are converted to training functional data by basis function expansions using B-splines (curve fitting), and then Functional Principal Component Analysis (FPCA) is performed on all the training data to determine a finite set of functional principal components (FPCs) that explain the modes of variation in the data.

Other sensors than those exploited in computer vision or in motion capture based approaches can be used: in [57], for example, accelerometers and multichannel electromyography (EMG) signals are used for segmentation.

As we want to validate our gesture characterization in an off-line situation, we decided to exploit a simple but robust boundary method for gesture segmentation based on arm movement considerations (see Sec. 5).

3. The dataset

The dataset examined is composed of 91 isolated coverbal symbolic gestures. These coverbal gestures are semantically autonomous and cover all DOF of the upper limb (see 4.1). Some gestures are performed using the entire upper limb, while others employ a subpart only (for example only the fingers) for a total of 150 different gestures reproduced by one person.

3.1. The marker-set and the biomechanical model

Gestures are collected using a 3D motion-capture system of digital infra-red cameras, which ensures the reliability of our understanding of the gesture and the effectiveness of the characterization of avatar motion. The system uses hemispherical reflectors glued to the skin and records their trajectory.

A list of cutaneous markers is established, in order to model the body segments in three dimensions (marker-set of 90 points, Fig. 1). This list references the anatomical positions that should be used in modelling each segment as a rigid body.

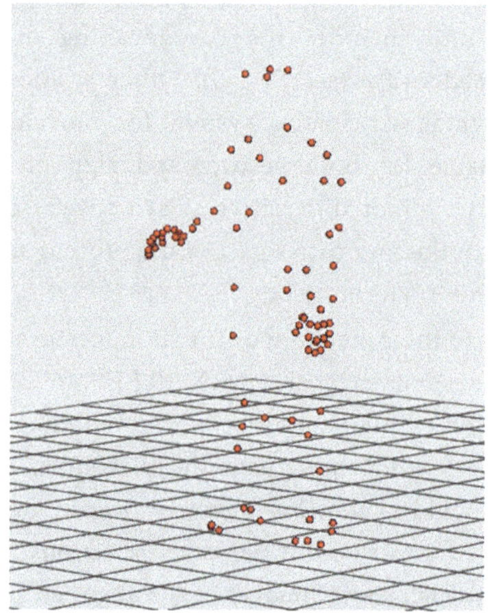

Figure 1. Marker-set visualization.

Generally, three non-aligned anatomical reference points are sufficient to define a segment. In our model (Fig. 2), the torso, arm, forearm, and hand segments have been defined based on coordinates of spatial perspective using a standardized method. This method enables the creation of three orthogonal axes for each system of segment coordinates [15, 55]. It involves calculation of the centers of the wrist, elbow, and shoulder joints, as well as those of the cervical and lumbar regions [15]. For the hand, the origin of the coordinate system is the centre of the wrist joint, Y is the unitary vector connecting the centre of the 2nd and 5th metacarpal heads to the origin, X is the normal unitary vector containing the origin and the 2nd and 5th metacarpal heads, Z is the vector result of axes X and Y. For the forearm, the origin of the coordinate system is the centre of the elbow joint, Y is the unitary vector connecting the centre of the wrist joint to the origin, X is the unitary vector normal to the plane containing the origin and the styloid processes of the ulna and radius, Z is the vector result of axes X and Y. For the arm, the origin of the coordinate system is the centre of the shoulder joint, Y is the unitary vector connecting the centre of the elbow joint to the origin, X is the unitary vector normal to the plane containing the origin, the epicondyle and the epitrochlea, Z is the vector result of axes X and Y. For the torso, the origin of the coordinate system is the centre of the cervical joint, Y is the unitary vector connecting the lumbar joint to the origin, Z is the unitary vector normal to the plane connecting the origin, the lumbar joint and the suprasternal space, X is the vector result of axes Y and Z.

The coordinate system of each joint is defined through sets of adjacent segment coordinates, allowing the description of the three-dimensional articulation of the shoulder, the elbow, and the wrist at every moment of the gesture. To establish the kinematics of the joints, we used a sequence of successive rotations around the mobile axes, using Euler angles [55]. The dynamic sequence of rotations enables the definition of joint coordinates through the axes of two adjacent segments: one axis for the proximal segment and another for the distal segment; and a floating axis, perpendicular to the other two.

The various joint movements of the wrist, elbow and shoulder are calculated thanks to this biomechanical model, as are the palmar/dorsal flexion and the adduction/abduction of the wrist. These correspond to the flexion/extension and adduction/abduction of the hand as described in the action schemas (Sec. 4). The extension/flexion and supination/pronation of the elbow correspond to the extension/flexion of the forearm and supination/pronation of the hand respectively for the action schemas (see 4.2). Finally, shoulder motion is measured in retropulsion/forward flexion, abduction/adduction and internal/external rotation. These correspond, respectively, to the extension/flexion, abduction/adduction and external/internal rotation of the arm for action schemas.

4. Linguistic redefinition in light of human-avatar interaction

The recorded coverbal gestures match emblematic quote gestures [27, 42], i.e. semantically autonomous gestures, whose significance is independent of the surrounding discourse. The 91 gestures can be

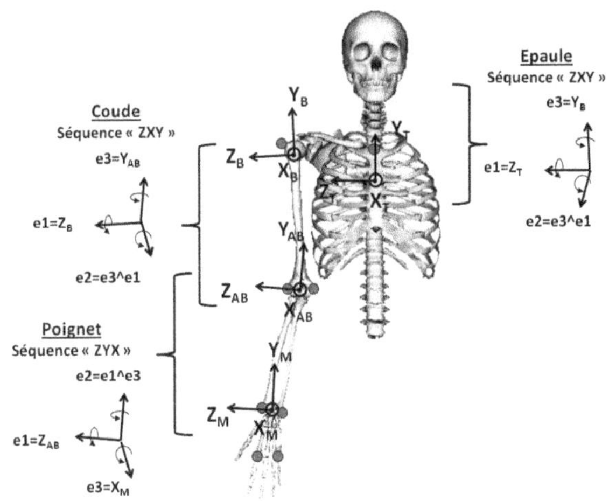

Figure 2. The adopted biomechanical model.

divided into a dozen GUs, with the following senses: reject, refuse, despise, discredit, pass, accept, consider something, consider someone, offer, not care, commit, revere. These semantic labels have been tested and validated with a French-speaking population in a previous study [6].

Each GU corresponds to a particular action schema implementing some (or all) of the segments of the upper limb. Action schemas are based on the motion of various DOF of the segments of the upper limb in a specific order. This order emerges from the difference in the range of motion of each DOF involved in the schema according to its range of motion. Motion is transferred through moments of inertia attached to each DOF and as a function of (involuntary) conjoint movement of the longitudinal axis (exterior/interior rotation or pronation/supination) associated with any joint with two DOF [9, 10, 35]. Thus, for the GU "refuse", for example (Fig. 3), the action diagram shows the hand motion towards the forearm.

4.1. Flow of motion propagation

In the action diagram, the position of the pole of adduction (motion towards the joint on the plane of the palm) determines the direction of motion propagation. If a movement of adduction is in first or second position, the flow of motion is distal-proximal, going from the hand towards the forearm. If adduction is in third position, then it is the result of the first two, and so does not present significant motion.

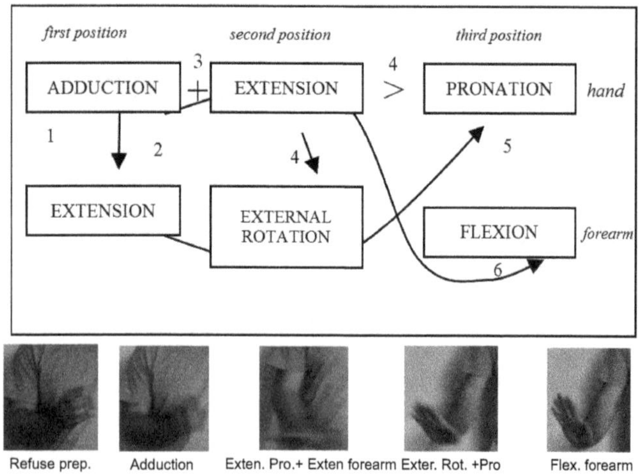

Figure 3. Action schema for the GU "refuse" . This gesture begins with a movement of the hand (Adduction, 1). The order of the motion is numbered. The photograms illustrate the execution of the gesture at different moments of what is gathered in the action schema. The first photogram capture the preparation phase before the stroke.

Therefore, the gesture is initiated on the forearm and spreads towards the hand in a distal-proximal flux. We thus define two types of GUs. The first 8 GUs in the list above are built on the hand while the last 4 (offer, not care, protect, revere) are built on the arm.

4.2. Action schema of hand motions

The sequence of hand motions is based on a structure such that the motion or position of the first two DOF cause involuntary motion of the third DOF. This third motion is either the result of a biomechanical constraint related to motion around the longitudinal axis (pronation/supination), or to a sequence based on the moment of inertia. In both cases, the poles of motion in third position are completely determinable and follow the first two movements such that their sequence affects the pole of the third motion. So, the sequence ADD.EXTEN leads to involuntary PRONATION, while the reverse order, EXTEN.ADD leads to SUPINATION [5, 6].

4.3. Grouping GUs by direction

Tracking the order of the poles in motion is affected by the range of motion, the temporal sequence of the emergence of motion, the initial position and the acceleration, but these criteria, which vary even among themselves, are difficult to hierarchise. On the other hand, it is possible to classify GUs on a formal basis by semantic field (Fig. 4).

Initially, it is necessary to determine the spread of motion; either the gesture starts from the hand and motion goes up the forearm, or it starts in the arm and spreads towards the hand (hand and arm in the diagram). For the hand (Fig. 4, left), the initial prono-supination of the gesture may be marked or unmarked. At the next level, we examine prono-supination with respect to the initial position. This gives us 8 manual action schemas. For the arm (Fig. 4, right), we examine the ADD/ABD position or motion of the arm. Subsequently, the 4 GUs of the arm can be distinguished through prono-supination.

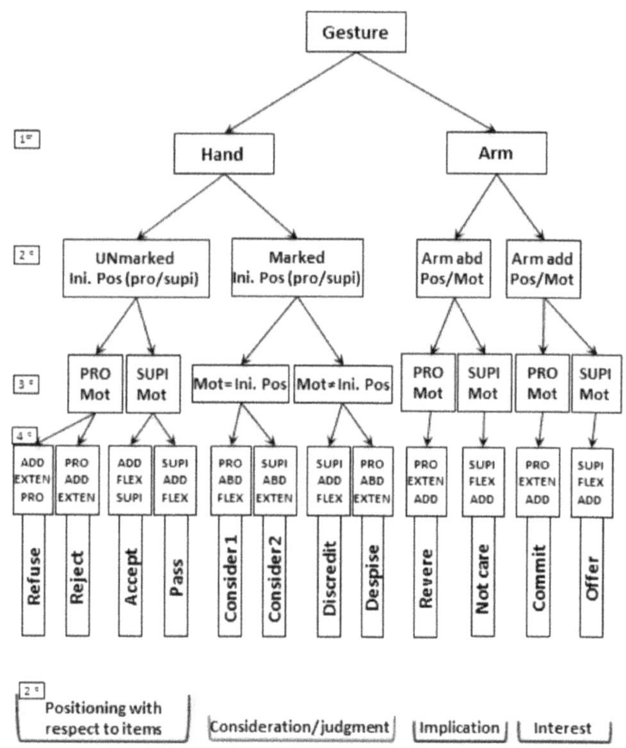

Figure 4. Diagram presenting the formal presentation of gestures according to semantic characterization.

Each of these GUs has a semantic label. A first level of hyperonymic grouping includes 4 semantic sets: i) positioning with respect to items, ii) consideration or judgment, iii) implication, and iv) interest. This semantic level corresponds to the 2nd level of formal disjunction in the diagram. Another possibility is a two part semantic grouping corresponding to the first formal disjunction (hand or arm): positioning in the world *versus* relational positioning. Therefore, various levels of formal differentiation correspond to specific semantic labeling.

5. Segmentation of gestural signals

Each gestural signal in our database is composed of a sequence: T-pose, gesture, T-pose. Extraction of the gesture requires automatic segmentation. The generally accepted sequence includes 4 phases [11, 36]: 1. resting position; 2. preparation (pre-stroke); 3. core (stroke); 4. retraction (post-stroke). This sequence describes the structure of a gesture. However, it is impossible to find an automatic, objective criterion to extract the stroke, the semantically significant part. This is a complex operation even for a human, and remains uncertain [47].

In our case, segmentation is carried out on the basis of morpho-kinetic properties (as defined by Kendon [29]). Indeed, the preparation of motion consists of a ballistic motion that brings the arm(s) to the core of the motion [8]. This ballistic motion involves acceleration followed by deceleration as the final position is approached, then symmetrical acceleration and deceleration to the first set, and a return to the resting position. T-poses are also characterized by acceleration and deceleration of movement.

In order to extract the stroke of each gesture, we consider the absolute value of the derivative of the Y index positions (seen in all cases as the body part that moves the most): the minimum of this signal represents the transition between acceleration and deceleration. So for automatic segmentation the stroke considered is the part between the minimal phase that precedes the second maximum (property of the beginning of a stroke) and the minimal phase that follows the penultimate maximum (end of a stroke) (Fig. 5). A threshold is set up to avoid minimal and maximal phases due to noise (small adjustment or preparatory motions) from being considered.

5.1. Segmentation evaluation

To evaluate automatic segmentation methods, it is necessary to compare an automatic segmenter's performance against the segmentations produced by human judges (coders). In general, methods for performing this comparison designate as comparison reference only the segmentation of a single coder [44]. However, this approach assumes that the only coder is unbiased and able to provide a perfect segmentation. Indeed, previous works, e.g. [22], showed that inter-annotator agreement between human coders can be rather poor. Thus, an automatic segmenter should be compared directly against different coders [19] to ensure that it does not over-fit to the preference and bias of one particular coder.

Given our dual aim, to evaluate inter-annotator agreement on the one hand and automatic segmentation on the other, we decided to adopt two methods: Accurate Temporal Segmentation Rate (ATSR) [45] and F-score [49]. ATSR is a time-based metric that measures performance in terms of accurately detecting the beginning and end of the stroke for each gesture signal. F-score provides more information than accuracy and enables individuated errors typologies. Three different cases are evaluated with both methods:

1. automatic segmentation is compared with the annotator considered as ground truth (case 1);
2. automatic segmentation is compared to a second annotator (the ground truth) (case 2);
3. the two annotators are compared against one another (case 3).

For each considered gesture, the ATSR was computed as follows: the Absolute Temporal Segmentation Error (ATSE) is evaluated by summing the absolute temporal error between the ground truth and the result of the algorithm for the start and stop event and dividing this sum by the total length of the gesture occurrence measured from the ground truth as formalized in Equation 1. Once the ATSE are calculated, ATSR metrics are computed by subtracting the average ATSE to 1 in order to obtain the accuracy rate as shown in Equation 2. A perfectly accurate segmentation produces an ATSR of 1.

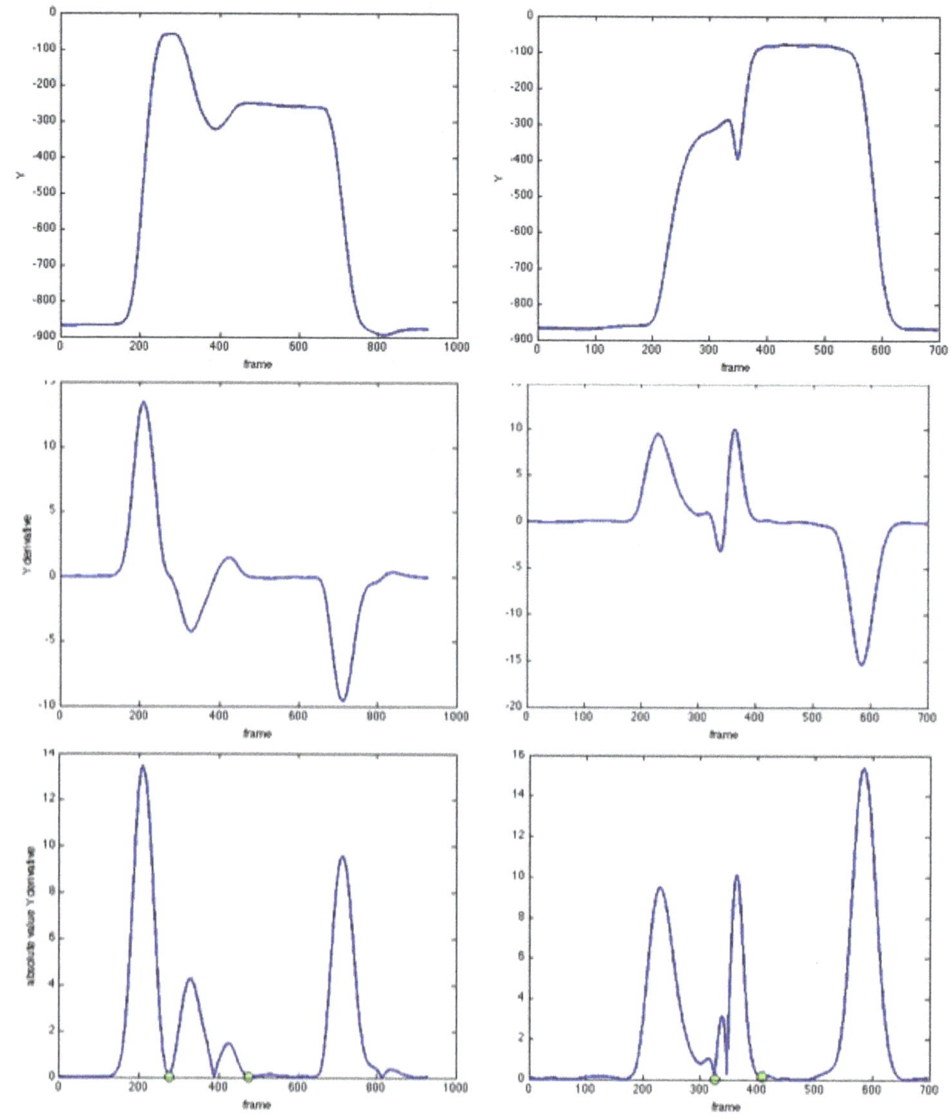

Figure 5. On the left, the signals of the gesture "revere". On the right, the signals for "accept". From top to bottom: positions of the index for Y, speed (derivative) and absolute value of the speed with automatic individuation of the stroke between two green points.

$$ATSE = \frac{| Start_{GT} - Start_{Alg} | + | Stop_{GT} - Stop_{Alg} |}{| Stop_{GT} - Start_{GT} |} \tag{1}$$

$$ATSR = 1 - \frac{1}{n} \sum_{i=1}^{n} ATSE(i) . \tag{2}$$

Equation 1 counts differences that occur frame by frame, so an error is taken into account even when annotations differ for just a few frames. To limit this effect and so to avoid small ground truth timing errors producing irrelevant penalties during the computation of the ATSE [45], it is possible to fix a toleration value α so that

$$\text{if } ATSE(i) < \alpha \text{ then } ATSE(i) = 0 . \tag{3}$$

As a stroke is in general of about 100 frames, we took $\alpha = 0.2$. This corresponds to a global difference of $\alpha * 100 = 20$ frames (around $0.17s$, given the acquisition rate is $120f/s$) compared to the duration of the ground truth, which is an adequate choice considering that, on average, it is easy to have 10-frame-errors for each start or stop. In case 1 we obtain $ATSR = 0.6038$, in case 2 $ATSR = 0.5857$ and in case 3 $ATSR = 0.8707$.

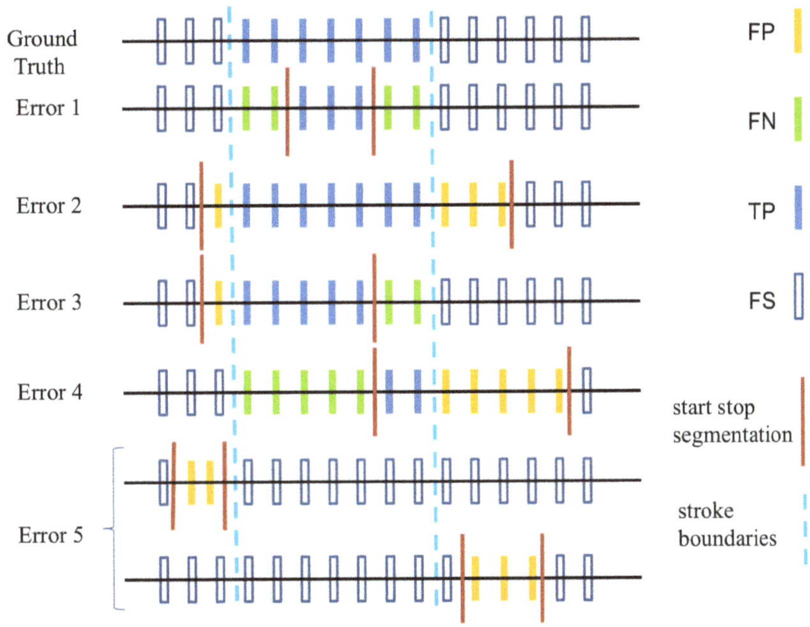

Figure 6. Different possible segmentation errors. Here, FP = False Positive, FN= False Negative, TP= True Positive and FS=outside segmentation.

This kind of method lacks completeness as it does not categorize the errors [52]. In fact, the errors can be categorized into 5 types, as shown in Figure 6.

It is, of course, quite important to know whether the automatic segmentation is wrong but the stroke is preserved (Error 2 in Figure 6) or cut out (all other cases in the figure). In order to assess the quality of our segmentation and of inter-annotator agreement, let us consider precision (p) and recall (r) [16, 41]: precision is the fraction of detections that are true positives rather than false positives (Equation 4), while recall is the fraction of true positives that are detected rather than missed (Equation 5). In probabilistic terms, precision is the probability that detection is valid, and recall is the probability that ground truth data was detected:

$$p = \frac{TP}{TP + FP} \quad (4) \qquad r = \frac{TP}{TP + FN}. \quad (5)$$

Precision and Recall can be combined in the F-score as follows:

$$F_\beta = (1 + \beta^2) * \frac{p * r}{\beta^2 p + r}. \quad (6)$$

When the parameter $\beta = 1$, F-score is said to be balanced and written as F_1:

$$F_1 = 2 * \frac{p * r}{p + r}. \quad (7)$$

The F_1 score can be seen as a weighted average of precision and recall; F_1 score reaches its best value at 1 and worst one at 0. The obtained results are summed up in Table 1. In general, high F_1 values are obtained; r is higher than p in the comparison with automatic segmentation meaning that the algorithm returned most of the relevant results, while p is higher for the inter-annotator agreement: they obtained more relevant than irrelevant agreement. Results concerning the kinds of errors are presented in Table 2.

It is worth highlighting that in Cases 1 and 2, Error 2 is the most frequent. This means that the segmentation method preserves the strokes despite the error. Moreover, the lower error is the cut stroke (Error 1). We can therefore assume that the presented segmentation method is robust to analyze the presented gestures model. For the inter-annotator agreement, we note that most errors stem from one annotator cutting the stroke or because one anticipated the other (Error 3).

Table 1. Results obtained for the three cases of study.

	p	r	F_1
Case 1	0.7430	0.92162	0.8227
Case 2	0.7358	0.9151	0.8157
Case 3	0.9077	0.9053	0.9065

Table 2. Errors occurred in the cases of study.

	Case 1	Case 2	Case 3
Error 1	4	5	17
Error 2	53	49	15
Error 3	24	16	54
Error 4	10	21	5
Error 5	0	0	0

6. Action schema component properties

To characterize the action schemas for each of the coverbal gestures recorded, the segmented signals are transformed into kinematic data in accordance with the biomechanical model (Sec. 3.1). The motion of different degrees of freedom for each joint in the right upper limb (shoulder, elbow, and wrist) is taken into account and normalized temporally on 101 points [14]. Our starting point for this characterization is the assumption that in human-human communication, the motion of DOF is most visible in prono-supination regardless of the type of gesture (performed on the entire upper limb or only one of its segments). We therefore decided to focus the analysis, initially, on the signals that were temporally aligned with the prono-supination zone, which contains the widest variation. Biomechanical parameters, such as initial and final positions of each DOF and their maximal range, were taken into account (Fig.7).

We analyzed gestures involving the motion of all segments of the upper limb (33 of the 91 gestures captured). The stages considered in the decision tree (diagram in Fig. 4) move from i) the 1st node (manual or brachial flow) to ii) the 4th node (separation of gestures by prono-supination).

The first stage involves determination of the flow of motion from the arm (proximal-distal) or from the hand (distal-proximal). We therefore calculated: 1) for the initial position, the moment in which the min. and max. of each DOF appear within the automatically cut stroke; 2) the temporal difference between the

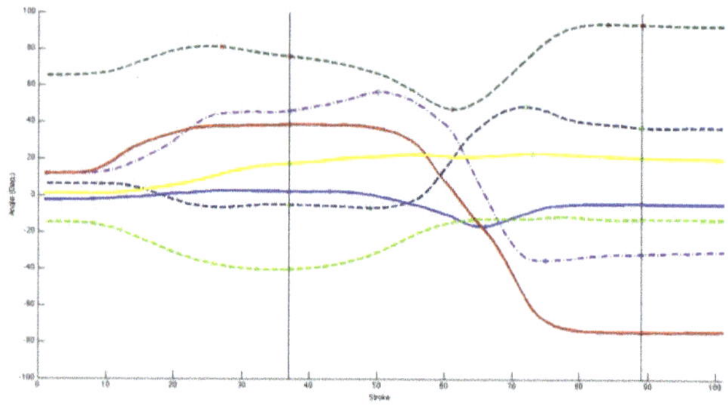

Figure 7. The signals for different segments of the upper limb. For the hand, supination/pronation in red, abduction/adduction in sky blue and dotted purple for flexion/extension. For the forearm, extension/flexion in dotted dark green. For the arm, internal/external rotation in dotted dark blue, abduction/adduction in dotted green and extension/flexion in yellow. Vertical lines indicate the highest variation in prono-supination.

min. and the max. of the DOF from one segment to another (arm [offer, not care, commit and revere], forearm and hand [for all the other gestures]). Within the latter calculation, the choice of the min. or max. value for one DOF or another correspond to the initial position, therefore a priori opposed to the pole seen in motion during the stroke. If hand motion is EXTEN (positive value), then the initial position corresponds to a minimum (flexion, negative value). Thus, for example, for the top line of diagram in Fig. 3, which illustrates the poles in motion in the "refuse" gesture, ADD.EXTEN >PRO, the initial position chosen was the max. value of the ADD/ABD, the min. value of the FLEX/EXTEN and the min. value of the SUPI/PRO. We set a minimal threshold of 10 frames, corresponding to 2 running video frames at 25f/s, for a temporal difference that enables the determination of the flow.

6.1. Discussion

Determination of flow using this method (for the 33 gestures tested, covering the 12 GUs presented in Sec. 4; we underline that each GU occurred between 2 and 3 times), is conformed to expectations in 87.88% of cases. Three of the four cases not validated were below the 10-frame threshold and therefore meet no determinable flow; a further case (a realization of "revere") shows a flow reverse to expectations.

The fourth stage of characterization – wherein poles in motion are used to determine the action schema – was conducted with two types of data (a. and b. below).

Initial calculations concern the average of two or three realizations by GUs (33 gestures in total), thus covering the 12 averaged GUs for which moving poles and semantic labels are already known. We then determine the maximum range for each DOF, either a. within the boundary of prono-supination as shown in Fig. 7, or b. a wider range, starting from the stroke and calculating the difference between the final and initial position of each DOF. We thus obtained the motion poles for all DOF that characterize the GUs, that is 60 DOF for 12 GUs. Results for the first type of data (a. in the demarcation of prono-supination) show a recognition rate of 76.67%. The other option (b. in the stroke with the difference in initial and final position) shows much better characterization ratios 90%. Out of 60 DOF, the opposite pole appears only for 6 expected poles. In both options, an average of 6 DOF measured per GU, the pole that was most prone to error was the hand in ABD/ADD (36% error in a., 67% error in b.). This pole also shows the smallest range (25° and 35°).

Intermediary stages of the characterization (2 and 3 diagram Fig. 4) involve - marking of the initial positions of prono-supination and ABD vs. ADD motion of the arm (stage 2); - determination of the initial position and the identical vs. opposed motion of prono-supination and the motion pole between PRO and SUPI (stage 3).

The characterization of ABD vs ADD of the arm and PRO vs SUPI is unproblematic. In contrast, marking the initial positions of prono-supination (stage 2) does not give the expected results. In this case, only the range difference of the interior/exterior rotation between the beginning and end of the stroke is significant. For a confidence interval of 95%, there is no overlap between "reject/refuse" on the one hand and "despise" on the other. For the trio "pass/accept/discredit" non-overlap was also checked. Thus, the DOF marking criterion (PRO/SUPI) of stage 2 should be modified into a differential of the range of rotation in relatively marked EXT/INT. For step 3, the identity or opposition between the initial position and prono-supination is a good criterion, since, with a confidence interval of 95%, there is no overlap between "reject/refuse/despise" on the one hand, and "consider something" on the other. The same is true between "pass/accept/discredit" on the one hand, and "consider someone" on the other.

All in all, the only phases which are not fully satisfactory are phases 1 (with a single case of inversion for "revere") and 4 (with 90% of expected poles). Intermediary steps are 100% reliable.

7. Conclusion

In this study, we have presented a method for the characterization of 12 gestural units involving the upper limb. A motion-capture system was used to build a reliable gesture database to prove that the meaning of different Gestural Units can be defined on the basis of forms along the upper limb, using multiple reference points, that are not limited to body-orientation, but oriented via each of the segments (hand, forearm, arm). For this purpose an automatic segmentation was exploited. Tests of the segmentation protocol demonstrate its robustness in the individuation of the stroke necessary for the characterization of gestures.

Simple characterization methods fulfill the requirement to associate each stage with formal semantic tagging. This is so since GUs that share the same poles differ only in the sequence in which these poles appear in the action schema. However, we have yet to characterize 4 of these. In this study, we cannot separate 'refuse' from 'reject' and 'accept' from 'pass'. Still both groups can be labeled: on the one hand, negative positioning with respect to things, and on the other, the same type of positioning, only positive. Consequently, all gestures are semantically associated with a variable granularity.

As this characterization has been conducted in light of a human-avatar interaction with an off-line system, the next step is to test the presented method in an on-line set-up using a simpler capture system, namely a kinect, on different subjects reproducing the presented kind of gesture also in other form as, for example, involving only a single moving segment (e.g., the hand).

References

[1] Abdelkader, M.F., Abd-Almageed, W., Srivastava, A. and Chellappa, R. (2011) Silhouette-based gesture and action recognition via modeling trajectories on riemannian shape manifolds. *Computer Vision and Image Understanding* **115**(3): 439–455.

[2] Aleotti, J., Cionini, A., Fontanili, L. and Caselli, S. (2013) Arm gesture recognition and humanoid imitation using functional principal component analysis. In *Intelligent Robots and Systems (IROS), 2013 IEEE/RSJ International Conference on* (IEEE): 3752–3758.

[3] Alon, J., Athitsos, V., Yuan, Q. and Sclaroff, S. (2009) A unified framework for gesture recognition and spatiotemporal gesture segmentation. *Pattern Analysis and Machine Intelligence, IEEE Transactions on* **31**(9): 1685–1699.

[4] Bobick, A.F. and Ivanov, Y.A. (1998) Action recognition using probabilistic parsing. In *Computer Vision and Pattern Recognition, 1998. Proceedings. 1998 IEEE Computer Society Conference on* (IEEE): 196–202.

[5] Boutet, D. (2008) Une morphologie de la gestualité: structuration articulaire. *Cahiers de linguistique analogique* (5): 81–115.

[6] Boutet, D. (2010) Structuration physiologique de la gestuelle: modèle et tests. *Lidil. Revue de linguistique et de didactique des langues* (42): 77–96.

[7] Brand, M., Oliver, N. and Pentland, A. (1997) Coupled hidden markov models for complex action recognition. In *Computer Vision and Pattern Recognition, 1997. Proceedings., 1997 IEEE Computer Society Conference on* (IEEE): 994–999.

[8] Chellali, R. and Renna, I. (2012) Emblematic gestures recognition. In *ASME 2012 11th Biennial Conference on Engineering Systems Design and Analysis* (American Society of Mechanical Engineers): 755–761.

[9] Cheng, P.L. (2006) Simulation of codman's paradox reveals a general law of motion. *Journal of biomechanics* **39**(7): 1201–1207.

[10] Codman, E.A. (1984) *The shoulder: rupture of the supraspinatus tendon and other lesions in or about the subacromial bursa* (RE Kreiger).

[11] Corradini, A. et Cohen, P.R. (2002) Speech-gesture interface for handfree painting on a virtual paper suing partial neural networks as gesture recognizer. In *Proceedings IJCNN'02*: 2293–2298.

[12] Cutler, R. and Turk, M. (1998) View-based interpretation of real-time optical flow for gesture recognition. In *2013 10th IEEE International Conference and Workshops on Automatic Face and Gesture Recognition (FG)* (IEEE Computer Society): 416–416.

[13] Darrell, T. and Pentland, A. (1993) Space-time gestures. In *Computer Vision and Pattern Recognition, 1993. Proceedings CVPR'93., 1993 IEEE Computer Society Conference on* (IEEE): 335–340.

[14] Desroches, G., Dumas, R., Pradon, D., Vaslin, P., Lepoutre, F.X. and Chèze, L. (2010) Upper limb joint dynamics during manual wheelchair propulsion. *Clinical Biomechanics* **25**(4): 299–306.

[15] Dumas, R., Cheze, L. and Verriest, J.P. (2007) Adjustments to mcconville et al. and young et al. body segment inertial parameters. *Journal of biomechanics* **40**(3): 543–553.

[16] Fawcett, T. (2006) An introduction to roc analysis. *Pattern recognition letters* **27**(8): 861–874.

[17] Feng, Z. and Cham, T.J. (2005) Video-based human action classification with ambiguous correspondences. In *Computer Vision and Pattern Recognition - Workshops, 2005. CVPR Workshops. IEEE Computer Society Conference on*: 82–82. doi:10.1109/CVPR.2005.549.

[18] Fod, A., Matarić, M. and Jenkins, O. (2002) Automated derivation of primitives for movement classification. *Autonomous Robots* **12**(1): 39–54. doi:10.1023/A:1013254724861, URL `http://dx.doi.org/10.1023/A%3A1013254724861`.

[19] Fournier, C. and Inkpen, D. (2012) Segmentation similarity and agreement. In *Proceedings of the 2012 Conference of the North American Chapter of the Association for Computational Linguistics: Human Language Technologies* (Montréal, Canada: Association for Computational Linguistics): 152–161. URL `http://www.aclweb.org/anthology/N/N12/N12-1016`.

[20] Françoise, J. (2011) *Realtime Segmentation and Recognition of Gestures using Hierarchical Markov Models*. Master atiam, UPMC - Ircam, Paris. URL `http://articles.ircam.fr/textes/Francoise11a/index.pdf`.

[21] Gritai, A., Sheikh, Y. and Shah, M. (2004) On the use of anthropometry in the invariant analysis of human actions. In *Proceedings of the Pattern Recognition, 17th International Conference on (ICPR'04) Volume 2 - Volume 02*, ICPR '04 (Washington, DC, USA: IEEE Computer Society): 923–926. doi:10.1109/ICPR.2004.652, URL `http://dx.doi.org/10.1109/ICPR.2004.652`.

[22] Hearst, M.A. (1997) Texttiling: Segmenting text into multi-paragraph subtopic passages. *Comput. Linguist.* **23**(1): 33–64. URL `http://dl.acm.org/citation.cfm?id=972684.972687`.

[23] Jenkins, O.C. and Mataric, M.J. (2002) Deriving action and behavior primitives from human motion. In *In International Conference on Intelligent Robots and Systems*: 2551–2556.

[24] Kahol, K., Tripathi, P. and Panchanathan, S. (2004) Automated gesture segmentation from dance sequences. In *Automatic Face and Gesture Recognition, 2004. Proceedings. Sixth IEEE International Conference on*: 883–888. doi:10.1109/AFGR.2004.1301645.

[25] Kendon, A. (1972) Some relationships between body motion and speech. an analysis of an example. *Studies in dyadic communication* .

[26] Kendon, A. (1980) Gesticulation and speech: Two aspects of the process of utterance. *The Relation Between Verbal and Nonverbal Communication* .

[27] Kendon, A. (1988) How gestures can become like words. *Cross-Cultural Perspective in Nonverbal Communication* .

[28] Kendon, A. (1990) Gesticulation, quotable gestures, and signs. *Senri ethnological studies* .

[29] Kendon, A. (1996) An agenda for gesture studies. *Semiotic Review of Books* .

[30] Kendon, A. (2004) *Gesture: Visible Action as Utterance* (Cambridge University Press).

[31] Kita, S., Van Gijn, I. and Van der Hulst, H. (1998) *Gesture and Sign Language in Human-Computer Interaction* (Springer).

[32] Lim, B., Ra, S. and Park, F. (2005) Movement primitives, principal component analysis, and the efficient generation of natural motions. In *Robotics and Automation, 2005. ICRA 2005. Proceedings of the 2005 IEEE International Conference on*: 4630–4635. doi:10.1109/ROBOT.2005.1570834.

[33] Lv, F. and Nevatia, R. (2007) Single view human action recognition using key pose matching and viterbi path searching. In *Computer Vision and Pattern Recognition, 2007. CVPR '07. IEEE Conference on*: 1–8. doi:10.1109/CVPR.2007.383131.

[34] Lv, F. and Nevatia, R. (2006) Recognition and segmentation of 3-d human action using hmm and multi-class adaboost. In *Proceedings of the 9th European Conference on Computer Vision - Volume Part IV*, ECCV'06 (Berlin, Heidelberg: Springer-Verlag): 359–372. doi:10.1007/11744085_28, URL http://dx.doi.org/10.1007/11744085_28.

[35] MacConaill, M.A. (1948) The movements of bones and joints. *Journal of Bone & Joint Surgery* **30-B**(2): 322–326.

[36] McNeill, D. (1992) *Hand and Mind: What Gestures Reveal about Thought* (University of Chicago Press). URL http://books.google.fr/books?id=3ZZAfNumLvwC.

[37] McNeill, D. (2008) *Gesture and Thought*, Phoenix Poets Series (University of Chicago Press). URL http://books.google.fr/books?id=N0SmyU4TKRwC.

[38] Morency, L.P., Quattoni, A. and Darrell, T. (2007) Latent-dynamic discriminative models for continuous gesture recognition. In *CVPR* (IEEE Computer Society). URL http://dblp.uni-trier.de/db/conf/cvpr/cvpr2007.html#MorencyQD07.

[39] Morguet, P. and Lang, M. (1998) Spotting dynamic hand gestures in video image sequences using hidden markov models. In *Image Processing, 1998. ICIP 98. Proceedings. 1998 International Conference on*: 193–197 vol.3. doi:10.1109/ICIP.1998.999009.

[40] Ogale, A.S., Karapurkar, A., Guerra-filho, G. and Aloimonos, Y. (2004) View-invariant identification of pose sequences for action recognition. In *In VACE*.

[41] Olson, D.L. and Delen, D. (2008) *Advanced Data Mining Techniques* (Springer Publishing Company, Incorporated), 1st ed.

[42] Payrató, L. (1993) A pragmatic view on autonomous gestures: A first repertoire of catalan emblems. *Journal of Pragmatics* **20**(3): 193 – 216. doi: http://dx.doi.org/10.1016/0378-2166(93)90046-R, URL http://www.sciencedirect.com/science/article/pii/037821669390046R.

[43] Peursum, P., Bui, H., Venkatesh, S. and West, G. (2004) Human action segmentation via controlled use of missing data in hmms. In *Pattern Recognition, 2004. ICPR 2004. Proceedings of the 17th International Conference on*, **4**: 440–445 Vol.4. doi:10.1109/ICPR.2004.1333797.

[44] Preciado, M. (2012) *Computer Vision Methods for Unconstrained Gesture Recognition in the Context of Sign Language Annotation*. URL http://books.google.fr/books?id=7DvyngEACAAJ.

[45] Ruffieux, S., Lalanne, D. and Mugellini, E. (2013) Chairgest: a challenge for multimodal mid-air gesture recognition for close HCI. In *2013 International Conference on Multimodal Interaction, ICMI '13, Sydney, NSW, Australia, December 9-13, 2013*: 483–488.

[46] Shi, Q., Wang, L., Cheng, L. and Smola, A. (2008) Discriminative human action segmentation and recognition using semi-markov model. In *Computer Vision and Pattern Recognition, 2008. CVPR 2008. IEEE Conference on*: 1–8. doi:10.1109/CVPR.2008.4587557.

[47] Sigalas, M., Baltzakis, H. and Trahanias, P. (2010) Gesture recognition based on arm tracking for human-robot interaction. In *Intelligent Robots and Systems (IROS), 2010 IEEE/RSJ International Conference on*: 5424–5429. doi:10.1109/IROS.2010.5648870.

[48] Sminchisescu, C., Kanaujia, A., Li, Z. and Metaxas, D. (2005) Conditional models for contextual human motion recognition. In *Computer Vision, 2005. ICCV 2005. Tenth IEEE International Conference on*, **2**: 1808–1815 Vol. 2. doi:10.1109/ICCV.2005.59.

[49] Van Rijsbergen, C.J. (1979) Information retrieval. *Butterworth* .

[50] Vogler, C. and Metaxas, D.N. (1999) Parallel hidden markov models for american sign language recognition. In *ICCV*: 116–122. URL http://dblp.uni-trier.de/db/conf/iccv/iccv1999-1.html#VoglerM99.

[51] Wang, T., Shum, H.Y., Xu, Y.Q. and Zheng, N. (2001) Unsupervised analysis of human gestures. In Shum, H.Y., Liao, M. and Chang, S.F. [eds.] *IEEE Pacific Rim Conference on Multimedia* (Springer), *Lecture Notes in Computer Science* **2195**: 174–181. URL http://dblp.uni-trier.de/db/conf/pcm/pcm2001.html#WangSXZ01.

[52] Ward, J.A., Lukowicz, P. and Gellersen, H.W. (2011) Performance metrics for activity recognition. *ACM Trans. Intell. Syst. Technol.* 2(1): 6:1–6:23. doi:10.1145/1889681.1889687, URL http://doi.acm.org/10.1145/1889681.1889687.

[53] Weinland, D., Ronfard, R. and Boyer, E. (2011) A survey of vision-based methods for action representation, segmentation and recognition. *Comput. Vis. Image Underst.* **115**(2): 224–241. doi:10.1016/j.cviu.2010.10.002, URL http://dx.doi.org/10.1016/j.cviu.2010.10.002.

[54] Wilson, A. and Bobick, A. (1999) Parametric hidden markov models for gesture recognition. *Pattern Analysis and Machine Intelligence, IEEE Transactions on* **21**(9): 884–900. doi:10.1109/34.790429.

[55] Wu, G., van der Helm, F.C., Veeger, H.D., Makhsous, M., Roy, P.V., Anglin, C., Nagels, J. *et al.* (2005) Isb recommendation on definitions of joint coordinate systems of various joints for the reporting of human joint motion—part ii: shoulder, elbow, wrist and hand. *Journal of Biomechanics* **38**(5): 981 – 992. doi:10.1016/j.jbiomech.2004.05.042, URL http://www.sciencedirect.com/science/article/pii/S002192900400301X.

[56] Yacoob, Y. and Black, M.J. (1998) Parameterized modeling and recognition of activities. In *Proceedings of the Sixth International Conference on Computer Vision*, ICCV '98 (Washington, DC, USA: IEEE Computer Society): 120–. URL http://dl.acm.org/citation.cfm?id=938978.939136.

[57] Zhang, X., Chen, X., Li, Y., Lantz, V., Wang, K. and Yang, J. (2011) A framework for hand gesture recognition based on accelerometer and emg sensors. *IEEE Transactions on Systems, Man, and Cybernetics, Part A* **41**(6): 1064–1076. URL http://dblp.uni-trier.de/db/journals/tsmc/tsmca41.html#ZhangCLLWY11.

[58] Zhao, L. (2001) *Synthesis and Acquisition of Laban Movement Analysis Qualitative Parameters for Communicative Gestures*. Master's thesis, University of Pennsylvania.

[59] Zhong, H., Shi, J. and Visontai, M. (2004) Detecting unusual activity in video. In *Proceedings of the 2004 IEEE Computer Society Conference on Computer Vision and Pattern Recognition*, CVPR'04 (Washington, DC, USA: IEEE Computer Society): 819–826. URL http://dl.acm.org/citation.cfm?id=1896300.1896418.

6

Mobile Agent Communication in Highly Dynamic Networks: A Self-Adaptive Architecture inspired by the Honey Bee Colony

Phuong T. Nguyen[1,*], Volkmar Schau[2], Wilhelm R. Rossak[2]

[1]Research and Development Center, Duy Tan University, Da Nang, Vietnam
[2]Department of Computer Science, Friedrich-Schiller-Univerität Jena, Ernst-Abbe-Platz 2-4, D-07743 Jena, Germany

Abstract

Communication is considered as a building block for mobile agent systems. In highly dynamic networks, thanks to environmental stimuli such as changes in connection quality and network topology, performance of communication between mobile agents may be degraded considerably. With focus on attaining fault tolerance and reliability, we propose a context-aware architecture for agent communication model inspired by the honey bee colony. To validate the hypothesis, a software prototype has been designed and implemented according to the proposed mechanism. Encouraging experimental results on a test system show that our approach brings benefits to a colony of agent platforms.

Keywords: Context-Awareness, Mobile Agents, Agent Communication

1. Introduction

Communication between mobile agents has been identified as an important issue in mobile agent technology, alongside other technical aspects of mobile agent technology, such as migrations and control mechanisms [9],[10],[11]. Under certain circumstances mobile agents can benefit from sending communication messages instead of migrating to remote platforms [9],[13].

For systems working in highly dynamic networks, agent communication is expected to adapt to environmental stimuli which appear only at execution time. Communication should deal not only with interoperability and location-transparency, but also with adaptability. Several studies have been published so far which address the design of an appropriate, highly flexible model for mobile agent communication. However, to the best of our knowledge none of the suggested solutions has been able to achieve the necessary performance and quality attributes to count as a practical solution. In most cases, these existing approaches seem to neglect the inherent dynamics of modern networks. Thus, in our view adaptive mobile agent communication remains a challenging topic. We are interested in developing an adaptive communication model for mobile agents which is suitable for tasks in highly dynamic networks.

In this paper we present an adaptive communication for mobile agents in highly dynamic networks. The main objective is to develop a model for agent communication that supports fault tolerance and reliability as the key qualities. Being inspired by the exotic behaviours of the honey bee colony, we suggested building a self-organizing network in the agent platform colony. Based on the proposed theory, we design, implement a software prototype for supporting mobile agent communication in highly dynamic networks. To validate our hypothesis, we also perform some evaluations for the software prototype in a laboratory scale test system.

The paper is organized as follows. In Section 2 we present the motivations as well as the aims of our work. Background on swarm intelligence is going

*Corresponding author. Email: phuong.nguyen@duytan.edu.vn

to be introduced in Section 3. Section 4 provides a biological background concerning the honey bee colony. Our proposed approach is highlighted in Section 5. Afterwards, we present our method for improving an existing consensus algorithm in Section 6. Section 7 deals with the implementation of the software prototype. Experimental evaluation results are introduced in Section 8. Finally, Section 9 concludes the paper.

2. Motivations and Objectives

2.1. Motivations

For the support of mass casualty incidents (MCI) rescue, a project has been executed by Friedrich Schiller University Jena and its partners. The project, named SpeedUp[1], consists of two main areas: SpeedUp Practice and SpeedUp Technology. SpeedUp Technology is responsible for the IT support of rescue forces in MCI situations so that in disaster events, rescue actions can be performed in a more effective way. The IT solution supports the usual rescue measures by providing a framework for improved information handling, additional preprocessed sensor data, and a basis for a highly flexible communication infrastructure [12]. In SpeedUp the mobile agent concept has been selected as one of the main technologies on the communication infrastructure level.

Our work was motivated by a system working in a communication infrastructure for emergency rescue scenarios [6]. Figure 1 illustrates the so-called mass casualty incident (MCI). In this scenario work conditions change dramatically as new tasks appear, rescue staff are unlikely to process incoming information in an effective way. Moreover, since each team performs tasks with its own criteria and under control of its own authority, conflicts in their goals might arise at anytime. As a consequence, performing rescue operations based on experience without any additional technical supports seems to be an impossible task. In an MCI rescue scenario, there are multiple regions of working staff being scattered over a wide area. In a region, the movement and relocation of rescue personnel make the availability of their devices unstable. Changes in network topology trigger other stimuli, such as changes in connection quality, in routing strategy, and in communication pattern. These stimuli generally adversely affect communication performance. A question arises: How to organize the logical connection among platforms in an operationally feasible manner to cope with these stimuli? Given that communication between mobile agents is important, it is expected that agent platforms can self-manage in order to be adaptive to perturbations, thereby attaining fault tolerance and reliability in message transferring. We consider the issues as our research problem.

Figure 1. Communication in MCI rescue scenarios

In our work, we propose a model for agent communication that focuses on the cooperation aspect of agent interaction and supports reliability and fault tolerance as the key qualities, while keeping up an acceptable overall performance at the same time. We design, implement, and evaluate in this project environment a prototype for adaptive mobile agent communication in highly dynamic networks.

2.2. Aims

We have been participating actively in designing and implementing a communication infrastructure for the support of rescue personnel in MCIs. The IT solution supports the usual rescue measures by providing a framework for improved information handling, additional preprocessed sensor data, and a basis for a highly flexible communication infrastructure. Essentially, we are aiming at providing necessary information to the right personnel so that in disaster events, rescue actions can be performed in a more effective way. The main contributions of our work are summarized as follows:

- Introduction of a self-organizing mechanism for the management of working regions in highly dynamic networks.
- Development of a communication model that is able to adapt in a mostly autonomous fashion to topological changes of the underlying network.
- Implementation of a software prototype for the proposed communication model that is adaptive to environmental stimuli.

[1]The project was funded within the Federal Government's program "Research for Civil Security" by the German Federal Ministry of Education and Research (BMBF), http://www.speedup-projekt.de

- Improvement of overall qualities of mobile agent collaboration concerning fault tolerance and reliability.
- General speaking, the proposed model is suitable for mobile agents working in highly dynamic networks of the SpeedUp-Type.

3. Swarm Intelligence

Swarm intelligence is considered as a branch of biomimicry which covers a wide range of applications, including architecture, mechanical engineering, nano technology, and computer science. The main idea of swarm intelligence in computer systems is to emulate activities in nature so that they can have alike features. As a result, they can react to environmental changes and therefore be resilient to perturbations.

In MCI rescue scenarios, rescue forces may be distributed widely. Each geographical location forms most likely a technological region in which different forces work together to do rescue tasks. The key point is how to manage the dynamics of mobile agent platforms in these regions efficiently, as they join and quit spontaneously, or they swarm onto a new location. We expect a technique that can help the system fulfil the following requirements:

- The control of the whole system does not rely on any central component, it is distributed to all constituent members.
- The system is able to adapt automatically to the surrounding environment.
- The system is resilient to errors, it can get to a stable status after changes or damages through autonomous interaction of its members.

By bearing on these points, our analysis converges on the concept of self-organization [18]. Self-organization is the way a system adjusts its behaviour to adapt to the surrounding environment. Through interaction among the constituent members on a local scale, the system reaches a globally sensible pattern. The control of a self-organizing system does not rely on any specific member, it is distributed to all of them [15].

Self-organization has been proven to help a system to alleviate "The complexity problem" [18], it is a common phenomenon in nature. Throughout thousands of years of evolution, biological systems have adapted themselves to a changing environment. Plants' leaves always lean to the direction where they can get sunlight, roots grow towards water sources and nutrients. In the animal world, flocks of bird and fireflies, schools of fishes, swarms of honey bees are typical examples of self-organization [15]. The common property for these examples is that there is no central control or external intervention to the control of the community, the final decision is made based on the communication among individuals. Robustness, flexibility and adaptability are the outstanding characteristics of natural systems [17]. Through the interaction among its members a system can gain stability, and it is able to recover when error occurs. The whole system fits itself "bottom-up" to the environmental stimuli. It becomes more robust since the control of the whole system does not rely on any individual component. New members can be integrated seamlessly into the system. Furthermore, the disappearance of members does not cause the whole system to stop working, it can scale flexibly to the number of members.

4. Biological Background

4.1. Overview

The bee colony is a typical example of self-organization in nature. A bee population consists of a queen bee who is reproductive, some male drones, and thousands of female workers. Worker bees undertake the most important tasks of the whole colony, they build the hive, keep it clean, regulate the temperature, search for food sources, and find a new resident location when the bee colony needs to split up.

Teamwork is the distinctive character of the colony, honey bees are a well-organized team and they work in strict disciplines. The control of the whole colony is not dependent on any individual, even the queen bee. This is done through the interaction among thousands of worker bees, when and how many bees the queen bee must create is actually controlled by the workers. The queen bee has only an indirect role in organizing the colony, she gives birth to bees. The self-organizing model of honey bees has inspired much work in computer science and shown to improve considerably computational performance in particular circumstances [3],[4],[5].

Regarding the network model in MCI rescue scenarios, we found that the honey bee community and the community of agent platforms have a substantial coincidence of behaviour. If we call honey bees and agent platforms simply *members*, then the following clauses hold for both communities:

- Members work in a distributed environment.
- Members cooperate to perform common tasks.
- Members work to maximize the global work performance.
- Congestion occurs when the population grows.

Furthermore, honey bees have efficient mechanisms to deal with their daily routine as follows:

- Control of the colony takes place without any external intervention.
- The colony can respond dynamically to the surrounding environment.
- The colony is still able to work, if some of the workers are missing.
- The colony can adjust its population to avoid conflict and maintain reproduction.

The fact that the two phenomena have a lot in common and honey bees possess a good mechanism to deal with their daily routines encourages us to apply honey bee behaviour to manage the dynamics of a networked environment that reflects a SpeedUp-Type of system. In the next sections, we are going to provide the reader with an overview of the honey bee colony, and then introduce our approach.

4.2. Bee Communication

Bees can produce honey only after they collect enough nectar and pollen. Since food sources are normally distributed far away from their hive, foraging is a vital skill of honey bees. Foraging takes place in the summer, bees spread out to search for food sources. Each individual bee may find different food sources, but as time goes by, the colony tends to head for rich food source in which they can get more nectar and pollen. It is exciting to know that a honey bee has a mechanism to notify other bees. A worker bee can communicate with her colleagues to inform about the food sources she has found. A scout bee, after finding a new food source, returns to the hive and performs a *waggle dance* to notify other honey bees about the food source [25]. It has been shown that, with the waggle dance, a forager provides the following information to her colleagues [26]:

- Direction to the food source compared to the sun.
- Distance to the food source.
- The amount of food: the more exciting and the longer she dances, the richer the food source is [2],[16].

With this information, other honey bees are able to know where and how far they must fly in order to get to the food source. Rich food sources attract more honey bees to come and get pollen than other sources that contain lower amounts of food. Thanks to this information sharing, the honey bee colony can maximize the amount and quality of food. The discovery of bee communication brought the biologist Karl von Frisch the Nobel Prize for Physiology or Medicine in 1973 [27].

4.3. Swarming

Swarming in bee colony occurs when the bee population increases and causes congestion and difficulties in maintaining good hygiene in the beehive. Swarming is also one of the honey bees' instincts, they maintain reproduction by splitting.

Before swarming, worker bees build a cup in which the old queen lays eggs, in order to raise a new queen bee. When the new queen bee arrives, the old queen bee takes a number of workers with her and they fly to a new temporary location, e.g. a tree branch, that is near to their hive. The swarm remains on the temporary residence for a short time while waiting for a new residence. Scout bees, which are the most experienced foragers in the swarm, are then deployed to find new suitable locations. When the exploration is finished, every scout bee returns to the bee cluster to inform the whole colony about the place she has found. This is done by the same waggle dance the scout bee does when she informs the community about food sources.

The scouts fly back and forth until the number of honey bees that gather at a site constitutes a quorum, then the scout bees return to the swarm and the swarm decides to leave for the new location [1]. The relocation may prolong, if the colony finds the new location does not satisfy expected requirements regarding available space, light intensity and hygiene conditions. The swarm will continue to depart for the next location. Eventually, it is summoned to the new residence.

5. A Context-Aware Architecture for Mobile Agent Communication

5.1. Network Model

Figure 2 depicts the SpeedUp communication infrastructure. The components are described as follows:

- Agent Platform: A software platform which provides executing environment for mobile agents.
- Lightweight Agent Platform: A special type of agent platform which is energy limited, such as PDAs and smart phones.
- Region: A group of platforms that work in the same geographical location.
- RegionMaster: A platform that has enough computing power to perform self-organizing tasks, manages agent's location information for the whole region.
- Global System: The highest logical view and consists of all regions of the system.

Based on the self-organizing behaviours of the honey bee colony, we proposed a model for agent communication which is able to cope with environmental stimuli

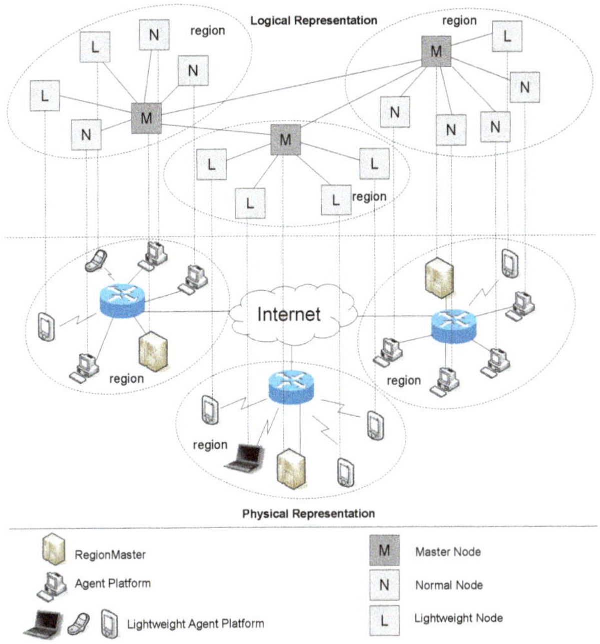

Figure 2. SpeedUp communication infrastructure

[8]. In the algorithm, two mappings from the honey bee colony to the rescue scenario network model will be employed. In the first mapping, RegionMaster is considered as the beehive and mobile agents are scout bees. All other platforms of the region are viewed as potential beehives. In the second mapping, we consider all agent platforms as honey bees, and RegionMaster as the queen bee. Our solution is going to bear on the following issues:

- The distributed nature of mobile agents.
- The parameters related to fault tolerance and reliability of communication between mobile agents.
- The scalability of the system with regards to the availability of heterogeneous end devices.
- The dynamics of the network.

In the SpeedUp architecture, regions of working devices are the prime elements which make up the global system. In our view, these facets should be handled in an effective way. Environmental stimuli affect the overall performance, therefore they need to be monitored so that the system can self-adjust accordingly. Since mobile agent communication is communication between individual agents, issues related to direct agent communication must be managed adequately from the start. Communication for mobile agents should reuse existing results from agent communication research. The main objective is, therefore, a model for agent communication that focuses on the cooperation aspect of agent interaction and supports fault tolerance and reliability as the key qualities.

5.2. Getting the Blueprint of a Region

The first mapping is inspired by bee communication. It is used for measuring the connection quality between an agent platform and all the remaining platforms of a region at a point in time. The connectivity map of the region can be constructed once a calculating process has been performed for all nodes of a region.

In this view, RegionMaster plays the role of the beehive and mobile agents are scout bees. Figure 3 illustrates how agents are sent to scout network. At regular intervals, RegionMaster deploys scout agents to all platforms of the region to get information about them. A RegionMaster is responsible for monitoring its region and regions are surveyed independently. Information related to connectivity and and platform's performance will be collected by scout agents.

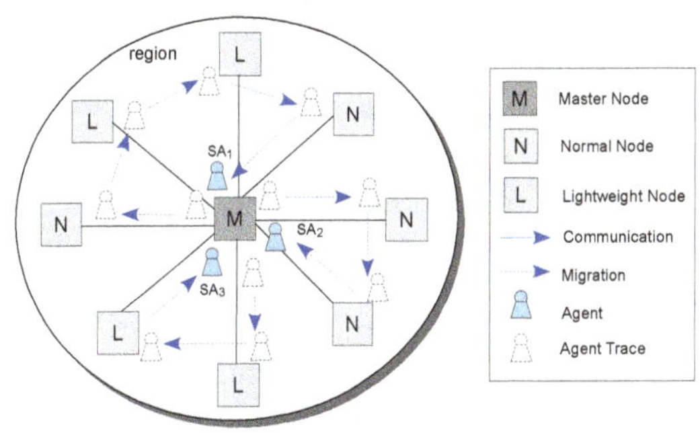

Figure 3. Scout agents collect network's information

When a scout agent arrives at a platform, it sends ping messages to all nodes of the region, including RegionMaster to measure the transfer time between the current platform and every other platform. After collecting the node's information, the agent migrates to the next platform. At the new platform, the scout agent performs the same routine. The process repeats until the last node of the itinerary has been visited.

A platform holds a boolean value *split* to indicate whether it expects its region to split or not. At each platform once a scout agent has obtained the transfer time to all neighbours by sending and receiving ping messages, it calculates the average transfer time τ_{avg}.

$$\tau_{avg} = \frac{1}{n}\sum_{i=1}^{n-1} t(i) \quad (1)$$

In which n is the number of nodes in the region, $t(i)$ is the transfer time between the current platform and the ith neighbour platform. Since every node in a region has the same number of neighbours, this value is the measure of the closeness between a node and other nodes in the region. Once the value is calculated, the platform compares the τ_{avg} with the transfer time to RegionMaster τ_{RM}. If $\tau_{RM} >> \tau_{avg}$ then it records this state. If the state is observed for a number of times then RegionMaster expects that the region will be splitted; it sets the value *split* to *true*. The two values τ_{avg} and *split* will then be given to the scout agent. Eventually, the agent migrates back to RegionMaster and submits all the information it has collected. RegionMaster then performs self-organizing tasks based on information fetched from all the scout agents it deployed.

5.3. Re-Organizing a Region

The second mapping is inspired by bee swarming. In this view, RegionMaster plays the role of the queen bee, and other platforms of the region are worker bees. This mapping is used to impose a self-orgarnizing mechanism on the platforms.

For the rescue scenario network model, in a region RegionMaster is responsible for performing self-organizing tasks and handling location information of all mobile agents working in the region. If the number of agent platforms in a region is large and too many requests are sent simultaneously, a bottleneck may occur at RegionMaster. The phenomenon resembles the necessity for swarming in the honey bee community. The platform population exceeds the region's capacity thus causing degradation in connection quality. As a result, it needs to be adjusted, the colony requires some kind of "swarming." Like in the bee colony, the region will be splitted into two regions. One of the nodes will be promoted to be a new RegionMaster. The new RegionMaster organizes to form a new region from the nodes it inherits. The two regions are then independent from each other, but logically connected.

Each agent platform plays a contributory role in organizing the region. A final decision of splitting the region is made based on a consensus of the platforms in the region. In mapping 1, at every platform, the scout agents collect the boolean value *split*. RegionMaster counts the number of platforms that want the region to be segregated. If the number constitutes a quorum, the region is about to be splitted. Like the queen bee chooses her successor, RegionMaster promotes a new RegionMaster from the candidates nominated the by scout agents. Afterwards, RegionMaster broadcasts the name of the new RegionMaster to all platforms. Every platform decides itself, which region it belongs to by sending a joining message to the new RegionMaster.

5.4. Managing Mobile Agent's Location

The mobility of agents may cause message losses when communication executes, especially in large scale and dynamic networks. The management of mobile agents' location should also be adaptive to the network topology changes. With respect to the dynamic region architecture established by the self-organizing algorithm, we introduce a technique for the management of mobile agent's location.

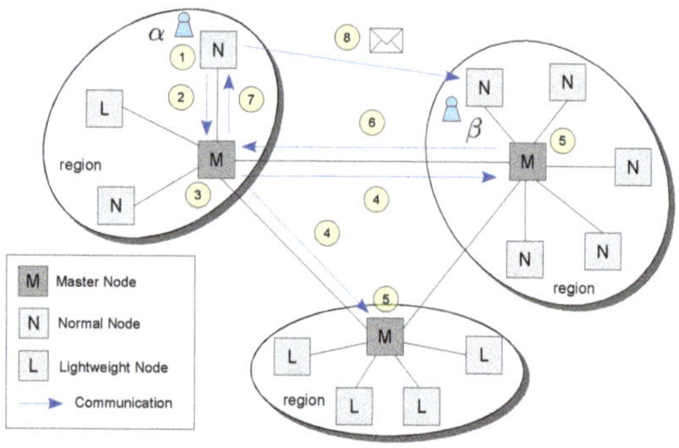

Figure 4. Location query

Figure 4 depicts an example of location searching where agent α wants to communicate with agent β whose location is unknown. The circled numbers depict the steps that are to be executed. Agent α asks the platform where it is currently in for the ID of agent β. This platform checks its cache to see if there is any information regarding the location of the requested agent (1). When the address is already in the cache, the platform returns the address to the agent. Otherwise, the platform queries its RegionMaster (2). RegionMaster then looks into its cache for the address (3). When the address is found, RegionMaster returns it to the requesting party (7); if not, RegionMaster broadcasts a query to all RegionMasters of the global system (4). Each RegionMaster of the global network checks its cache for the address (5). Whenever the address is found by one of the RegionMasters, it will be sent back to the requesting RegionMaster (6). This RegionMaster in turn forwards the address to the requesting platform (7). After the location of the agent β has been determined, the agent sends message to the targeted platform (8).

If a region is splitted, RegionMaster advertises the change to all other RegionMasters to notify other RegionMasters of the new region's appearance.

6. Fault Tolerance: A Honey Bee inspired Approach for Building Consensus

6.1. Finding consensus in distributed systems

The proposed mechanism seems to rely much on RegionMasters, which may generate a single point of failure. In a region, RegionMaster acts as the beehive, where scout agents unload information they collect. It also performs self-organizing tasks for the whole region. If RegionMaster goes haywire or disconnects suddenly, the node community has no information about the network and, therefore, cannot re-organize wherever necessary. As a result, a constructive approach to fulfil the fault tolerance requirement is needed. This leads to the problem of finding consensus in distributed systems, all nodes promote a new RegionMaster by exchanging messages.

The problem of building consensus is described in detail as follows: A set of $n \geq 2$ nodes $P = \{p_1, p_2, .., p_n\}$ has to negotiate by using asynchronous message passing and decide on the same value from a common set of inputs. In the scope of this paper, we discuss binary consensus, that means the input values $v \in \{0, 1\}$ [21],[22],[23],[24].

The properties of consensus [23]:

- At most f nodes may fail. The failed nodes are called faulty nodes; the working nodes are called non-faulty or correct nodes. Once fails a node remains faulty for ever.
- Each node p broadcasts its proposal as a message to all other nodes.
- A node makes a decision based on the set of messages it received.
- A message never fails once it has been sent. It arrives eventually at the targeted receivers.
- A node p can send its report or proposal value v to some nodes, and it may crash before sending the message to the remaining non-faulty nodes. As a result, some nodes have v and some not.

A consensus algorithm needs to satisfy three key requirements as follows [22]:

- Termination: Every node must ultimately decide.
- Agreement: All non-faulty nodes decide on the same value.
- Validity: The chosen value must be the input of at least one of the nodes.

No matter how simple the definition is, the solution for the problem remains a challenge in distributed computing. In [14] Fischer, Lynch and Paterson prove that with an asynchronous message passing system, there exists no deterministic algorithm for solving the problem of consensus in presence of node failures.

6.2. Ben-Or's algorithm for finding consensus

Ben-Or proposes a practical solution to solve consensus using randomization. In his approach, it is assumed that at most $f < n/2$ nodes may fail during execution. The algorithm is described in the pseudo code as follows [21],[31]:

Algorithm 1 Ben-Or's Consensus Algorithm

```
procedure BenOrConsensus(v_p)
    x ← v_p
    k ← 0
    while true do
        k ← k + 1
        Report(k,x)
        WaitFor(k,*) from n-f nodes        ▷ * is either 0 or 1
        if there are more than n/2 messages containing v then
            Propose(k,v)
        else
            Propose(k,abstention)
        end if
        WaitFor(k,*) from n-f nodes
        if there are at least f+1 messages containing v then
            Decide(v)
        else if there is at least 1 message containing v then
            x ← v
        else
            x ← ChooseRandom(0,1)
        end if
    end while
end procedure
```

- Procedure *Report(k, x)* broadcasts a message containing information about the current round k and the proposed value to all other non-faulty nodes.
- Procedure *WaitFor(k, *)* waits for all incoming messages containing k and a report/proposal value.
- Procedure *Propose(k, v)* broadcasts a proposal value v in round k to all non-faulty nodes.
- Procedure *Decide(v)* makes a decision on the value v.
- Procedure *ChooseRandom*(0, 1) returns either 0 or 1 with equal probability.

A comprehensive proof for the correctness of Ben-Or's algorithm can be found in [31].

6.3. A proposal for improving the convergence speed

The original Ben-Or's algorithm may reach a pretty slow convergence given that nodes do not choose the same value in the randomization phase. Regarding the problem of reaching an eventual agreement among nodes in distributed computing, we witness a substantial coincidence between consensus in distributed systems and consensus in the honey bee colony as shown in Table 1.

Bee Colony	Distributed Systems
Bees	Nodes
Sites	Values
Scout bees nominate a site by dancing	Nodes propose a value by broadcasting messages
Bees select a site if the number of bees around the site exceeds a threshold	Nodes decide on a value v if they receive more than f report messages containing v
All bees reach an eventual agreement for a site	All nodes eventually decide on the same value

Table 1. Analogies

The fact motivates us to propose some amendments to the original consensus algorithm. We call $M_p(k)$ is the set of messages received by node p at round k; $C_p(k,0)$ is the cardinality of the set of messages that contain 0 and $C_p(k,1)$ is the cardinality of the set of messages that contain 1. When $C_p(k,0) = C_p(k,1)$ we say it is a tie for the two sets.

As we have seen in Algorithm 1, it is expected that the majority of nodes will report the same value in Line 6 and then a quorum can be reached in Line 14. However, a node proposes a value only when it receives more than $n/2$ messages containing the same report value, that means only when the majority requirement is satisfied. Otherwise it proposes the value *abstention*, which does not help to build a consensus.

We may "waste" the chance that nodes report the same value in round $k+1$ given that almost $n/2$ messages containing the same value v are sent in Line 13. This should be considered as a room for improvement [7].

It should be noted that from rounds $k > 1$, the value x reported in Line 6 is the result from either Line 15 or 17 or 19 in round $k-1$. We propose a way to *guide* the nodes to opt for the most sensible value v instead of choosing a random value. We consider the case that the majority requirement is not satisfied: $C_p(k,v) < n/2$; but there is a *plurality* of nodes that report v, that means: $C_p(k,1-v) < C_p(k,v) < n/2$.

The honey bee inspired algorithm for building a consensus is illustrated in the pseudo code Algorithm 2. The procedures and functions of the pseudo code are explained as follows:

- Function $Plurality(0,1)$ returns 0 if $C_p(k,0) > C_p(k,1)$ and returns 1 if $C_p(k,0) < C_p(k,1)$.
- Procedure $Follow(v)$ indicates that a node prefers a value v.
- Function $Evaluate(0,1)$ calculates a correlative value between the cardinalities $C_p(k,0)$ and $C_p(k,1)$; a proposal is $Evaluate(0,1) = abs(C_p(k,0) - C_p(k,1))$.
- Procedure *Decrease(strength)* subtracts a specified number from *strength*.

Algorithm 2 Honey Bee Consensus Algorithm

```
1: procedure HoneyBeeConsensus(v_p)
2:     x ← v_p
3:     k ← 0
4:     strength ← 0
5:     while true do
6:         k ← k + 1
7:         Report(k,x)
8:         WaitFor(k,*) from n-f nodes
9:         if there are more than n/2 messages containing v then
10:            Propose(k,v)
11:        else
12:            if strength ≤0 and C_p(k,0) ≠ C_p(k,1) then
13:                v ← Plurality(0,1)
14:                Follow(v)
15:                strength ← Evaluate(0,1)
16:            end if
17:            Propose(k,abstention)
18:        end if
19:        WaitFor(k,*) from n-f nodes
20:        if there are more than f messages containing v then
21:            Decide(v)
22:        else if there is at least 1 message containing v then
23:            x ← v
24:        else
25:            if strength >0 and Follow(v) then
26:                x ← v
27:                Decrease(strength)
28:            else
29:                x ← ChooseRandom(0,1)
30:                strength ← 0
31:            end if
32:        end if
33:    end while
34: end procedure
```

7. Implementation

This section deals with the implementation of the software prototype for an adaptive communication framework for mobile agents in highly dynamic networks of the SpeedUp-Type. The implementation of the software prototype will bear on the honey bee inspired alogrithm with the aims and objectives of the targeted qualites: fault tolerance and reliability in transferring agent messages.

Ellipsis is an open source mobile agent system developed by the Chair of Software Engineering of Friedrich Schiller University Jena. Ellipsis provides a framework for hosting and executing mobile agents and runs on any operating system that supports Java, thereby adhering to the slogan "write once, run anywhere." Ellipsis can be deployed on any device as long as the device is equipped with Java Virtual Machine (JVM). We base our design and implementation on Ellipsis and the software architecture in conjunction with the mobile agent system is depicted in Figure 5. The following sections provide an overview of the software components.

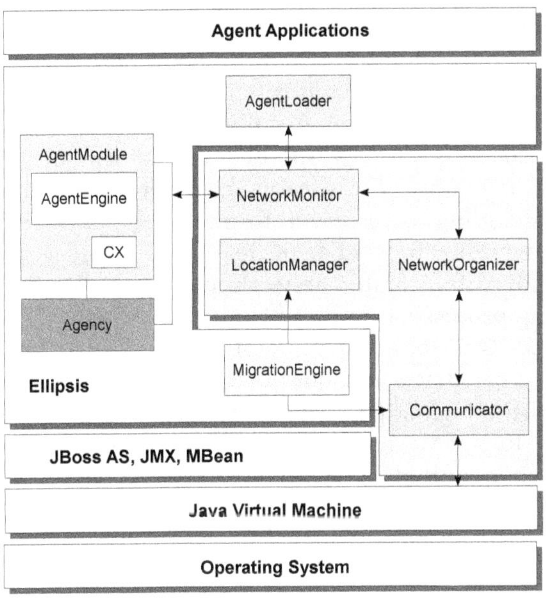

Figure 5. Software Architecture

7.1. NetworkMonitor

NetworkMonitor observes the agent platforms of a region. It collects connectivity information between a platform and its neighbours and information of the platforms. NetworkMonitor undertakes the following tasks:

- Managing a dynamic list of agent platforms.
- Deploying scout agents periodically to get update on the network situation.
- Receiving and replying ping messages.
- Calculating the values τ_{avg} and $split$.
- Making a decision of splitting region by counting the boolean values $split$.

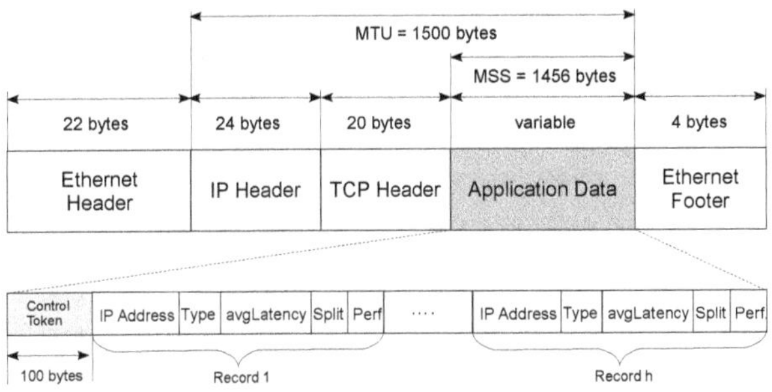

Figure 6. Packet Format

7.2. NetworkOrganizer

NetworkOrganizer re-organizes the logical network based on the conditions percepted by NetworkMonitor. This component is present at every platform of a region. However its role is dependent on the type of the platform. In a RegionMaster, NetworkOrganizer has the following functions:

- NetworkOrganizer receives the list of platforms from NetworkMonitor.
- NetworkOrganizer selects a new RegionMaster from the candidates nominated by NetworkMonitor. It broadcasts the ID of the new RegionMaster to all platforms of the region to conduct a poll.
- NetworkOrganizer activates a platform when the platform has been promoted to the post of a RegionMaster.
- NetworkOrganizer notifies other NetworkOrganizers of the changes by sending broadcast messages.

In a normal platform, NetworkOrganizer performs the succeeding tasks:

- It receives a proposal for a node to be the new RegionMaster from RegionMaster.
- It chooses which region the corresponding platform should belong to by measuring the latencies to the proposed platform and to RegionMaster.
- It sends the ID representing the chosen platform back to RegionMaster.

7.3. LocationManager

LocationManager manages location information of mobile agents, it updates location of mobile agents when they migrate, answers location query and queries agent's location information. LocationManager holds a cache for storing location information. Essentially, the component performs its function to answer the question: *Which agent locates where?*

The following tasks are undertaken by LocationManager:

- Interfacing to Ellipsis to manage a dynamic list of location information for mobile agents.
- Receiving and answering location query from other nodes as well as from agents.
- Sending location information query to other LocationManagers.

7.4. Communicator

The software entity Communicator is responsible for initializing connection between platforms and transferring messages between agents. It encodes, decodes, and transfers agent messages over the network. This component maintains two caches of agent messages at every platform, one for incoming messages and one for outgoing messages.

Communicator has the following functions:

- Establishing connection between two agent platforms.
- Encoding the envelope and the payload of an ACL message at the sender side.
- Sending byte streams containing ACL messages over the network.
- Decoding the envelope and the payload of ACL messages at the receiver side.

8. Experimental Evaluation

8.1. Setup

We deploy a laboratory scale test network with the presence of eight computers connected through a local Gigabit Ethernet LAN 1000 Mbps network. Each computer corresponds to an agent platform and is equipped with the software framework as well as other essential softwares. The system configuration is specified in Table 2.

The agent platforms representing by their alias are going to be used throughout the succeeding test scenarios.

Alias	OS	RAM	Processor
D1	Fedora 12	2.0 GB	AMD 2.2 GHz
D2	Debian 6.0.4	4.0 GB	Intel 2*2.0 GHz
D3	Debian 6.0.4	4.0 GB	Intel 2*2.0 GHz
D4	Debian 6.0.4	4.0 GB	Intel 2*2.0 GHz
FJ	Fedora 12	3.0 GB	Intel 2*2.8 GHz
PT	Windows 7	4.0 GB	Intel 2*2.4 GHz
T1	Fedora 12	2.4 GB	Intel 2*2.4 GHz
T2	Windows XP	1.0 GB	Intel 1.7 GHz

Table 2. Hardware configuration of the experiments

8.2. Feasibility

In a dynamic network environment the cost for monitoring network might be considerably high. Monitoring activities may place a burden on the network traffic, thereby reducing overall system's performance. The proposed mechanism is beneficial only if network monitoring gains a good performance whilst keeping a reasonable running cost. The issues need to be addressed throughtfully.

The feasibility of the proposed mechanism means that it maintains a reasonable cost for monitoring while providing necessary information for the self-organizing tasks. To validate the feasibility, scout agents are deployed to monitor network and the parameters regarding processing time and exploration time will be measured. One round of exploration happens when a scout is created at a RegionMaster, travels through all nodes of the region, performs its routine and migrates back to RegionMaster. The following information is going to be acquired:

- The time needed for a scout agent to accomplish its tasks at a platform: $t_{processing}$.
- The time needed for an agent to perform one round of exploration: $t_{exploration}$.

The processing time at an agent platform is the duration from when an agent arrives until it completes its tasks and leaves for the next node. It is calculated as follows:

$$t_{processing} = t_l - t_a \quad (2)$$

where t_a is the time when a scout arrives at a platform and t_l is the time when it leaves for the next node of its itinerary.

The processing time is made of:

$$t_{processing} = t_{RTT} + t_s + t_d \quad (3)$$

in which

- t_{RTT} is the period of time from the first ping message sent until the last response received.

- t_d is the time to deserialize a scout agent from an incoming stream of byte.
- t_s is the time to serialize a scout agent into a byte stream to be sent over the network.

By measuring the processing time, we examine how fast a scout agent performs its tasks at a specific platform.

The average exploration time is computed:

$$t_{exploration} = t_{end} - t_{begin} \tag{4}$$

where t_{begin} and t_{end} are the time when a scout starts and completes one round of exploration, respectively. This paramater indicates how often a scout agent supplies a RegionMaster with up-to-dated information for the self-organizing activities.

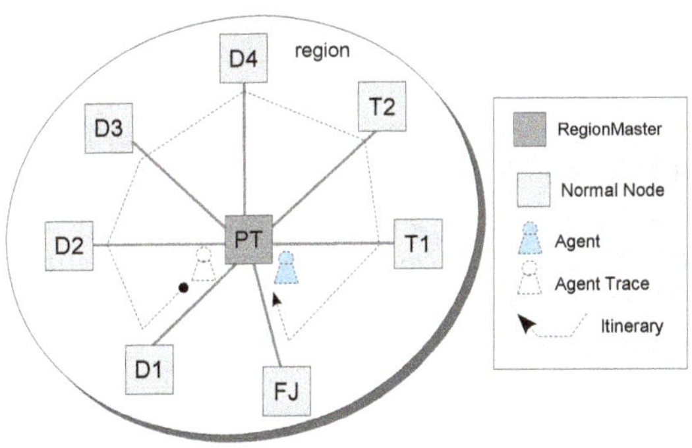

Figure 7. Network Setup

Figure 7 shows the logical connection for the first evaluation test. In this scenario, one scout agent is created at RegionMaster. The RegionMaster assigns the list of agent platforms $\{D1, D2, D3, D4, FJ, PT, T1, T2\}$ to the scout agent. One round of surveillance takes place when the agent is sent by RegionMaster, travels around the nodes of the region, performs its routine at every node and then migrates back to RegionMaster.

To get a more reliable result, by every platform a measurement is done for 80 times. The box plot diagram in Figure 8 depicts the processing time $t_{processing}$ of the platforms. This parameter represents the time that the agent needs to perform its tasks at that platform. It is dependent on the processing power of agent platforms as well as the latencies to the other platforms, which are in turn dependent on the network connection speed. It can be seen that, except platform $T2$ that has a higher processing time because of its limited processing power (as specified in Table 2), the processing time for the other platforms is considerably low. It guarantees that the processing activities place comparatively little burden on the system performance.

To measure $t_{exploration}$, the agent is sent around the network for different numbers of rounds r. In

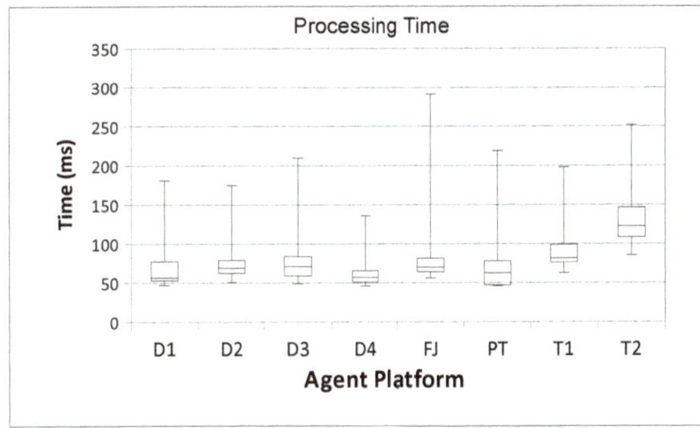

Figure 8. Processing time

the second experiment r is set to different values, i.e. $r = \{100; 200; 300; 500; 1000; 2000; 3000; 5000\}$. The average exploration time is given by:

$$t_{exploration} = \frac{t_{end} - t_{begin}}{r} \tag{5}$$

This parameter demonstrates how fast the scout agent can supply RegionMaster with up-to-date information about the network. If the agent needs a long time to perform its tasks at the platforms or to hop from platform to platform, the information submitted to RegionMaster might be out-of-date. As a consequence, the reactions produced by RegionMaster would not be adequate.

However, the measurement results show that this is not the case. In the diagram in Figure 9, the curve represents the accumulated exploration time. The straight line which depicts the average time for finishing one round of surveillance (second/round) provides evidence that the parameter is stable, no matter how many rounds the agent has migrated. This indicates that the scout agent produces no overhead when it works in the long run.

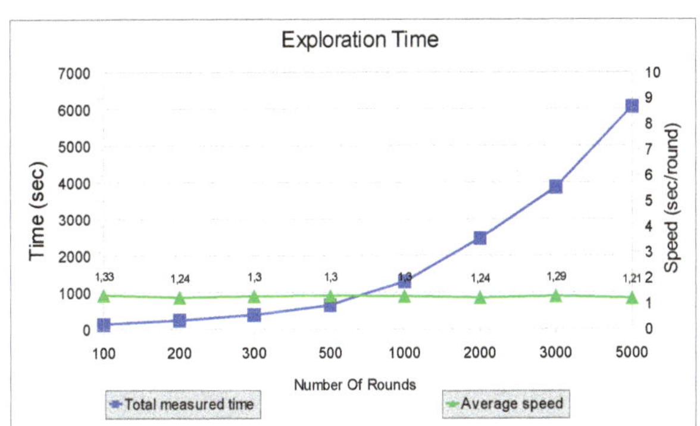

Figure 9. Average time for a scout agent to explore the region

8.3. Self-Organizing

In a dynamic network it is a common occurrence that a group of platforms leaves other platforms of the same region and swarms onto a new location, causing the connectivity to change correspondingly. Under the circumstances, maintaining a fixed logical connection between the abandoning platforms and the stationary platforms is not an optimal solution. It is more sensible to re-organize the colony to adapt to the new structure. The abandoning platforms should form a new region while the stationary platforms group into a second region. A reasonable re-organizing manoeuvre helps facilitate a harmonious correlation between the platforms, thereby balancing network load and saving processing time.

In network applications, a service is identified by the combination of an IP address and a network port. A traffic shaper controls network bandwidth by delaying traffic to meet pre-defined requirements. A traffic can be precisely allotted to a service by referring to the service's identification. By traffic shaping, two important factors for a service are determined. A minimum usable bandwidth ($MiUB$) which is the guaranteed bandwidth that the service can consume and a maximum usable bandwidth ($MxUB$) which is the bandwidth that the service can reach as long as there is free bandwidth available.

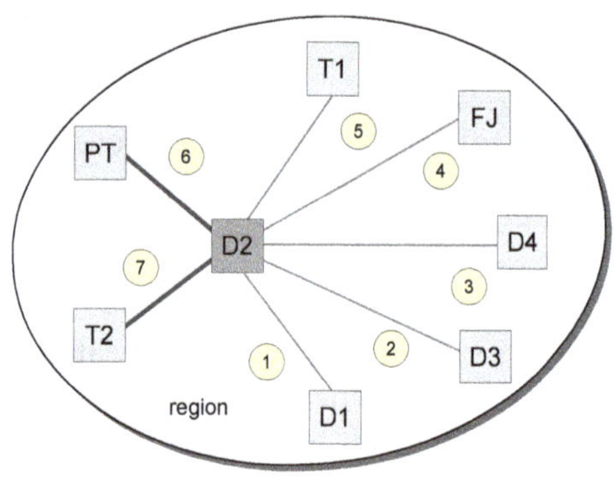

Figure 10. The network layout and bandwidth shaping imposed on ping services

The logical network layout for this test scenario is depicted in Figure 10. In the beginning, platform $D2$ is configured as RegionMaster. A scout agent is deployed to survey the region. We use traffic shaping to simulate the occurrence when the platforms, together, move far away from RegionMaster and swarm onto a new location, resulting in a decrease in the corresponding bandwidths.

In this test scenario, the traffic shaper which is comprised of software packages *tc* (traffic control), *iptables* and *iproute* in Linux distributions is used to regulate the traffic of the ping messages exchanging service between RegionMaster and the platforms $\{D1, D3, D4, FJ, T1\}$ on TCP port 3242. Table 3 lists the corresponding values $MiUB$ and $MxUB$ for every connection representing by the circled numbers depicted in Figure 10.

Con.	(1)	(2)	(3)	(4)	(5)	(6)	(7)
$MiUB$ (Mbps)	0,512	1	2	20	54	1000	1000
$MxUB$ (Mbps)	1	2	3	30	60	1000	1000

Table 3. Bandwidth allotted to each link

The bandwidth degradation will have a certain effect on the communication between agent platforms. While bandwidth shaping is being activated, there are changes in network connectivity within the region. The platform colony re-organizes to react to the changes in network bandwidth. Every platform sets the value $split$ based on the ratio $\tau = \tau_{avg}/\tau_{RM}$. In this experiment the splitting threshold is set to $\tau < 50\%$. Since more than a half of the agent plaforms set $split = true$, the region is bisected. Two regions emerge from the original region. Figure 11 depicts the splitting process, those platforms with a ratio $\tau \geq 50\%$ namely $\{D2, PT, T2\}$ group into Region 1; Region 2 is made of the platforms with $\tau < 50\%$ i.e. $\{D1, D3, D4, FJ, T1\}$.

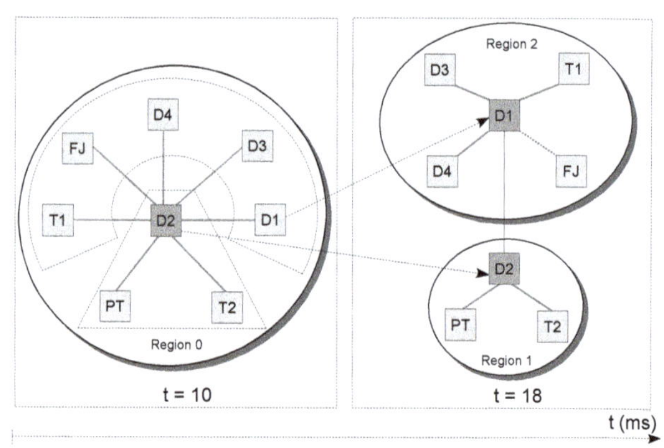

Figure 11. The self-organizing process of the platform colony

Table 4 displays the metrics measured for Region 1 and Region 2 before and after splitting. The measurement unit for the ratio τ is %. According to the metrics shown in the tables, both Region 1 and Region 2 benefit from the splitting, the ratio τ grows markedly and exceeds the threshold 50%. An increase in the ratio τ implies that the connection from each platform to RegionMaster has been improved. All the platforms reset $split$ to $false$.

The metrics show that after network connectivity degrades, the splitting helps the platforms improve connection quality to RegionMaster significantly. Since

	Before splitting		After splitting	
Platform	τ(%)	**Split**	τ(%)	**Split**
Region 1				
D2	—	*false*	—	*false*
PT	75,0	*false*	75,0	*false*
T2	120	*false*	83,3	*false*
Region 2				
D1	14,8	*true*	—	*false*
D3	16,8	*true*	75,0	*false*
D4	15,7	*true*	100	*false*
FJ	38,9	*true*	87,5	*false*
T1	29,4	*true*	83,3	*false*

Table 4. Metrics measured for Region 1

RegionMaster is responsible for performing the self-organizing tasks of the whole region, an improvement in connectivity will lead to an increase in performance of the system. This test supports the hypothesis that the software framework can assist the system to react appropriately to changes happening in the surrounding environment. In our judgement, the features demonstrate that the software framework complies with one of the requirements: the ability to self-organize in a changing environment.

8.4. Context-Awareness

In the next experiment, platform *FJ* is configured as RegionMaster, a scout agent is deployed to survey the region. In this scenario, the connectivity between RegionMaster and the remaining platforms is going to be degraded using software. This aims to investigate the countermeasures of the system given that the quality of the connection to RegionMaster has declined. Since RegionMaster is responsible for the management of the whole region, the deterioration adversely affects system performance, processing speed may considerably slow down. Given the circumstances, it is expected that the software prototype helps the platform colony to re-organize and recover from the degradation.

To conduct this experiment, a simulation of connectivity degradation is required. A certain network traffic between RegionMaster and the other platforms will be created, resulting in a smaller bandwidth left to the remaining platforms. For the purpose of creating network traffic, we utilize the open source software Iperf [28]. This tool is used for measuring throughtput and performance of a network. It can also be used to produce both TCP and UDP data streams over networks; data sent by the client will be received and eventually discarded by the server.

Since Iperf consumes a certain bandwidth on the connection between RegionMaster and the rest of the region, there is smaller bandwidth left for other applications. As a result, each agent platform will experience effects from the traffic generator. The latencies between the platforms and RegionMaster grow sharply. These changes are sensed by the scout agent every time it visits the platforms. Every platform sets the value *split* based on the ratio $\tau = \tau_{avg}/\tau_{RM}$; where τ_{avg} is the average latency and τ_{RM} is the latency to RegionMaster, respectively. In this experiment the splitting threshold is set to $\tau < 50\%$.

Table 5 shows the ratio and the corresponding value *split* for every platform. At the end of a round of exploration, the scout agent goes back to RegionMaster and submits information it collected. Based on this information, RegionMaster makes a decision.

Platform	τ_{avg}/τ_{RM} (%)	Split
D1	47,8	*true*
D2	47,6	*true*
D3	37,5	*true*
D4	27,5	*true*
FJ	—	*false*
PT	37,5	*true*
T1	44,4	*true*
T2	29,7	*true*

Table 5. Metrics measured at the time of self-organizing

Since most of the platforms set the value *split* to *true* the region is splitted. The new region structure is depicted in the central part of Figure 12. There is only one platform staying at the old region, this is RegionMaster. The remaining nodes join the new region with *D*2 promoted to be the new RegionMaster. In this case, an adaptation has been made, the old RegionMaster relinquishes its leadership in the only-one platform region and becomes an inferior node of the new region. Eventually, a new region emerges from the original region.

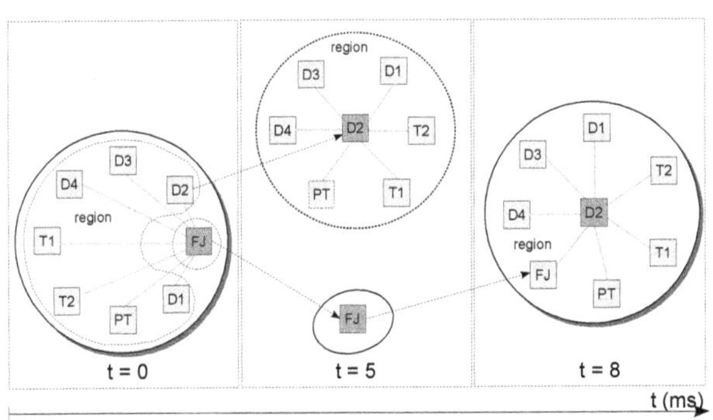

Figure 12. Exchange in role of the command node

Figure 13 displays the latency to RegionMaster τ_{RM} of every platform in three phases. For a platform, the left column is the latency before Iperf is activated;

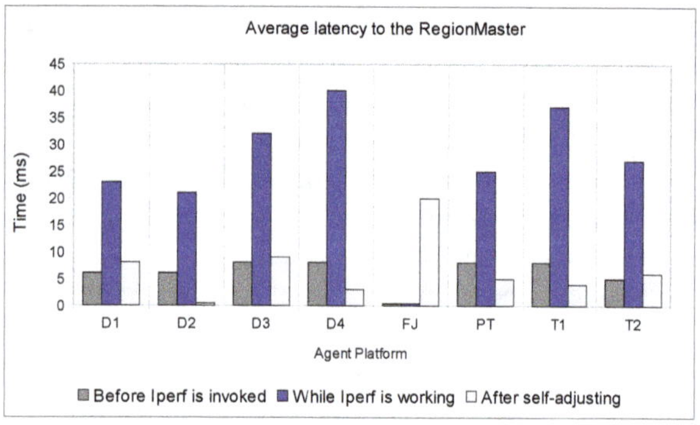

Figure 13. Average latencies to RegionMaster

the middle column represents the time while Iperf is operating and the right column is the latency of the platform after the self-adapting process has occurred. In the beginning and while Iperf was working *FJ* was RegionMaster so $\tau_{RM}(FJ) = 0$. Similarly, after *D2* has taken its job as RegionMaster $\tau_{RM}(D2) = 0$. As usual expected, while Iperf is being executed, the latencies to RegionMaster measured for every platform increase significantly. However, once *D2* takes over as RegionMaster, the latencies decrease proportionally.

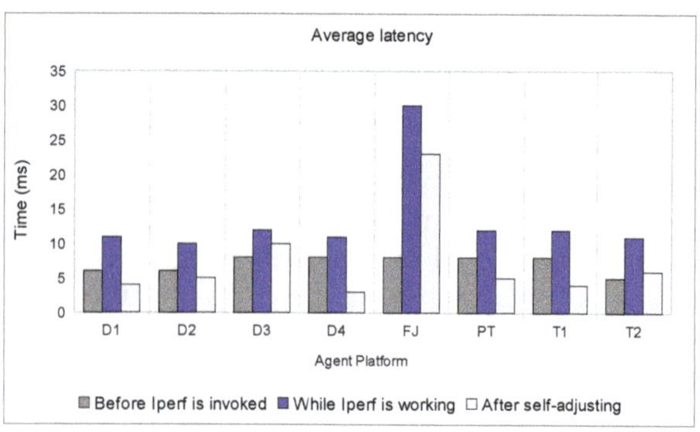

Figure 14. Avarage latencies of all platforms

Figure 14 shows the average latencies τ_{avg} of every platform in three phases: Before Iperf is executed, while Iperf is being executed and after the adaptation has been made. These latencies also shift in the same pattern as by τ_{RM}. Before additional bandwidth was produced, the average latencies had been at a normal level. While Iperf was operating the latencies rose markedly. After the region has been restructured, these values resume to a normal level. It can be seen that the network monitoring activities supply the framework with up-to-date information about network situation, thereby facilitating the decision making process. The measurement results suggest that the swap in role of RegionMaster from *FJ* to *D2* brings a more stable connectivity to every platform compared to that of the old arrangement, right after Iperf started producing bandwidth.

8.5. Fault Tolerance

We decided to export the test program to run on simulation environment, since a more precise evaluation result can be obtained if performance tests are performed with presence of several network nodes. The software Network Simulator NS2 is a discrete event simulation framework. It was written in the programming languages C++ and Python and has been used widely for simulating applications running in wired and wireless networks. NS2 supports various routing protocols, including TCP, UDP, multicast. NS2 is normally utilized for testing distributed network applications on a laboratory scale. Tcl (Tool Command Language) is used as the language for scripting simulation scenarios in NS2 [30].

AgentJ is a software tool for embedding real applications written in Java in the NS2 simulation environment. AgentJ allows Java source code to execute on NS-2 with minor modification. Essentially, AgentJ works as a bridge between Java source code and NS2 [32]. The use of AgentJ is practical given that there is a need for simulating program written in Java with solicitation of a large number of nodes. The combination NS2 and AgentJ provides us with a convenient way to simulate evaluation tests without needing to setup a real network with multiple computers.

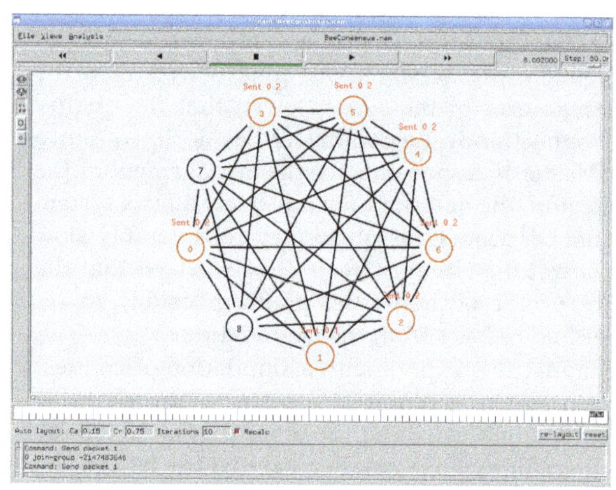

Figure 15. NS2 Simulation with a handful of network nodes

Network Animator (NAM) is a tool embedded in NS2 to visualize simulation. Figure 15 depicts an example of multicast network nodes in NAM, a circle represents a network node and a line between one pair of nodes corresponds to a link between them. In a multicast scenario, every node is connected to all other nodes. For a clear representation, in the figure we depict only a network with a handful of nodes, however in simulation, we can increase the number of network nodes to meet our requirement.

We setup NS2 and AgentJ in the Linux Distribution Fedora 12. Network crash is simulated by shutting down some nodes randomly during execution given that at most $f < n/2$ nodes may fail. A node fails with a probability of a randomized value ranging from 0 to 1.

In the evaluation, the number of nodes is set to the following values $n = 5, 10, 15, 20, 25, 30$ and the number of failed nodes $f \in \{0, \lfloor n/2 \rfloor\}$. We ran the tests with different sets of input values and with repetition. This aims to cover a wide range of possibilities of execution.

For the implementation, Function $Evaluate(0, 1)$ in Line 15 is determined as $Evaluate(0, 1) = abs(C_p(k, 0) - C_p(k, 1))$. Once a node starts to follow a value v, it assigns the difference to the strength value $strength = Evaluate(0, 1)$. Every time it is called, Procedure $Decrease(strength)$ subtracts 1 from $strength$ until it is smaller than 0.

The outcomes of the execution are shown in Table 6 and Table 7. For each table, the first row represents the number of nodes taking part in building consensus n. The second row is the number of nodes that fail during execution f. The third and fourth rows r_1 and r_2 are the corresponding numbers of rounds that Ben-Or's algorithm and the honey bee inspired version reach a consensus, respectively.

n	5			10			15		
f	0	1	2	0	3	4	4	6	7
r_1	1	1	12	1	10	20	4	23	60
r_2	1	1	2	1	2	2	2	2	2

Table 6. Performance Comparison for $n = 5, 10, 15$

n	20			25			30		
f	6	8	9	8	10	12	10	12	14
r_1	17	32	189	32	291	1319	112	1531	7554
r_2	2	2	2	2	3	2	4	2	2

Table 7. Performance Comparison for $n = 20, 25, 30$

At first sight, it can be seen that, the larger the number of nodes is, the more difficult a consensus can be reached. If no node fails, that means $f = 0$ then both approaches can gain a swift consensus. When failure is present, there are differences in performance. By the original Ben-Or's algorithm, if the number of failed nodes increases, the number of rounds that it gains a consensus increases correspondingly. Especially, when f is nearly approaching $n/2$ a large number of rounds can be seen.

Both tables show that the honey bee inspired algorithm has a considerably better computational performance, especially when network nodes fail en masse. Compared to Ben-Or's algorithm, it can gain a consensus after a small number of rounds of exchanging messages. In addition, its performance is stable towards the number of failed nodes. We witness an improvement in performance when applying the honey inspired mechanism.

9. Conclusion

In this paper, we have introduced our proposed architecture for mobile agent communication in highly dynamic networks. Our research was motivated by a mobile agent system working in networks for rescue forces in a mass casualty incident. In this type of networks, connection instability militates against successful transmission of agent messages. To facilitate message transmission, it is necessary to organize the logical connection among agent platforms in an operationally feasible manner. Our work aims to develop a communication model for mobile agent systems that is able to deal with the inherent dynamics of modern networks.

We found inspiration from the honey bee colony where honey bees have a special mechanism to reach self-organization. The organization of the whole colony is done through the interaction among thousands of bees. The behaviours of honey bees inspire us to employ the self-organizing model on the colony of agent platforms. We proposed a solution to a network management mechanism where network organizing tasks are performed in an autonomous manner.

Mobile agents are sent over the network to gather network and platform information. Information collected by mobile agents gives a base for managing the network. Each agent platform plays a contributory role in organizing the region. A final decision to reorganize the network is made based on a consensus of the platforms. In a platform colony, though the master node plays an important role, it is replaceable. Thanks to a failure detector, whenever it fails a second platform can substitute for it and system's operation is not interrupted. Based on the existing mobile agent system Ellipsis, the software prototype for an adaptive model for mobile agent communication has been designed and implemented.

The software prototype was deployed to run in a system of a laboratory scale where conditions of a real dynamic network had been simulated using various software tools. The experiments demonstrated that the software prototype provides the system with the ability of being self-adaptive. It helps the system to react adequately to environmental stimuli as well as to take suitable measures to counteract unexpected network degradation. In addition, given that network failure happened, the system can achieve fault tolerance and maintain a reasonable processing cost for message coding and transferring. From our perspective, the software framework fulfils the basic requirements.

It is our firm belief that the honey bee inspired model cannot only be utilized for mobile agent systems. It can also be applied in other types of P2P network, where it is necessary to network member nodes in a flexible manner. The model can be used to solve other types of problems in network management, such as for discovering resources in P2P networks [29].

References

[1] Seeley, T.D. and Kirk Visscher, P. (2003) *Choosing a home: How the scouts in a honey bee swarm perceive the completion of their group decision making* Journal of Behavioral Ecology and Sociobiology, vol. 54, pp. 511–520

[2] Seeley, T.D. and Kirk Visscher, P. (2004) *Group decision making in nest-site selection by honey bees* Journal of Apidologie, vol. 35, pp. 101–116

[3] Nakrani, S. and Craig, T. (2004) *On Honey Bees and Dynamic Server Allocation in Internet Hosting Centers* Journal Adaptive Behavior - Animals, Animats, Software Agents, Robots, Adaptive Systems , pp. 223–240

[4] Karaboga, D. and Bahriye, A. (2009) *A survey: algorithms simulating bee swarm intelligence* Journal Artif. Intell. Rev., pp. 61–85

[5] Pham, T. and Afify, A. and Koc, E. (2007) *Manufacturing Cell Information using the Bees Algorithm* In Proceedings Innovative Production Machines and Systems Virtual Conference

[6] Nguyen P.T. and Schau V. and Rossak W.R. (2011) *Towards an Adaptive Communication Model for Mobile Agents in Highly Dynamic Networks based on Swarming Behaviour.* The 9th European Workshop on Multi-agent Systems (EUMAS)

[7] Nguyen P.T. (2013) *Building Consensus in Context-Aware Systems Using Ben-Or's Algorithm: Some Proposals for Improving the Convergence Speed.* In Proceedings of the Second International Conference on Context-Aware Systems and Applications, ICCASA 2013, pp. 87–96

[8] Nguyen P.T. and Schau V. and Rossak W.R. (2013) *A Context-Aware Model for the Management of Agent Platforms in Dynamic Networks.* In Proceedings of the Second International Conference on Context-Aware Systems and Applications, ICCASA 2013, pp. 77–86

[9] Finin, T. and Labrou, Y. and Peng, Y. (1998) *Mobile Agents can Benefit from Standards Efforts in Inter-agent Communication* IEEE Communications Magazine, vol. 36, pp. 50-56

[10] Braun, P. (2003) *The Migration Process of Mobile Agents* PhD Dissertation, Friedrich Schiller University Jena.

[11] Baumann, J. (2000) *Control algorithms for mobile agents* PhD Dissertation, the University of Stuttgart.

[12] FSU Jena (2011) *The SpeedUp Project (2011)* http://www.speedup-projekt.de

[13] Baumann, J. and Hohl, F. and Radouniklis, N. and Rothermel, K. and Strasser, M. (1997) *Communication Concepts for Mobile Agent Systems* In Proceedings of the First International Workshop on Mobile Agents

[14] Fischer, M.J. and Lynch, N.A. and Paterson, M.S. (1983) *Impossibility of Distributed Consensus with One Faulty Process* Proceedings of the 2nd ACM SIGACT-SIGMOD symposium on Principles of database systems, pp. 1–7

[15] Heylighen, F. (1999) *The Science Of Self-Organization And Adaptivity* The Encyclopedia of Life Support Systems, vol. 62, pp. 253–280

[16] Seeley, T.D. and Buhrman, S.C (2001) *Nest-site selection in honey bees: how well do swarms implement the best-of-N decision rule?* Journal of Behavioral Ecology and Sociobiology, vol. 49, pp. 416–427

[17] Yeom, K. (2010) *Bio-inspired self-organization for supporting dynamic reconfiguration of modular agents* Journal of Mathematical and Computer Modelling, vol. 52, pp. 2097–2117

[18] Heylighen, F. and Gershenson, C. (2003) *The meaning of self-organization in computing* IEEE Intelligent Systems.

[19] FIPA (2002) *ACL Message Representation in String Specification*

[20] FIPA (2002) *ACL Message Representation in Bit-Efficient Specification*

[21] Ben-Or, M. (1983) *Another Advantage of Free Choice (Extended Abstract): Completely asynchronous agreement protocols* Proceedings of the Second Annual ACM Symposium on Principles of Distributed Computing

[22] Vavala, B. and Neves, N and Moniz, H. and Veríssimo, P. (2010) *Randomized Consensus in Wireless Environments: A Case Where More is Better* Proceedings of the 2010 Third International Conference on Dependability, pp. 7–12

[23] Aspnes, J. (2003) *Randomized Protocols for Asynchronous Consensus* Journal of Distributed Computing, vol. 16, pp. 165–175

[24] Aspnes J. (2002) *Fast deterministic consensus in a noisy environment* Journal of Algorithms, vol. 45, no. 1, pp. 16–39

[25] Thom, C., Gilley, D.C., Hooper, D., Esch H.E.: The Scent of the Waggle Dance. PLoS Biology (2007)

[26] Hayes, J.: Pleasure chemical controls bee dance. Cosmos Online (2007)

[27] Gadagkar,R.: The Honeybee Dance-Language Controversy. Resonance: Journal of Science Education, vol. 1, pp. 63–70 (1996)

[28] NLANR/DAST: Network Performance Measurement. http://sourceforge.net/projects/iperf/, (2012)

[29] Mittelbach F. and Goossens M (2004) *The LaTeX Companion* (Addison-Wesley), 2nd ed.

[30] The Network Simulator NS2, http://www.cs.virginia.edu/ cs757/slidespdf/cs757-ns2-tutorial1.pdf

[31] Aguilera, M.K., Toueg,S.: The correctness proof of Ben-OrŠs randomised consensus algorithm (2011)

[32] Taylor, I., Downard, I., Adamson, B., and Macker, J. AgentJ: Enabling java NS-2 simulations for large scale distributed multimedia applications. In Second International Conference on Distributed Frameworks for Multimedia DFMA 2006, Penang, Malaysia, pp.1-7, (2006)

7

Highlighted Activities of ICCASA 2015

Nguyen Thanh Tung[1,*]

[1]International School, Hanoi Vietnam National University , Vietnam

Abstract

The Fourth International Conference on Context-Aware Systems and Applications (ICCASA 2015) [1] is jointly organized by EAI, Nguyen Tat Thanh University (NTTU), and Ba Ria-Vung Tau University (BRVTU) and endorsed by the European Alliance for Innovation (EAI), a leading community-based organization devoted to the advancement of innovation in the field of ICT. The event consists of a main track on context-aware systems and applications, and two special tracks on contextual recommendation systems and on computational aspects of context in natural language processing, respectively. Especially, there are three keynote speeches, which will be presented at ICCASA 2015 by prominent invited speakers

Keywords: Context-awareness, Context-aware applications, Context-aware systems

1. Introduction

The Fourth International Conference on Context-Aware Systems and Applications (ICCASA 2015) [1] jointly organized by EAI, Nguyen Tat Thanh University (NTTU), and Ba Ria-Vung Tau University (BRVTU) is a place for highly original ideas about how context-awareness is going to shape networked computing systems of the future. Hence, it focuses on rigorous approaches and cutting-edge solutions which break new ground in dealing with the properties of context-awareness. Its purpose is to make a formal basis more accessible to researchers, scientists, professionals and students as well as developers and practitioners in ICT by providing them with state-of-the-art research results, applications, opportunities and future trends. For this fourth edition, we hope to repeat the success of previous year, when the conference received many papers and the participation of a large number of students, researchers, and professionals from all over the world.

Relevant topics include the following:

- Fundamentals of Context-aware systems
- Context-aware Systems:
- Context-aware Technologies:

The event is endorsed by the European Alliance for Innovation (EAI), a leading community-based organization devoted to the advancement of innovation in the field of ICT. All accepted papers will be published by Springer and made available through SpringerLink Digital Library, one of the world's largest scientific libraries. The proceedings are submitted for inclusion to the leading indexing services: DBLP, Google Scholar, Thomson Scientific ISI Proceedings, EI Elsevier Engineering Index, CrossRef, Scopus, as well as ICST's own EU Digital Library (EUDL).

2. Special Tracks

Special tracks in ICCASA 2015 will provide a meeting for presenting novel ideas in a possibly more focused way than the conference itself. Its aim is to engage in active exchange, interaction, and comparison of approaches, methods, and ideas related to specific topics, both theoretical and applied, in the general area of context-awareness.

*Corresponding author. Email: tungnt@isvnu.vn

2.1. Track on Contextual Recommendation Systems (CtxRecSys)

Recommendation system is an interesting topic for a large number of researchers over recent years. There are several approaches, models and techniques that can deal with some issues in recommendation systems. Most of existing approaches have focused on recommending the most relevant items to users without considering any additional contextual information, such as time, location, and expertise. In many applications, it contains several useful data to extract more information about the users. For example, using spatial context, a restaurant recommendation system can provide different lists of places for the same user, depending on his/her current locations. As another example, in music recommendation system, user may listen to romantic songs at the weekend but he/she may like jazz before sleeping. Many issues such as contextual information integration, context-aware recommendation techniques combination and evaluation approaches still remain open. Therefore, this track tries to be meeting point where researchers and practitioners in areas of Knowledge Management and Social Sciences, Collective Intelligence could have the opportunity to present current research results, to share their research achievements and solutions, and to look for new idea of context-aware recommendation systems.

Relevant topics include the following:

- Context-aware information retrieval
- Context-aware profiling, clustering and collaborative filtering
- Machine learning for context-aware information retrieval and ontology learning
- Context-aware e-learning/tutoring
- Ubiquitous and context-aware computing
- Context-aware advertising
- Recommendations for mobile users
- Context-awareness in portable devices
- Context-aware services
- Social and Conversational Agents
- Social Behavior
- Understanding Human Activity
- Memory and Long-term Interaction
- Cases studies from Collective intelligence applied to CtxRecSys
- Cases studies from Social Sciences applied to CtxRecSys

2.2. Track on Computational Aspects of Context in Natural Language Processing (CACNLP)

The vision of this special session is focused on computational aspects of context in natural language processing. The emergent issues of context-awareness include methods of context modeling and computing, impacts of context on meaning, natural language understanding and argumentation, interpretation of contextual information in theoretic and empirical models, and so on.

Relevant topics include the following:

- Context modeling and computing
- Context in natural language and speech processing
- Context in natural language understanding
- Context in natural language argumentation
- Context in discourse
- Context in dialog
- Context in information retrieval
- Context-awareness in (multidisciplinary) multimedia data analysis and mining, machine learning, knowledge representation and discovering

3. Keynote Speeches

The keynote speeches establish the framework for the technical program of ICCASA 2015. This year we choose three prominent invited speakers, Prof.Vangalur Alagar from Concordia University in Canada; Prof.Chintan Bhatt from Charotar University of Science And Technology in India and Prof. Phan Cong Vinh from Nguyen Tat Thanh University in Vietnam

3.1. Autonomic Urban Traffic Controller System - A Cyber-physical Systems Approach

Prof.Vangalur Alagar will give a talk at ICCASA 2015 on a hot topic of Cyber-Physical Systems (CPS) with the title of "Autonomic Urban Traffic Controller System - A Cyber-physical Systems Approach"

In almost every country in the world transportation infrastructure is severely overstressed. The traditional traffic control infrastructures have become awfully inadequate to handle the modern-day vehicular traffic, which can be characterized by density of traffic, speed of vehicles, behavior of human drivers, and constraints imposed by the traffic regulation policies laid down by urban administrators. Consequently, roadway corridors regularly experience increased level of traffic congestion, not only in developed countries such as United States, Canada and Europe, but also in all developing countries. Several reports have quantified the loss of productivity and wasted fuel usage resulting from traffic congestions. According to one report, in 2013 traffic congestion in U.S.A has cost Americans 124 billion US Dollars in direct and indirect losses, and this amount is estimated to rise 50 percent by 2030. Wasted fuel in traffic congestion is not only contributing to monetary loss but also a threat to health because of the steep increase in environmental pollution.

3.2. Toward IoT Universe

Prof.Chintan Bhatt will give a talk at ICCASA 2015 on a hot topic of Internet of Things (IoT) with the title of "Toward IoT Universe"

Ambient intelligence and autonomous control are not part of the original concept of the Internet of Things. Ambient intelligence and autonomous control do not necessarily require Internet structures, either. However, there is a shift in research to integrate the concepts of the Internet of Things and autonomous control, with initial outcomes towards this direction considering objects as the driving force for autonomous IoT. In the future the Internet of Things may be a non-deterministic and open network in which auto-organized or intelligent entities (Web services, SOA components), virtual objects (avatars) will be interoperable and able to act independently (pursuing their own objectives or shared ones) depending on the context, circumstances or environments. Autonomous behavior through collecting and reasoning context information plays a significant role in IoT. Modern IoT products and solutions in the marketplace use variety of different technologies to support such context-aware automation.

3.3. Recent Advances in Autonomic Computing

Prof.Phan Cong Vinh will give a talk at ICCASA 2015 on a hot topic of autonomic computing with the title of "Recent Advances in Autonomic Computing"

A common way to interpret self-* in autonomic systems (ASs) is to say that self-* actions are running on ASs. Triggers of self-* actions from self-* can be performed concurrently to transform one AS state into another. A first rule for self-* actions is this: the performance of a sequence of several self-* actions is itself the performance of a self-* action - a more complex self-* action, but a self-* action nonetheless. Algebraic objects called monoids are tasked with encoding the self-* actions perspective in all this, i.e. what the self-* action can do, and what happens when different self-* actions are done in succession. A monoid can be construed as a set of self-* actions, together with a formula that encodes how a sequence of self-* actions is itself considered a self-* action.

References

[1] ICCASA 2015: HTTP://ICCASA.ORG/2015/SHOW/HOME

8

Evaluation of a Facial Animation Authoring Pipeline Seamlessly Supporting Performance Capture and Manual Key-pose Editing

Fabrizio Nunnari[1,*], Alexis Heloir[1,2]

[1]Deutsches Forschungszentrum für Künstliche Intelligenz (DFKI), CAMPUS D3.2, 66123, Saarbrücken, Germany
[2]LAMIH-UMR CNRS 8201, Le Mont Houy, Univ. Valenciennes, France

Abstract

In this paper, we present an architecture following a novel animation authoring pipeline seamlessly supporting performance capture and manual editing of key-frames animation. This pipeline allows novice users to record and author sophisticated facial animations in a fraction of the time that would be required using traditional animation tools. This approach paves the way towards novel animation pipelines which seamlessly merge the roles of the animator and the actor. The second contribution is a method assessing a facial retargeting system, we conducted a user study where participants assessed the emotions conveyed by the facial expression displayed in the control and the authored animation. Contrary to existing evaluation methods, it factors out possible misinterpretations of the intended emotion and focuses on assessing the retargeting quality.

Keywords: facial animation, performance capture, retargeting, evaluation, animation authoring.

1. Introduction

The face is a very expressive and complex non-verbal communication channel for humans. It is the preferred mean of expressing emotions and share feelings. The importance of facial expression is not only the privileged location for emotion expression [1], it is also a paramount articulator for signed language where lack of facial animation prevents the content to be understood [2]. Depending on how researchers agree on categorizing them, the human face counts between 42 and 52 facial muscles. These muscles combines to activate and move the above skin to produce a surfacing expression.

In computer graphics, A facial *mesh*, is defined by set of triangles, stitched to each other in a three dimensional space, small enough to approximate the smoothness of the curvatures of the face. The underlying facial muscles are, in most cases not simulated. Instead, their surfacing influence is reenacted using a combination of so called blendshapes [3]. A blendshape is a vector of offsets to be applied to the 3D vertices describing the topology and the morphology of the face in a neutral configuration. Each blendshape is configured to move a small subset of the face triangles, as a single facial muscle would do. An animator can configure complex facial expressions by modulating the influence of a blendshape (i.e., by applying a scale to the vector of offsets) and by using multiple blendshapes at once.

The most straightforward metaphor depicting the craft of animation would be the puppetry metaphor: one could think of an animated virtual character as a digital marionette that might be actuated by an animator. In both cases, the limbs and other body parts are not usually directly manipulated, there is an intermediate interface offering higher-levels handles [4] that give an abstracted control to the puppeteer. This higher set of controls is called *control rig* [5] and its purpose is to ease the work of the animator. Professional animators manually author key-framed animations using elaborated control rigs, which often present more than two hundreds controllers. Twenty or more of these controllers are dedicated to the animation of the face. Despite of the presence of the control rig, animating the face of a 3D avatar is an activity which requires a significant amount of time, passion, and dedication. Fortunately, recent consumer-range technology [6] has proved to be capable of enabling users authoring animation of human-like bodies or interactively controlling physical or digital puppets.

Past research already proposed solutions to increase the productivity of animation through the performance

*Corresponding Author. Email: fabrizio.nunnari@dfki.de

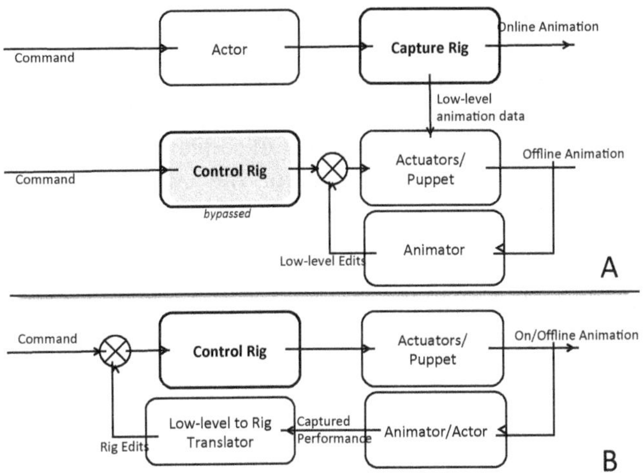

Figure 1. A: In existing workflows, the captured motions are recorded only once by an actor, then edited using low level controls.
B: In our approach, the animator is also the actor, and he is able to seamlessly switch between motion capture and manual edit during the animation authoring.

capture of facial expressions [7, 8]. In performance capture the animator can use his own face to drive the animation of an arbitrary avatar. So far, no performance capture system is able to guarantee the same expressive quality attained by manual animation. Hence, for real productions, the animation captured via performance capture must be furtherly post-processed and edited, by the animator. However, the output of the performance capture is expressed as a low-level animation data, which drive the low-level mesh deformers, i.e. the blendshapes, that are available on the animated figure. As already stated, it is not a convenient method for manual animation. The animator needs a control rig in order to refine captured performances.

In existing animation production workflows, the performance capture produces low-level animation data which are then transferred to the manual animation pipeline bypassing the control rig. Further edit operations will be a hard task for the human operator, and are for this reason relegated to data cleanup operations. If a major editing is needed, a new capture is required, as depicted in Figure 1A.

We propose to map the performance of the actor onto the control rig. As such, the animator can later refine the animation using the sliders of the control rig. In this use-case, the roles of actor and animator thus overlap. Experienced animators can seamlessly switch between traditional space-time constraint edit and interactive performance capture recording, as shown in Figure 1B.

In this paper, we introduce a pipeline driving the animation of arbitrary faces using performance capture and manual animation in an integrated approach. The animation pipeline includes an innovative stage which maps a captured facial performance onto a control rig. In so doing, the animator, after the capture phase, is able to perform further edits using the same control rig used in pure manual animation.

A facial retargeting method must be evaluated in order to prove its effectiveness. The evaluation can be either qualitative or quantitative. A quantitative evaluation based on inter-surface comparison can be applied only when the source and the target faces match, which happens when the target face is a 3D reconstruction of a real face. Since the source (perfomer's facial morphology) and the target faces (animated character's morphology) differ, a qualitative method must be used. A qualitative evaluation involves a user study. So far, there are no de facto standards to conduct the qualitative evaluation of a facial retargeting method. The second contribution of this paper is a qualitative evaluation procedure based on the recognition of emotions conveyed through facial expressions.

The assessment of a retargeting system is generally based on the judgment of facial expressions conveying specific "pure" emotions (e.g., happiness, anger, ...). But emotions are subject to misinterpretation; even when performed by professional actors. If the evaluation includes only correct interpretations, important cues about how subjects perceive the intended emotion are lost. Our procedure takes into account misinterpretations and allows for the evaluation of mixed facial emotions. This method is thus also suited to casual users without acting background.

The evaluation procedure starts with the selection of the users who will perform a set of reference facial animations. Each animation intends to convey a pure emotion. The performances of the users are later retargeted to the face of a virtual character and rendered as animated video clips. Another set of users, called respondents, will judge the reference videos, the original live performances, and the rendered 3D face. For each video, the judgment consists of expressing a level of perception for all the emotions involved in the study.

The judgment is implemented as an online survey asking to fill a form. The output of the online survey is a set of contingency tables. Contingency tables are a data structure commonly used in statistics to describe the frequency distribution of variables [9]. We measure the quality of the mapping method by comparing the contingency tables of the live performances and of the rendered videos.We show, in a user study, that the retargeting system proposed in this paper allows for users with no experience in computer animation and computer graphics to effectively use our tool

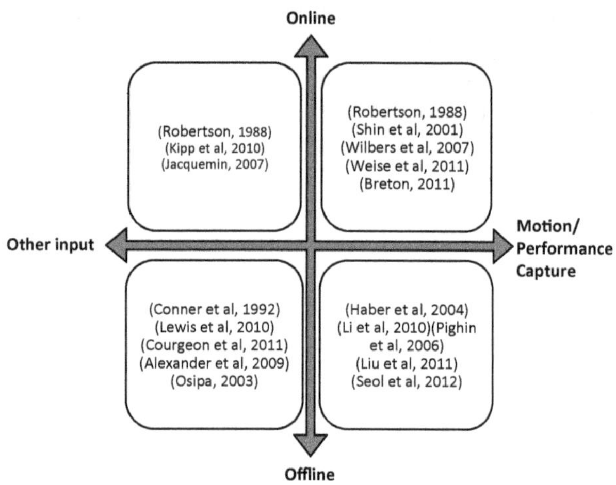

Figure 2. A comprehensive classification of the related work.

to produce convincing animations. Our evaluation method allows for the measurement of the quality of the mapping method even if the involved performers are not professional actors.

This paper is organized as follows. Section 2 presents related work in the field of facial performance capture. Section 3 describes in detail the architecture proposed in this paper, which is depicted in Figure 1B. Section 4 describes the method which maps the low-level capture information to the control-rig. Section 5 presents the user study that we conducted to evalutate the proposed workflow. Finally, Section 6 concludes the paper.

2. Related Work

In computer animation, the animation control interface and the way it is connected to the underlying body elements is called *control rig* [5]. As we will see in the following, rigs provide the interface between the manual inputs of an animator who authors an animation on a key-frame basis [3, 5]. Alternatively, the interface between an interactive input method and a virtual or real puppet generates low-level animation data [8, 10]. In fact, most animation authoring pipelines can be classified according to two dimensions: input type (manual edit vs captured motion) and usage (online puppetry vs offline animation). Figure 2 gives an overview of the related work we are going to present and discuss.

2.1. Offline Animation

In traditional animation, rigging is the process of setting up a group of controls to operate a 3D model, analogous to the strings of a puppet. It plays a fundamental role in the animation process as it eases the manipulation and editing of expressions. Control rigs are created to facilitate and speed up the task of animation. Their purpose is to reduce the degrees of freedom (DOFs) needed to control the character. They provide the animator with a simple but convenient layer of abstraction [4]. It is essential for the animator to have a rig that allows a right expression palette, that is convenient to use, and at the same time compact (not too many handles) and expressive (the handles provide a good factorization of the controllers). In the industry, rigs are created by a dedicated professional, the rigger, who sets it up in tight collaboration with the animator. A rig is therefore a complex mixture of science, art, and craft. As a consequence, rigs are very often unique, tailored for a single character, and tuned for a particular animator.

Following the conventions introduced in the early 90's [11], a manual control rig is a set of 3D objects, known as *controllers*, or *rig-handles*, which are used to manipulate a character's body or facial configuration. For example, a sliding controller (one degree of freedom) for the *smile* expression is generally associated to the movement of four or more blend shapes [3] which operate at the level of facial muscles. Another example, at skeletal level [12], is the rig-handle associated to the *hand opening and closing* action, where a single degree of freedom is associated to the rotation of 14 hand joints. Finally, the *wrist* rig-handle, very commonly used in full body animation, allows the animator to position and rotate a hand in space (6 DOFs) without having to manually edit the rotation of the joints of the arm, the clavicle and the spine (potentially more than 15 DOFs). This last handle generally invokes *Inverse Kinematic* (IK) algorithms [13], an approach inherited from robotics to calculate the rotations of a chain of joints given the desired position and rotation in space of a terminal end-effector.

Traditionally, the handles that compose the control rig are exposed to the animator in two ways: either in a different window than the viewport window displaying a 3D view of the character being edited, in which case handles are represented by simple 2D controllers constrained to squares [5, 14, 15], or more complex anthropomorphic controllers [16].

Sketching techniques have also been used as UI control systems [17]. Kipp et Nguyen [18] proposed a multi-touch interface allowing an animator to interactively control an IK-driven human arm and record poses or captured motions, whereas Sanna et al. [6] achieved a similar task for the whole body using the Kinect[1] depth-camera. Like Sanna et al., the proposed animation workflow uses a depth sensor, but for recording facial animation instead of the body. Like

[1] http://www.xbox.com/en-US/kinect/ - 7 Jan. 2015

Kipp et al., our system is also capable of storing single poses or recorded performances.

2.2. Online Puppetry and Performance Capture

When performing online puppetry, the puppeteer interactively manipulates a dedicated interface that allows the live control of a *puppet*. This puppet might be physical, made out of rag and wood, in which case the animation rig is a wooden frame actuating the puppet using strings. The physical puppet might as well be a way more complex Geminoid [10], with an elaborate control rig capable of converting real time live 3D data collected from a set of facial markers glued onto a performer's face into dynamic commands sent to the geminoid's actuators. In 1988, Robertson demoed the first interactive anthropomorphic computer puppet: Mike [19]. Mike was capable of performing simple facial expressions and moving the lips according to the puppeteer's speech. Mike's animation rig consisted of data glove and a speech recognition system.

Mike was later followed by systems capable of capturing in real time the animation of an actor's face without markers. For example, see [8] or the Dynamixyz[2] systems. When it includes face, Motion Capture [20] is often referred to as *performance capture*. In performance capture, the puppeteer's face usually does not match the puppet's face: they often have different topology and morphology. The mapping of the puppeteer's facial motion on the puppet face is called *retargeting* [7, 21–23].

All the surveyed retargeting approaches are conceived to map the performance to low-level animation controls. None of them maps directly to a control rig, ignoring the need of a post-production editing phase. Seol et al. [24] specifically address the problem of providing room for additional editing by an animator. But again, the target of their retargeting method is a set of blendshapes, which, for animators, is harder to use with respect to a control rig.

The animation workflow presented in this paper is based on the technology developed by Weise et al. [8], where a facial expression is described through a weights vector of a set of blendshapes. Section 4 describes the mapping method which translates the weights vector into a configuration of the control rig. The translation is computed via a linear cross-mapping [25]. This translation stage is a form of *facial motion retargeting* which, following the terms introduced by Pighin [7], is based on a *scattered data interpolation* approach (the mapping is driven by a matrix multiplication function) fed by an *art directed* input (the authors manually edited the correspondence values – matrix elements – between each source blend shape to a target control rig configuration.)

The effectiveness of a facial retargeting method must be evaluated. The evaluation can be qualitative or quantitative. A qualitative evaluation involves a user study. A quantitative evaluation is based on the computation of a numerical difference between the source and the target expressions. Previous results, whether qualitative [23] or quantitative [26], tell us that a perfect match between the source and the target is never achieved. The animation pipeline presented in this work is evaluated, qualitatively, by measuring the recognizability of the six basic facial emotions defined by Ekman [1]. In a previous work, Song et al. [26] used facial emotions like *expressing surprise*, *crying*, and *laughing* to train the retargeting routine. In our work, we exploit emotion transfer to evaluate the retargeting. Also Battocchi [27] and Costantini et al. [28] used emotion judgment as mean of evaluation. However, as shown in [27], even if a professional actor perform the emotions, subjects falls into misinterpretation. As great as an actor can be in conveying a certain emotion through a facial expression, there are no guarantees that the emotion is correctly perceived. The correctness of the perception decreases when the same emotion is conveyed by the target animated face.

In the above mentioned studies, subjects are asked to guess the emotion conveyed by a short video, or a static picture, which shows a facial expression. Subjects rate a video by selecting the perceived expression from a closed list. The goodness of a facial expression is evaluated as ratio between correct vs. incorrect classifications. In this way, information about the incorrect interpretation is lost. When the facial expression of a human is transferred, to a virtual character, this misinterpretation is expected to be found also in the target virtual face.

The evaluation method proposed in this paper retain misinterpretations and includes them in measuring the quality of the retargeting. This is accomplished by asking subjects to rate each video across all the emotions. Subjects rate a video (showing an expression) through several Likert scales, one for each emotion. The result of the evaluation of a video is a contingency table describing the distribution of the votes.

The quality of the facial retargeting is then measured by comparing the contingency tables associated to the source and the target expressions. In statistics, the similarity between contingency tables is generally computed using a Chi-square test for goodness of fit[3] [9]. However, the chi-square test is not applicable to the proposed method because the votes distribution is likely to leave many cells at 0. This breaks one of the

---

[2]http://www.dynamixyz.com/ - 7 Jan. 2015

[3]http://en.wikipedia.org/wiki/Chi-square_test - 7 Jan. 2015

pre-conditions for the applicability of the Chi-square test, which requires that all levels have a minimum value of 5.

Hence, our method relies on a different metrics to compute the similarity between contingency tables, namely, the generalized Jaccard similarity coefficient[4] [9]. The Jaccard similarity coefficient expresses the similarity between two vectors of real numbers by performing a pairwise comparison on the elements of the vectors. Section 5.3 describes in the detail how the Jaccard coefficient is computed and used.

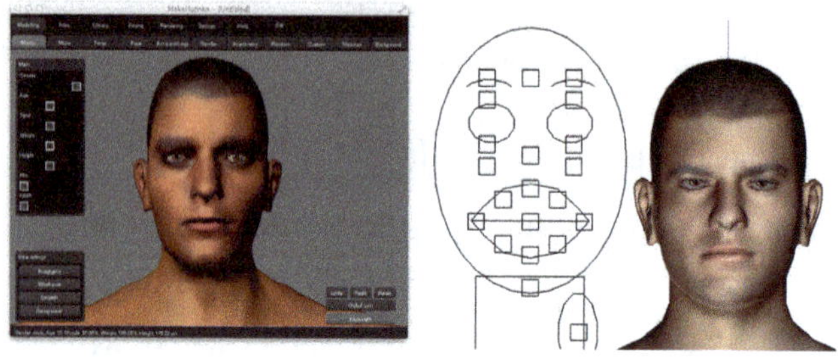

Figure 4. Left: the MakeHuman GUI. Right: the head mesh and the control rig imported in Blender

3. Architecture

This section describes the authoring pipeline proposed in this paper. The description gives details of all the stages composing the pipeline, the software therein used, and how the stages are connect with each other.

Four stages compose the authoring pipeline, as depicted in Figure 3: 1) Creation of the 3D character, 2) Animation of the character's face using a Natural User Interface, 3) Refinement of the animation using traditional mouse+keyboard approach, and 4) Exportation of the resulting animation as movie or as animation data.

Stage 1 of the proposed pipeline is based on MakeHuman[5]: an open-source tool for making 3D characters [15]. MakeHuman allows non-expert users for easily authoring a digital character in all its details, without the need to accomplish the complex actions required by full-fledged 3D animation tool (like Autodesk's 3D Studio and Maya, or the open-source Blender), such as vertex-based modelling, vertex weighting, UV mapping, and control rig construction. It offers users an easy-to-use Graphical User Interface (GUI) to edit a human model. The GUI of MakeHuman (Figure 4, left) is mainly composed of sliders. Each slider controls the deformation of an attribute of the human body: age, gender, height, weight, arms and legs length and size, distance between eyes and ears, torso depth and width, are only a few among the more than hundred controls available. Starting from a default androgen 3D character, the user operates the sliders to achieve the desired result.

The edited human can be exported as static 3D mesh in the OBJ and Collada formats. For the proposed pipeline, an editable version of the character can be imported in Blender through the dedicated MHX (Make Human Exchange) exchange format. Once imported in Blender, the character presents both a skeletal structure and an external 3D mesh. The 3D mesh includes a rich set of blend shapes for facial animation (70 in total). Finally, the character offers both a facial and a full-body control rig. The character is thus ready-to-use for a full featured animation work. The MakeHuman face animation rig is composed of 20 controllers. Each controller is a 3D box, sliding on the xz plane, which controls a particular element of the face plus the tongue. The latter has not been used in our work because FaceShift doesn't provide motion capture data of the tongue. Three additional anthropomorphic controllers allows for the rotation of the neck and the eyes. Figure 4, on the right, shows a screenshot of the control rig together with the associated mesh of the head and the face.

Stage 2 of the pipeline consists of animating the character via a performance capture, within the Blender software, using a Natural User Interface. The performance capture is based on the use of a depth sensor (like the Microsoft Kinect, or any equivalent on the market) together with the FaceShift[6] application.

FaceShift is an application able to use a 3D depth sensor to reconstruct the 3D mesh of a face an animate it in real-time according to the actual facial expression of the performer. The details of its underlying implementation can be seen in [8]. Faceshift is based on the OpenNI[7] middleware, which guarantees compatibility with a wide range of Kinect-like hardware available on the consumer market. FaceShift requires a calibration phase for each user. The calibration, lasting less than 10 minutes, consists of mimicking a set of facial expressions. While keeping an expression on his/her face, the user turns his/her head of about 30 degrees on both sides, for a couple of times, to let the system collect data about the shape of the face. The result of the calibration is a 3D reconstruction of the user head and face. The 3D face can be animated via blendshape deformations. A vector with the weights of the blendshapes represent the current state of the user facial expression. Each weight is a real number normalized in the range $[0, 1]$. After the calibration,

[4]http://en.wikipedia.org/wiki/Jaccard_index - 7 Jan. 2015
[5]http://www.makehuman.org/ - 7 Jan. 2015
[6]http://www.faceshift.com/ - 7 Jan. 2015
[7]http://structure.io/openni - 7 Jan. 2015

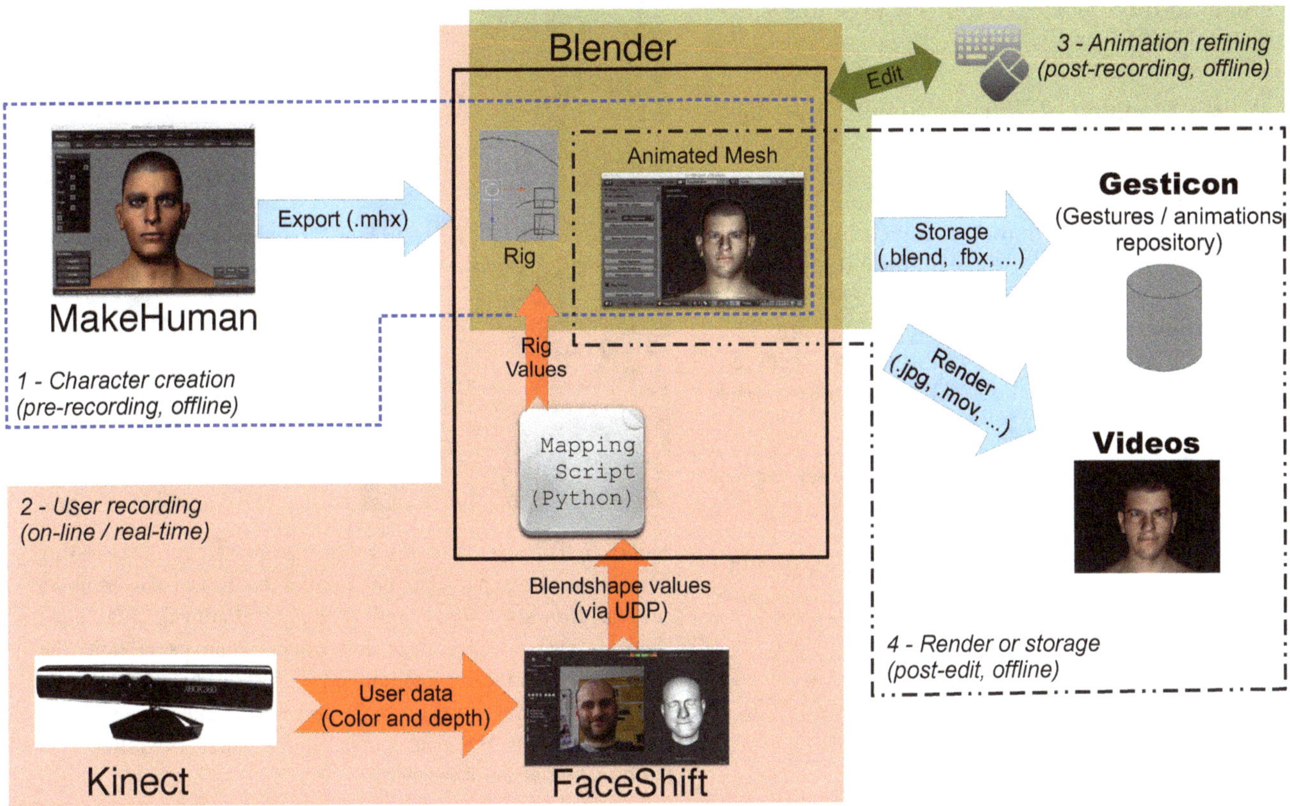

Figure 3. The architecture of the proposed workflow.

FaceShift can be used in real-time tracking mode. Here, the user face expression framed by the Kinect is continuously analyzed and used to pilot the 3D reconstruction of the user face.

The integration with Blender is performed thanks to the capability of FaceShift to output its tracking information to TCP or UDP network sockets. The captured data, which include the vector with the blendshape weights, is processed by a Blender addon, written in Python, developed by the authors. The Blender addon, described in Section 4, decodes the FaceShift binary network communication protocol and allows for the real-time recording of the facial expression and the head movement. More importantly, the addon maps the FaceShift blend shapes values to the control rig of the face. In so doing, the Blender user can have access to the control rig animation keyframes to further edit the facial animation.

Stage 3 of the pipeline allows for the user to manually operate on the control rig to refine facial animations. Obviously, the refinement of the facial and head animations can be associated to a synchronised full body animation for a more comprehensive production. The recording session generates animation keyframes at a rate of about 30 frames per seconds. As a result, animations present a very high data density which makes them hardly editable by hand. Thus, in our pipeline we suggest to perform a curve simplification through the *simplify curve* addon included by default in Blender.

Finally, in **Stage 4**, the edited animations can be rendered as videos, or image sequences, or stored in a separate Gesticon (gestures and animations repository). A Gesticon is intended to be a repository of classified animation data that can be reused across different scenes of the same project or across different projects. Common actions such as smiling, walking, running, or a full set of signs used in sign languages are all perfect candidates for a Gesticon.

The pipeline was developed and tested with the following software versions: MakeHuman v1.0 alpha7, FaceShift 1.1.05, Blender 2.66a. MakeHuman and Blender are both open-source projects and available for free. FaceShift is a commercial application, but facial performance capture is a feature of increasing popularity which might be integrated in some competing open-source project in the near future.

The Python scripts developed during this study are published online[8].

4. Real-time Mapping of Tracked Blendshapes to Control-rig

This section describes the retargeting method which converts the tracking information received by FaceShift into the MakeHuman control rig. The method has been developed as an addon of the Blender 3D authoring tool and is implemented in the Python language. The tracking information sent by FaceShift are received by the addon via UDP packets. The addon uses the following data:

1. a *track_ok* flag indicating whether the user face is being correctly tracked or not;
2. head rotation (yaw/pitch/roll angles as quaternion);
3. eyes rotation (yaw and pitch angles for each eye);
4. and a set of 48 float values, in the range [0,1], representing the weights of the 48 blend shapes animating the 3D model of the face.

The mapping script works as follows. If the *track_ok* flag is false, the received data are skipped and the MakeHuman rig isn't moved. Otherwise. The rotation of the head rotation is directly applied to controller *Neck*. The *theta* and *pi* rotation angles of the eyes are mapped as *-theta, 0, phi* euler angles (in XYZ order) to the controllers *Eye_L* and *Eye_R*. Finally, the blend shape values are mapped to offsets of the facial control rig, as described in the rest of this section.

Each FaceShift blend shape value consist of a float number in the range $[0,1]$. When all the blend shape values are at 0 the FaceShift face is in its neutral position. In Blender, each controller of the facial rig is associated to a 3D vector which acts as offset of the controller from its default position. The face is neutral when all offsets are at $< 0,0,0 >$. The main mapping procedure consist of applying the 48 FaceShift blend shape values (EyeBlink_L, EyeBlink_R, EyeSquint_L, ..., CheekSquint_R) to 3D vector offsets for the 20 controllers of the MakeHuman facial rig (PBrows, PBrow_R, PBrow_L, ..., PJaw).

The mapping is computed as a single multiplication between a matrix M and the blend shapes vector. Given the vector V_{bs} of 48 blend shape values, the resulting facial rig configuration vector V_r (20 3D vectors) is computed as $V_r = V_{bs} * M$, where M is a 48 lines by 20 columns matrix. Each column of the matrix is associated to a controller of the rig, while each line of the matrix is associated to a blendshape. Each line of the matrix represent how much the 20 controllers have to be offsetted when the corresponding blend shape is a maximum value (1.0). In other words, during the multiplication, the 48 offset vectors associated to a rig controller are scaled according to the actual blend shape values. The 48 scaled offsets are then accumulated (added) into a final offset. That's repeated for each controller.

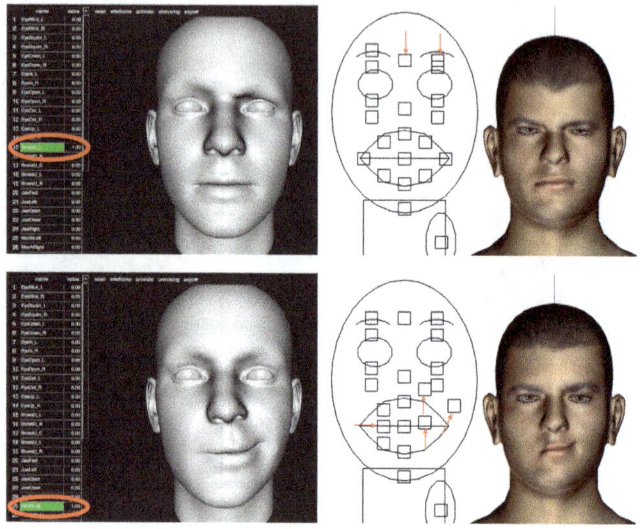

Figure 5. Examples from the calibration procedure of the matrix M. The picture shows the configuration for two of the 48 blend shapes: *BrowsD_L* (top) and *MouthLeft* (bottom). The blend shape is shown in the FaceShift configuration panel (left, the active blend shape is highlighted). The resulting expression is manually reproduced (as close as possible) in Blender using the control rig (right, the arrows indicate the moved controllers and the direction of the movement).

The matrix M has been manually configured by the authors, following an artistic-directed approach [7]. The FaceShift 3D model and the MakeHuman character in Blender have been put side-by-side. One by one, each of the 48 FaceShift blend shapes has been visualized at full weight. For each blend shape, the MakeHuman control rig has been configured to match (as much as possible) the Faceshift expressions. Each resulting control rig configuration is a list of 20 3D offsets. Each list has been inserted as line of the matrix M. Figure 5 shows examples of the side-by-side authoring process. Table 1 reports an extract of the resulting matrix.

After the multiplication, a logarithmic function is applied to each separate component (x,y,z) of the vectors in V_r. The logarithm limits the saturation resulting from the additive approach while leaving sensibility for low values. Given the control rig components offset range $[-0.25, 0.25]$, a log base of 3.5

[8]http://slsi.dfki.de - 7 Jan. 2015

Table 1. An extract of the matrix M mapping the blendshapes to the facial control rig.

		rig offsets (20)			
		PBrows	PBrow_R	...	PJaw
blend shapes (48)	EyeBlink_L	0, 0, 0	0, 0, 0	...	0, 0, 0
	EyeBlink_R	0, 0, 0	0, 0, 0	...	0, 0, 0
	...	...	...	...	...
	BrowsD_L	0, 0, 0.12	0, 0, 0	...	0, 0, 0
	BrowsD_R	0, 0, 0.12	0, 0, 0.12	...	0, 0, 0
	...	...	...	...	...
	CheekSquint_R	0, 0, 0	0, 0, 0	...	0, 0, 0

was empirically found to be satisfactory.

Actually, the rig controllers are constrained to move along 1 or 2 dimensions, yielding to a total of ca. 30 controlled degrees of freedom out of the 60 possible (20 vectors * 3 axes). Additionally, for most of the blend shapes, only one or two offsets out of 20 are used. As a consequence, most of the matrix elements are set to 0. Hence, there is room for optimizations. But this specific case is peculiar to the MakeHuman rig, while the algorithm is open to support more complex mapping configurations.

Two modification were applied to the original MaheHuman rig in order to increase its expressivity level. First, the range of movement of the Eyebrows was quite limited in the default MakeHuman rig. Hence, the range of movement of the PBrows, PBrow_L and PBrow_R controllers was doubled before calibrating the matrix M. Second, the corners of the mouth did not retract and widen enough. This is solved by controlling the value of the *mouth retraction* Expression Unit. Among the rich set of controllers in the MakeHuman rig, Expression Units are a set of facial expressions achievable through a set of dedicated sliders (range $[0, 1]$). Such expressions cannot in many cases be matched using the 3D controllers. The level of the mouth retraction control w_{mr} is calculated through an average of the offset applied to the mouth corners:

$$w_{mr} = \frac{PMouth\_L_z + PMouth\_R_z}{2} * 1.8$$

The Constant 1.8 has been again artistically determined.

Consider that the two above described modifications can be implemented by editing the control rig in Blender, rather than via explicit code, thus leaving the whole retargeting to a pure matrix multiplication. However, such option would have required a significant amount of 3D authoring which goes beyond the expertise of the authors.

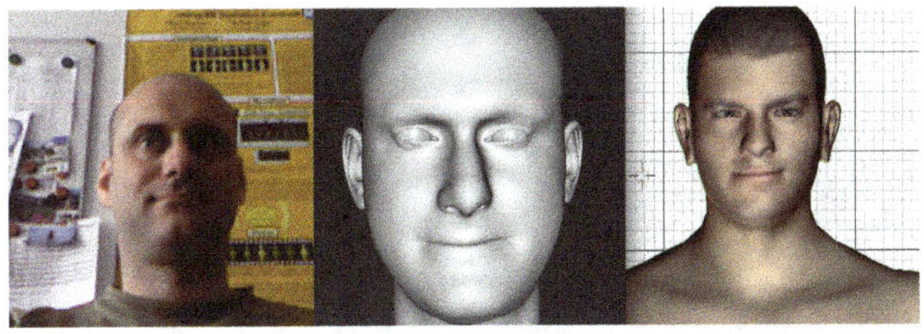

Figure 6. From left to right, screenshots of: the *live* Kinect color camera, the preview in FaceShift, the *rendered* character.

It is worth noticing that the resulting matrix M can be re-used for any FaceShift user and any additional character exported by MakeHuman. This is possible because both the blendshape values and the control rig offset spaces are normalized, respectively, in the ranges $[0, 1]$ and $[-0.25, 0.25]$.

5. Evaluation

This section describes the subjective user study which evaluates the animation workflow presented in the previous sections. The purpose of the user study was to give an estimation of the efficiency of the facial retargeting method and consisted of two stages. Firstly, the subjects used the system at the lab and recorded a set of dynamic facial expressions. The facial expressions were inspired by a set of *reference* videos performed by a professional actor. Each video showed one of the six Ekman [1] basic emotions: happiness, anger, disgust, surprise, fear, sadness. The recording stage produced two new set of videos: the *live* recordings of the subjects and the *renders* of the virtual character. Figure 6 shows an example. Secondly, another set of subjects (respondents, in the following) assessed the reference, the live, and the rendered videos via an online questionnaire.

The remaining of this section gives the details of the two evaluation stages and reports a discussion of the results.

5.1. Recording of Dynamic Expressions

Sixteen students from the campus participated to the recording stage. All of them were studying Computer Science or Communication Technologies. The call has been advertised in mailing lists, social networks, fliers, and posts on public billboards. Participants were rewarded with a 3D model of their head. The 3D model was acquired during the calibration of FaceShift. Two participants were female and two were male. They were informed of the purpose of the study, were instructed to watch the reference videos, and were invited to reproduce the emotions through their facial expression.

The performance capture has been conducted on a laptop and an additional external 22-inches monitor.

Figure 7. The user study setup.

The subjects were sitting in front of the monitor, without keyboard nor mouse. No written instruction was given to them since they were directed by an operator sitting next to them.

The Kinect camera was positioned in front of the monitor, under the screen (See Figure 7). Subjects's face distance from the Kinect was kept around 65cm, which is the optimal distance suggested by the FaceShift software. A mirror was provided to help users during the task.

The operator, using the laptop, assisted the subject during the calibration and provided guidance during the task. The operator was able to monitor the behaviour and the facial expression of the subject by looking the Kinect color camera preview on his screen.

Before starting the capture, subjects were assisted by the operator in configuring the FaceShift software. During the configuration, the FaceShift control window was shown on the screen of the subjects. The configuration includes a FaceShift training phase, which consists of holding a set of facial expressions while the Kinect capture depth and color information. After the training, users could play for one to two minutes with the FaceShift tracking interface, which showed the Kinect color camera output next to the 3D rendering of their head reconstruction.

After the configuration, the performance capture started. The FaceShift GUI was removed from the subjects' monitor, leaving space to the Blender GUI and the window of a video player. The Blender GUI was showing the virtual character exported by MakeHuman, which was moving its head and face, in real time, according to the subject's performance. Subjects were given one to two minutes to accommodate with the animated character. The task for the subject consisted of watching a reference movie of an actor performing a dynamic facial expression and reproducing the same facial expression on the virtual character. Starting from a neutral expression, after a short countdown, subjects had 15 seconds to reproduce the facial expression and go back to a neutral expression. The mirror was helping the subjects to fix the desired expression on their own face before applying it to the virtual character. Professional animators do the same: they explore different stances in front of a small mirror before authoring a facial pose using their traditional manual animation rig.

The six reference videos, displaying the six Ekman's emotions [1], belong to the DaFEx database of dynamic facial expressions [27] (Actor 1, high intensity). The videos show a professional female actress. High intensity can be considered as mild. For each subject the experiment lasted between 30 and 40 minutes: 8 to 10 minutes were needed for FaceShift configuration, the remaining time for the recordings.

5.2. Online Survey

The second part of the study was conducted as a public online questionnaire written in English, and advertised in mailing lists as well as social networks. The questionnaire was anonymous and consisted of 57 pages. The first and the third pages introduced the questionnaire and presented the context of the study. The second page gathered anonymous information about the respondents (age, experience with computers and video games). The following pages (54) displayed, in random order, the expression videos described below.

Four subjects ($U5$; $U6$; $U8$; $U10$) were selected among the 16 who participated to the recording stage. These subject were selected according to their "acting talent". The aim was to select users able to perform convincing facial expressions. The acting talent was first determined by the authors via a screening of the performances and later verified according to the rates of the online survey. The verification consisted in comparing the rates of the users to the ones of the professional actor.

As such, the 54 videos submitted for rating were composed as follows. Six *reference* videos, showing the six facial expressions performed by the professional actor; 24 *live* videos ($= 4 * 6$), showing the six facial expressions performed by the four selected subjects (these videos were recorder during the performance capture); another 24 *rendered* videos showing the virtual character animated by the performance capture.

Respondents of the online questionnaire were asked two questions per video. The first request was to rate each video along the six Ekman dimensions: happy, surprised, scared, sad, angry, disgusted. Each rating was done on a 6 point scale from 0 to 5.

The second question was: *"If you had to choose one word that qualifies the expression, what would it be?"*. Users had to select one among: happiness, surprise, fear, sadness, anger, disgust. This question was mainly used to detect possible inconsistent answers and, if necessary,

Value	Range
Happy	$[0 \ldots 5]$
Surprised	$[0 \ldots 5]$
Scared	$[0 \ldots 5]$
Sad	$[0 \ldots 5]$
Angry	$[0 \ldots 5]$
Disgusted	$[0 \ldots 5]$

discard respondents. Circa 30 minutes were needed to complete the on-line survey.

In total 96 participants answered the questionnaire. Among them, 30 completed it. The questionnaire was driven by the Limesurvey[9] software.

5.3. Similarity between contingency tables

This section describes the formula to calculate the similarity coefficient between two set of ratings (reference vs live, live vs rendered, ...).

Table 2 shows an example of the votes obtained from the video displaying a happy expression in the reference, live, rendered, and "ideal" version. The *ideal* table is an hypothetical distribution of votes where all the subjects give the maximum score to the intended emotion and zero to all the other emotions. The rates for the live and rendered videos are normalized to have the same total per line as the reference videos. The normalization is required to compute the similarity between types.

The similarity between two types of videos is computed using the generalized Jaccard similarity coefficient [9]. The coefficient J is defines as:

$$J(\mathbf{x}, \mathbf{y}) = \frac{\sum_i \min(x_i, y_i)}{\sum_i \max(x_i, y_i)}$$

where $\mathbf{x}$ and $\mathbf{y}$ are two vectors of the same length. The coefficient lies in the range $[0, 1]$. Comparing a vector with itself results in a coefficient value 1.0: a perfect match.

The coefficient introduced above can be used to measure the similarity between any video type. The tables are converted into vectors by serializing in the same order, line by line, the voting counts. Table 3 shows the results of the comparisons between some video types, divided by expression. The similarity between the *ideal* and the *reference* tables measures the performance of the actor. The similarity between the *ideal* and the *live* tables measures the performance of the subjects. The similarity between the *ideal* and the *rendered* tables assess the performance of the virtual character. Finally, the similarity between the *live* and the *rendered* tables assesses the goodness of the retargeting method.

The Jaccard coefficient is always positive.Hence, to evaluate which of the two compared type performs better, or worse, the coefficient must be compared with the averages of the input vectors. The coefficients shown in Figure 3 are discussed in detail in the next section.

[9]http://www.limesurvey.org/ - 7 Jan. 2015

5.4. Results and Discussion

Complete vs incomplete questionnaires. In total, 96 respondents participated to the survey. Only 30 respondents completely filled the questionnaire, probably due to the excessive time required (ca. 30 minutes). Before including the ratings of the incomplete questionnaires in the overall statistics, ANOVA tests between the ratings obtained in a dimension (0 to 5) vs. the questionnaire status (complete, incomplete) have been conducted. An ANOVA test has been conducted separately for each dimension (happiness, anger, ...). For all of the dimensions the p-value was > 0.1, meaning that, with at least 10% confidence, there is not significant difference between the scores obtained in the complete questionnaire vs. the scores obtained in the incomplete questionnaires. We therefore included the partially filled questionnaires into the statistical analysis.

Reliability of the respondents. The respondents of the questionnaires rated the videos similarly to previous published results. Figure 8 shows the ratings obtained by the six *reference* videos along each dimension.

These results can be compared with the first row of Table 3, showing the coefficients of the comparison between the ideal rating and the reference videos. Our subjects recognized pretty wrongly the *disgust* dimension (J=0.438). All the other results match with the results claimed in [27], who evaluated their reference videos on 80 subjects. Quoting [27]: "anger seems to be the emotion which is best recognized, [...]. On the other hand, fear seems to be the emotion which is worse recognized [...]. Happiness seems to be the most stable emotion".

The average of the coefficient (average J=0.656) can be taken as measurement of the talent of the actor. The standard deviation (0.122) given an estimation of the consistency of the acting across all the dimensions.

"Acting talent" of the performers. Figure 9 shows the results of the rates obtained by the four subjects selected as best performers. This graphs can be compared with the second row of Table 3, showing the coefficients of the comparison between the ideal condition and the live videos. The coefficients confirm that, as average, the subjects performed worse than the professional actors (average J=0.536 vs. 0.656). There is an exception is the disgust dimension, where subjects

Table 2. The rates distribution for the happiness expression: i) the rates obtained by the reference videos, ii) the rates for the live videos, iii) the rates for the rendered videos, and iv) the ideal distribution. The live and rendered rates have been scaled to match the same total per line of the reference rates. The ideal table is generated ad-hoc.

Reference	0	1	2	3	4	5	Tot
Hap	0	0	1	4	12	33	50
Ang	45	5	0	0	0	0	50
Dis	46	4	0	0	0	0	50
Sur	40	6	2	2	0	0	50
Fea	45	4	0	1	0	0	50
Sad	43	5	1	1	0	0	50

Live	0	1	2	3	4	5
Hap	0.00	0.63	2.22	7.91	15.82	23.42
Ang	47.78	2.22	0.00	0.00	0.00	0.00
Dis	47.15	2.85	0.00	0.00	0.00	0.00
Sur	36.71	8.54	3.48	1.27	0.00	0.00
Fea	48.10	1.90	0.00	0.00	0.00	0.00
Sad	45.25	3.48	0.32	0.95	0.00	0.00

Rendered	0	1	2	3	4	5
Hap	8.72	2.33	5.52	10.17	12.21	11.05
Ang	38.95	4.36	1.74	1.45	1.45	2.03
Dis	42.73	4.36	1.45	1.16	0.29	0.00
Sur	41.28	6.98	0.58	0.58	0.58	0.00
Fea	41.86	4.07	0.87	1.74	0.00	1.45
Sad	45.06	2.91	1.45	0.29	0.29	0.00

Ideal	0	1	2	3	4	5
Hap	0	0	0	0	0	50
Ang	50	0	0	0	0	0
Dis	50	0	0	0	0	0
Sur	50	0	0	0	0	0
Fea	50	0	0	0	0	0
Sad	50	0	0	0	0	0

Table 3. The Jaccard coefficient between selected couples of video types.

Groups	Meaning	Hap	Ang	Dis	Sur	Fea	Sad	Average	Std dev
ideal vs. reference	Actor performance	0.724	0.769	0.438	0.661	0.560	0.781	0.656	0.122
ideal vs. live	Subjects' acting perf.	0.707	0.475	0.599	0.540	0.445	0.449	0.536	0.094
ideal vs. rendered	Virtual character perf.	0.583	0.533	0.459	0.480	0.380	0.387	0.470	0.073
live vs. rendered	Retargeting quality	0.754	0.784	0.738	0.842	0.834	0.793	0.791	0.038

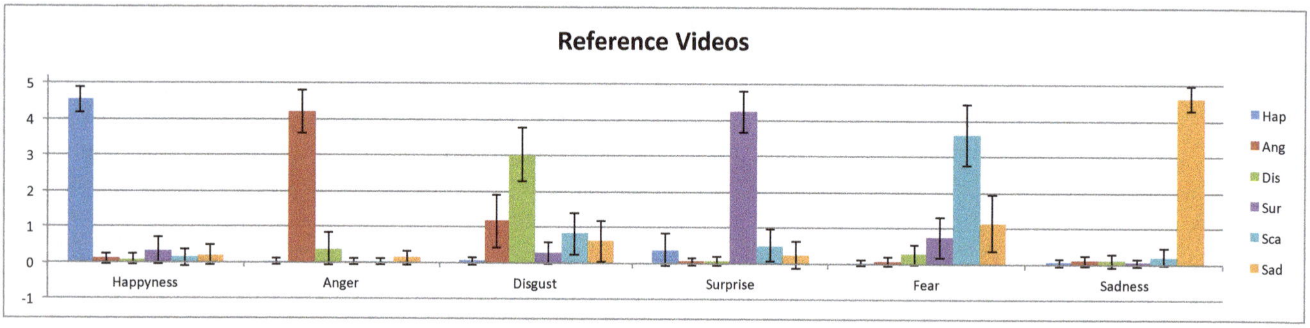

Figure 8. Rating obtained for the reference videos (actor) for each expression.

performed better than the actor (J=0.599 vs. 0.438). The fear dimension performed really bad (J=0.445), with the subjects misinterpreting the expression as *surprised* rather than *scared*.

Emotions conveyed by the virtual character. Figure 10 shows the rates obtained by the rendered videos. This graphs can be compared with the third row of Table 3, showing the coefficients of the comparison between the ideal condition and the rendered videos. These values indicate that that the 3D rendered videos have been rated worse than the live videos (average J=0.470 vs. 0.536), with the exception on anger (J=0.533 vs. 0.475).

Efficiency of the mapping method. Lastly, a comparison between the rates obtained by the live vs. the rendered videos gives an estimation of the quality of the facial retargeting method. The rates obtained by the live and the rendered videos can be visually compared by looking at Figures 9 and 10. It is possible to notice a similarity between the two performances. The similarity is numerically expressed by the coefficients reported in the fourth row of Table 3. The overall quality of the retargeting method is expressed by its average coefficient (average J=0.791). The low standard deviation value (0.038) indicates that the retargeting is consistent across all the six dimensions.

It is worth noticing that this evaluation metric allows for an estimation of the quality of the retargeting even if the performers do not present high acting talent. In fact, as can be seen in Figure 9, in the live videos the recognition of Anger, Fear and Sadness is pretty low, with the rates scattered through all other dimensions (except happiness). Also, Fear is misinterpreted as Surprise. Nevertheless, the same phenomenon can be observed for the rendered videos (Figure 10). The high values of the coefficients of the comparison (Respectively J=0.784, 0.834, and 0.793 for the anger, fear, and sadness dimensions) confirm this similarity.

6. Conclusion

To sum up, we presented an animation authoring workflow enabling unexperienced users to intuitively and quickly author facial animations. The workflow maps the facial performance to a control rig, allowing for convenient editing at any point of the animation pipeline. We suggest that this system might also be used by professional animators to increase their productivity and foster their creativity by allowing them to explore and test multiple alternative performance captured animations while maintaining the possibility to seamlessly refine and edit the captured animation using traditional animation rigs.

Allowing users to manually edit an animation after it has been recorded requires a retargeting method that maps the captured data – recorded as vectors of blend shapes values – onto the animation rig – described as a combination of sliders – that is eventually edited by the user.

We assessed the accuracy of our facial retargeting method by conducting a user study where participants assessed the emotions conveyed by the facial expression displayed in the control and the authored animation. Contrary to existing evaluation methods, the method we used accounts for possible misinterpretations of the intended emotion.

In the future, we would like to perform further study involving both professional actors, casual users, and experienced animators. This would give better evidence of the efficiency of the evaluation method. The evaluation will be performed on an updated version of the facial retargeting method, which will map to a bone-based facial control rig rather than to the slider-based one that we used in this work.

References

[1] Ekman, P., Friesen, W.V. and Ellsworth, P. (1972) *Emotion in the human face: guide-lines for research and an integration of findings*, no. PGPS-11 in Pergamon general psychology series (New York: Pergamon Press).

[2] Kipp, M., Heloir, A. and Nguyen, Q. (2011) Sign language avatars: animation and comprehensibility. In *Proceedings of the 10th international conference on Intelligent virtual agents*, IVA'11 (Berlin, Heidelberg: Springer-Verlag): 113–126.

[3] Lewis, J.P. and Anjyo, K.i. (2010) Direct manipulation blendshapes. *IEEE Computer Graphics and Applications* **30**(4): 42–50. doi:10.1109/MCG.2010.41.

[4] Witkin, A. and Kass, M. (1988) Spacetime constraints (ACM Press): 159–168. doi:10.1145/54852.378507.

[5] Osipa, J. (2003) *Stop staring: facial modeling and animation done right* (San Francisco: SYBEX).

[6] Sanna, A., Lamberti, F., Paravati, G. and Rocha, F.D. (2012) A kinect-based interface to animate virtual characters. *Journal on Multimodal User Interfaces* doi:10.1007/s12193-012-0113-9.

[7] Pighin, F. and Lewis, J.P. (2006) Facial motion retargeting. In *ACM SIGGRAPH 2006 Courses*, SIGGRAPH '06 (New York, NY, USA: ACM). doi:10.1145/1185657.1185842.

[8] Weise, T., Bouaziz, S., Li, H. and Pauly, M. (2011) Realtime performance-based facial animation. In *ACM SIGGRAPH 2011 papers*, SIGGRAPH '11 (New York, NY, USA: ACM): 77:1–77:10. doi:10.1145/1964921.1964972.

[9] Tan, P., Steinbach, M. and Kumar, V. (2006) *Introduction to Data Mining*, Pearson International Edition (Pearson Addison Wesley).

[10] Wilbers, F., Ishi, C. and Ishiguro, H. (2007) A blendshape model for mapping facial motions to an android. In *IEEE/RSJ International Conference on Intelligent Robots and Systems IROS* (IEEE): 542–547. doi:10.1109/IROS.2007.4399394.

[11] Conner, B.D., Snibbe, S.S., Herndon, K.P., Robbins, D.C., Zeleznik, R.C. and van Dam, A. (1992) Three-dimensional widgets (ACM Press): 183–188. doi:10.1145/147156.147199.

[12] Magnenat-Thalmann, N. and Thalmann, D. (2004) *Handbook of virtual humans* (Chichester, England ; Hoboken, NJ: Wiley).

[13] Tolani, D., Goswami, A. and Badler, N.I. (2000) Real-time inverse kinematics techniques for anthropomorphic limbs. *Graphical Models* **62**(5): 353–388. doi:10.1006/gmod.2000.0528.

[14] Villagrasa, S. and Susin, A. (2009) Face! 3d facial animation system based on facs. In *Proceedings of the IV Iberoamerican Symposium in Computer Graphics*.

[15] Bastioni, M., Re, S. and Misra, S. (2008) Ideas and methods for modeling 3D human figures: the principal algorithms used by MakeHuman and their implementation in a new approach to parametric modeling. In *Proceedings of the 1st Bangalore Annual Compute Conference*, COMPUTE '08 (New York, NY, USA: ACM): 10:1–10:6.

[16] Alexander, O., Rogers, M., Lambeth, W., Chi-Ang, M. and Debevec, P. (2009), Creating a photoreal digital

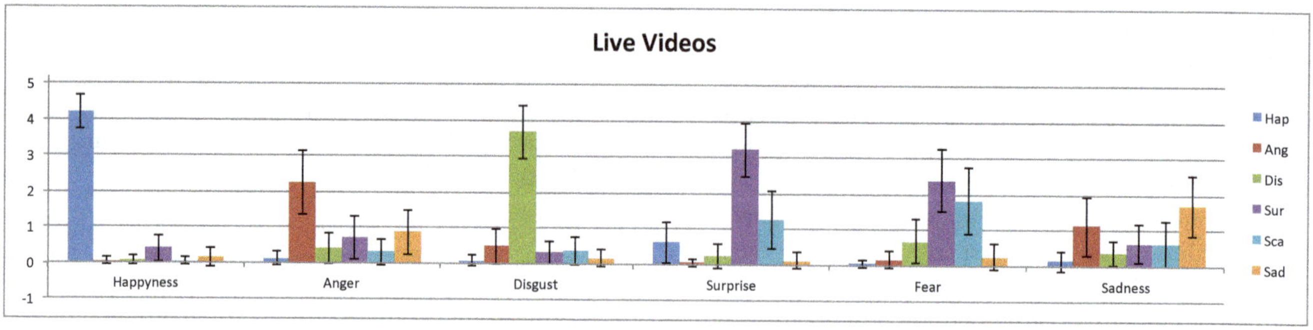

Figure 9. Rating obtained for the live videos (subjects) for each expression.

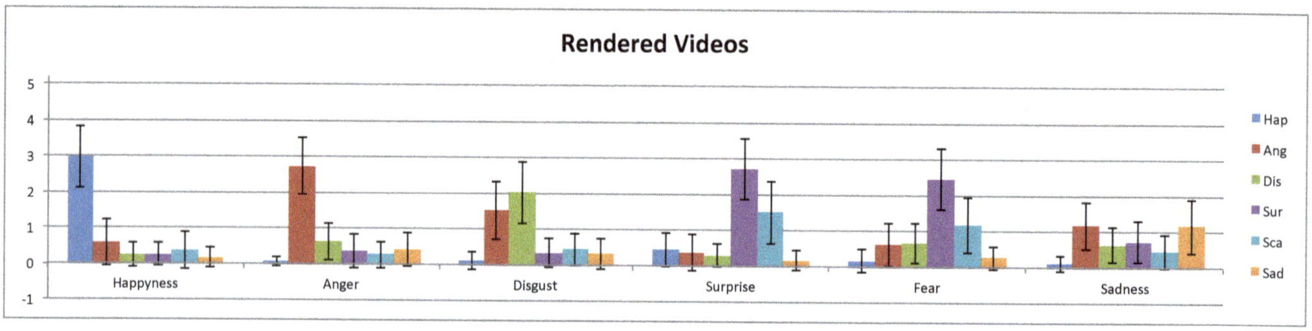

Figure 10. Rating obtained for the rendered videos (avatar) for each expression.

actor: The digital emily project.

[17] Chang, E. and Jenkins, O.C. (2008) Sketching articulation and pose for facial animation. In Deng, Z. and Neumann, U. [eds.] *Data-Driven 3D Facial Animation* (London: Springer London), 145–161.

[18] Kipp, M. and Nguyen, Q. (2010) Multitouch puppetry: creating coordinated 3D motion for an articulated arm (ACM Press): 147. doi:10.1145/1936652.1936682.

[19] Robertson, B. (1988) Mike the talking head. *Computer Graphics World* **11**(7).

[20] Menache, A. (1999) *Understanding Motion Capture for Computer Animation and Video Games* (San Francisco, CA, USA: Morgan Kaufmann Publishers Inc.).

[21] Shin, H.J., Lee, J., Shin, S.Y. and Gleicher, M. (2001) Computer puppetry: An importance-based approach. *ACM Transactions on Graphics* **20**(2): 67–94. doi:10.1145/502122.502123.

[22] Liu, K.Y., Ma, W.C., Chang, C.F., Wang, C.C. and Debevec, P. (2011) A framework for locally retargeting and rendering facial performance. *Computer Animation and Virtual Worlds* **22**(2-3): 159–167. doi:10.1002/cav.404.

[23] Curio, C., Breidt, M., Kleiner, M., Vuong, Q.C., Giese, M.A. and BÃijlthoff, H.H. (2006) Semantic 3D motion retargeting for facial animation. In *Proceedings of the 3rd Symposium on Applied Perception in Graphics and Visualization* (ACM Press): 77. doi:10.1145/1140491.1140508.

[24] Seol, Y., Lewis, J., Seo, J., Choi, B., Anjyo, K. and Noh, J. (2012) Spacetime expression cloning for blendshapes. *ACM Transactions on Graphics* **31**(2): 1–12. doi:10.1145/2159516.2159519.

[25] Lewis, J.P. and Pighin, F. (2005) Cross-mapping. In *ACM SIGGRAPH 2005 Courses* (ACM Press): 3. doi:10.1145/1198555.1198583.

[26] Song, J., Choi, B., Seol, Y. and Noh, J. (2011) Characteristic facial retargeting. *Computer Animation and Virtual Worlds* **22**(2-3): 187–194. doi:10.1002/cav.414.

[27] Battocchi, A., Pianesi, F. and Goren-Bar, D. (2005) A first evaluation study of a database of kinetic facial expressions (DaFEx). In *Proceedings of the 7th international conference on Multimodal interfaces*, ICMI '05 (New York, NY, USA: ACM): 214–221.

[28] Costantini, E., Pianesi, F. and Prete, M. (2005) Recognising emotions in human and synthetic faces: The role of the upper and lower parts of the face. In *Proceedings of the 10th International Conference on Intelligent User Interfaces*, IUI '05 (New York, NY, USA: ACM): 20–27. doi:10.1145/1040830.1040846.

9

An Introduction to the Book Titled *Nature of Computation and Communication*

Phan Cong Vinh[1]

[1]Nguyen Tat Thanh University, Ho Chi Minh City, Vietnam

Abstract

We introduce the book with the title of "Nature of Computation and Communication" edited by Prof. Phan Cong Vinh, Dr. Emil Vassev and Prof Mike Hinchey, and published by Springer in 2015 to scientists, researchers, professionals and students in the field of computing. Hopefully, they will find this book provoking for their research related to hot topics of computing and being useful to their future work.

Keywords: Collective adaptive system, Formal methods, Self-adaptive system

1. Introduction

The 1st **I**nternational **C**onference on na**T**ure of **C**omputation and **C**ommunication (ICTCC 2014) was a place for highly original ideas about how the nature is going to shape computing systems of the future. There, it focused on rigorous approaches and cutting-edge solutions, which were to make a formal basis more accessible to researchers, scientists, professionals and students as well as developers and practitioners in ICT by providing them with state-of-the-art research results, applications, opportunities and future trends.

The ICTCC 2014 was an international scientific conference research in the fields related to nature of computation and communication held during November 24-25, 2014 in Ho Chi Minh City of Vietnam. This conference provided an excellent opportunity for researchers to discuss natural approaches and techniques for computation and communication. The proceedings of ICTCC 2014 [1] were published by Springer in series Lecture Notes of the Institute for Computer Sciences, Social Informatics and Telecommunications Engineering (LNICST) (indexed by DBLP, EI, Google Scholar, Scopus, Thomson ISI).

*Corresponding author. Email: pcvinh@ntt.edu.vn

For this edition, the Program Committee received over hundred submissions from twenty countries and each paper was reviewed by at least three expert reviewers. Thirty four papers were selected after intensive discussions held among the Program Committee members. The excellent reviews and lively discussions of the Program Committee members and external reviewers were highly appreciated in the review process.

This book, with papers contributed by researchers from academia and industry, serves as a technical guide and reference material for researchers, scientists, professionals and students in computer science and computer engineering as well as developers and practitioners in computing and networking systems design by providing them with state-of-the-art research findings and future opportunities and trends. These contributions include formal approaches and applications in natural computing. In particular, the book covers existing and emerging research issues in the formal methods for natural computing.

2. Contents of the Book

The book has thirty four papers organized into two sections. First section contains five papers addressing formal and practical aspects of self-adaptive systems.

Figure 1. Cover of the book

Second section consists of twenty nine papers presenting formal and practical aspects of collective adaptation and others.

2.1. Formal Methods for Self-Adaptive Systems

A self-adaptive system changes its behavior in response to stimuli from its execution and operational environment. As software is used for more pervasive and critical applications, support for self-adaptation is increasingly seen as vital in avoiding costly disruptions for repair, maintenance and evolution of systems. Self-adaptation may result in changes to some functionality, algorithms, or system parameters, as well as to the system's structure or any other system aspect. Moreover, an autonomic self-adaptive system may have intrinsic intelligence that may help it reason about situations where autonomous decision making is required. The major challenges to overcome in order to achieve adequate self-adaptive behavior are related to monitoring, quality and quantity of knowledge and awareness. A successful self-adaptive system meets those challenges to emphasize features like adaptability, autonomy, mobility, robustness, and dynamicity.

Developing self-adaptive systems requires new approaches that can efficiently tackle the problems of expressing autonomy requirements, designing and implementing self-adaptive features, and efficiently testing self-adaptive behavior. This section of the book contains both theoretical and applied contributions related to formal methods in engineering self-adaptive systems as follows:

- Modular Design and Verification of Distributed Adaptive Real-Time Systems
- Modeling Swarm Robotics with KnowLang
- Reasoning on Data Streams: An Approach to Adaptation in Pervasive Systems
- Autonomic Computing Software for Autonomous Space Vehicles
- Logic-Based Modeling of Information Transfer in Cyber-Physical Multi-Agent Systems

2.2. Nature of Computation and Communication

The socio-technical structure of our community increasingly depends on systems, which are built as a collection of varied agents and are tightly coupled with humans and social interrelations. Their agents more and more need to be able to develop, cooperate and work all by themselves as a part of an artificial community. Hence, for such collective adaptive systems (CASs), one of major challenges is how to support self-adaptation in the face of changing interactions. In other words, how does a CAS understand relevant interrelations and then self-adapt to become better able to live in its interactions? Dealing with this grand challenge of CASs requires a well-founded modeling and in-depth analysis on the notions of self-adaptation and self-adaptive trait. With this aim, this section of the book covers some highlighted contents as follows.

- Categorical structures of self-adaptation in CASs where a collection of entities is able to self-adapt its configuration and self-optimize its performance in the face of changing interactions.
- Self-adaptation monoid to shape series of self-adaptive traits in CASs.
- Justifying equivalence between two series of self-adaptive traits based on a powerful method so-called proof principle of coinduction

3. Remarkable Features of the Book

This book has the following remarkable features:

- Provides a comprehensive reference on natural aspects of computation and communication.

- Presents state-of-the-art natural aspects of computation and communication.
- Formally specifies, develops and verifies natural aspects of computation and communication.
- Includes illustrative figures facilitating easy reading.
- Discusses emerging trends and open research problems on nature of computation and communication.

The book serves as a comprehensive and essential reference on natural aspects of computation and communication and is intended as a collection of revised selected papers for senior undergraduate and graduate-level courses. It can also be used as a supplementary book for graduate courses. The book is a useful resource for the students and researchers to learn natural aspects of computation and communication. In addition, it will be valuable to professionals from both the academia and industry and generally serves instant appeal to the people who would like to contribute to natural computing technologies.

References

[1] P.C. Vinh, E. Vassev, M. Hinchey(2015) *Nature of Computation and Communication*, Springer, 1st ed.

10

Multi-GPU based framework for real-time motion analysis and tracking in multi-user scenarios

Sidi Ahmed Mahmoudi[1,*]

[1]University of Mons, Faculty of Engineering, Computer Science Department. Place du Parc, 20. 7000, Mons, Belgium

Abstract

Video processing algorithms present a necessary tool for various domains related to computer vision such as motion tracking, event detection and localization in multi-user scenarios (crowd videos, mobile camera, scenes with noise, etc.). However, the new video standards, especially those in high definitions require more computation since their treatment is applied on large video frames. As result, the current implementations, even running on modern hardware, cannot provide a real-time processing (25 frames per second, fps). Several solutions have been proposed to overcome this constraint, by exploiting graphic processing units (GPUs). Although they exploit GPU platforms, they are not able to provide a real-time processing of high definition video sequences. In this work, we propose a new framework that enables an efficient exploitation of single and multiple GPUs, in order to achieve real-time processing of Full HD or even 4K video standards. Moreover, the framework includes several GPU based primitive functions related to motion analysis and tracking methods, such as silhouette extraction, contours extraction, corners detection and tracking using optical flow estimation. Based on this framework, we developed several real-time and GPU based video processing applications such as motion detection using moving camera, event detection and event localization

Keywords: Multi-GPU computing, camera motion estimation, event detection and event localization

1. Introduction

Recently, the CPU power has been capped, essentially for thermal reasons, to less than 4 GHz. A limitation that has been circumvented by the change of internal architecture, with multiplying the number of integrated computing units. This evolution is reflected in both general (CPU) and graphic (GPU) processors which present a large number of computing units, their power has far exceeded the CPUs ones.

Video processing and more particularly motion estimation algorithms present a very active research topic in computer vision domain. They can be used, for example, in surveillance systems tracking humans in public places, such as metro or airports, to identify possible abnormal behaviors and threats [1]. Motion estimation algorithms serve therefore as a common building block of some more complex routines and systems. However, these algorithms are hampered by their high consumption of both computing power and memory. The exploitation of graphic processors can present an efficient solution for their acceleration.

Unlike algorithms requiring a high dependency of computation between the input data and hence a complicated parallelization, most of image and video processing algorithms consist of similar computations over many pixels. This fact makes them well adapted for acceleration on GPU by exploiting its processing units in parallel. Nevertheless, the new standards, especially those in high resolutions require more computation since their treatment is applied on large video frames. Thus, the current implementations, even running on modern hardware, cannot provide a real-time processing. Moreover, the treatment of TV broadcast images, which cannot be down sampled, require an accelerated object detection and recognition. Therefore, a fast processing of videos is needed to ensure the treatment of 25 high definition frames per second (25 fps). To overcome these constraints, several GPU computing approaches have recently been proposed. Although they exploit GPU platforms, they are not able to provide a real-time processing of high definition video sequences. Moreover, they are not well adapted for exploiting multiple GPUs. Indeed, clusters and computers that contain multiple GPUs are becoming commonplace nowadays. However, their exploitation requires an efficient sharing of data and

*Corresponding author. Email: sidi.mahmoudi@umons.ac.be

computations between the available GPUs. Otherwise, the computation time (with many GPUs) can be even more elevated, compared to the case using one single GPU due to data transfers and memory allocation times.

In this paper, we propose a new framework enabling an effective exploitation of single and multiple GPUs for accelerating video processing algorithms, and hence achieving real-time treatment of high definition videos. This framework allows an efficient management of single and multiple GPUs memories and a fast visualization of results. The proposed framework includes several CUDA[1] based primitive functions related to motion analysis and tracking methods, such as silhouette extraction, contours extraction, corners detection and tracking using optical flow estimation. With this framework, we developed several real-time video processing applications such as motion detection using mobile camera, event detection and event localization.

The remainder of the paper is organized as follows: related works are discussed in the second section. Section 3 presents the proposed framework for real-time video processing using multiple GPUs. In sections 4, 5, 6, we present three Multi-GPU based use cases of our framework : motion detection using mobile camera, event detection and event localization, respectively. Section 8 compares and evaluates the related CPU, GPU and multi-GPU implementations. Finally, conclusions and future works are described in the last Section.

2. Related work

In general, event detection and localization methods consist of analyzing and modeling normal behaviors, and then detecting the difference between the normal behavior model and the observed behaviors. These variations can be labeled as abnormal or emergency events. In this category, Bilinski et al. [2] extract hog descriptors in order to calculate predefined models (i.e. crowd scenarios) that allow to recognize crowd events. Authors in [3] propose a context-aware method that allows to detect anomalies by tracking all moving objects in the video. There are also some work in [4] which addresses the problem of analyzing video events in crowded scenes. A novel manifold learning method was developed to achieve an effective modeling of video events in a low dimensional space. The general idea of these methods consist on estimating the displacement and velocity of features in a given video frame with respect to the previous one. In this work, we are more focused on optical flow methods since they present a promising solution for tracking even in noisy and crowded scenes or in case of small motions. Moreover, most of camera motion estimation techniques are based on optical flow methods such as shown in [5]. The latter presents a method that creates prototypes of optical flows before performing a linear decomposition of motion vectors, which enable to estimate the camera parameters.

Otherwise, several GPU based motion tracking methods have been proposed such as in [20], which developed a GPU version of motion detection technique from a moving platform. As result, they can build the background model and detect motion regions at around 18 fps on 320 × 240 video captured from a moving camera. Authors in [21] proposed an eye blink detector which can be used in dry eye prevention system. Their solution, exploiting GPU, is based on histogram back projection and optical flow techniques.

In case of optical flow-based motion tracking algorithms, one can distinguish two categories of related works. The first presents so called dense optical flow which tracks all frame pixels without selecting any features. In this context, Marzat et al. [6] proposed a GPU implementation of the Lucas-Kanade method for the optical flow estimation. The software was programmed using the CUDA library (Compute Unified Device Architecture) to compute dense and accurate velocity field at about 15 fps with 640x480 video resolution. Authors in [7] presented the CUDA implementation of the Horn-Shunck optical flow, that offered a real-time processing of 316×252 video resolution. Gwosdek et al. [8] developed a GPU implementation of the Euler-Lagrange (EL) framework for solving variational optical flow methods using sequences with 640x480 pixels in near-real-time.

The second category includes software tools tracking selected image features only. Sinha et al. [9] developed a GPU implementation of the popular KLT feature tracker [10] and the SIFT feature extraction algorithm [11]. This was developed with the OpenGL/Cg libraries allowing to extract about 800 features from 640x480 video at 10 fps which is approximately 10 times faster than the corresponding CPU implementation. There is also a work in [12] proposing a GPU based block matching technique using OpenGL. This implementation offered a real-time processing of 640x480 video. Sundaram et al. [13] developed a method for computing point trajectories based on a fast GPU implementation of the optical flow algorithm that tolerates fast motion. This parallel implementation runs at about 22 fps, which is 78 times faster than its CPU version. However, despite their great speedups, none of the abovementioned GPU based software tools can provide real-time processing of high definition videos. Moreover, they are not well adapted for exploiting multiple GPUs simultaneously.

Our contribution consists on proposing a new framework for video processing on single or multiple

[1] CUDA. https://developer.nvidia.com/cuda-zone

GPUs. This framework contributes within three main factors:

1. Efficient and adapted exploitation of GPU in case of video processing applications. Indeed, we apply several optimization techniques such as: the adapted selection of the number of GPU threads within blocks, the exploitation of GPU shared and texture memories, the overlapping of data transfers by kernels executions that allows reducing data transfer times, the OpenGL fast visualization of results.

2. Scalable exploitation of multiple GPUs that allows to benefit from the full power of each graphic cards. The number of selected GPUs depends directly from the computational intensity of the input application. As result, the framework allows to reduce significantly the computation time and the energy consumption.

3. Several primitive GPU based image and video processing functions are developed such as: silhouette extraction, contours and corners detection, features tracking, etc. These functions can be selected easily within our framework for accelerating video processing applications. As example, we developed three real-time HD/Full HD or even 4K video processing applications:

    - Camera motion estimation using optical flow vectors.
    - Event detection in crowd videos
    - Event localization in multi-user scenarios.

3. Multi-GPU based framework for real-time video processing

This section is presented in two parts: the first one describes our framework for video processing on single or multiple GPUs. The second part presents the related GPU based primitive functions : silhouette extraction, features detection and tracking, based on optical flow estimation. These primitive functions can be directly selected within our framework.

3.1. The proposed framework

Our framework is based upon CUDA for parallel constructs and OpenGL for visualization, using three steps: multiple GPUs selection, CUDA parallel processing and OpenGL visualization (Fig. 1).

1. **Multiple GPUs selection :** Video processing algorithms are known by their high intensity computation, and their well adaptation for parallel calculation. Therefore, apart from implementing our video processing algorithms on a single GPU, we propose a version taking advantage of multiple GPU systems that nowadays are becoming commonplace. The default setup of our framework selects a number of one GPU only. During the first 20 frames , we test the performance of our application. If the condition of real-time processing is achieved, the number of 1 GPU is maintained, else we increase the number of GPUs until achieving a treatment of 25 fps. This allows to select one GPU only in case of processing low resolution videos or applying low intensive treatments. As result, accelerated treatments are obtained with a reduced energy consumption. Once the number of GPUs selected, the program initializes all of them. Then, the input image frame is first uploaded to each GPU. This frame is virtually divided into equally sized subframes along y dimension and once the image data is available, each GPU is responsible for treating its part of the frame (subframe).

2. **CUDA parallel processing :** Before launching the parallel treatments of the current subframes. The number of CUDA threads, within each GPU, in the so called blocks and grid has to be defined, so that each thread can perform its processing on one or a group of pixels in parallel. This enables the program to process the image pixels in parallel. Note that the number of threads depends on the number of pixels. Once the number and the layout of threads is defined, different CUDA functions (kernels) are executed sequentially, but each of them in parallel using multiple CUDA threads. In our case, we selected a number of CUDA threads equal to the number of pixels, (within each subframe) which enables for each CUDA thread to treat its corresponding pixel.

    For a better exploitation of GPU, we propose to load the input image on GPU texture memory for a fast access to pixels. We have also loaded each pixel neighbors on GPU shared memory for a fast processing of pixels using their neighbors values. This optimization is so useful in case of applying image convolutions, the related filters are loaded in the shared memory. Moreover, we employ the streaming technique within multiple GPUs so that each GPU can overlap effectively data transfers by kernels executions.

3. **OpenGL visualization :** At the end of computations for each frame (the subframes). The results can be displayed on screen using the OpenGL graphics library that allows for fast visualization, as it can operate on the already existing buffers on GPU, and thus requires less data transfer between host and device memories. In case of Multi-GPU

treatments, each GPU result (subframe) need to be copied to the GPU which is charged of displaying. This, however, is a fast operation since contiguous memory space is always transferred. Once the visualization of the current image is completed, the program goes back to the first step to load and process the next frames of the video.

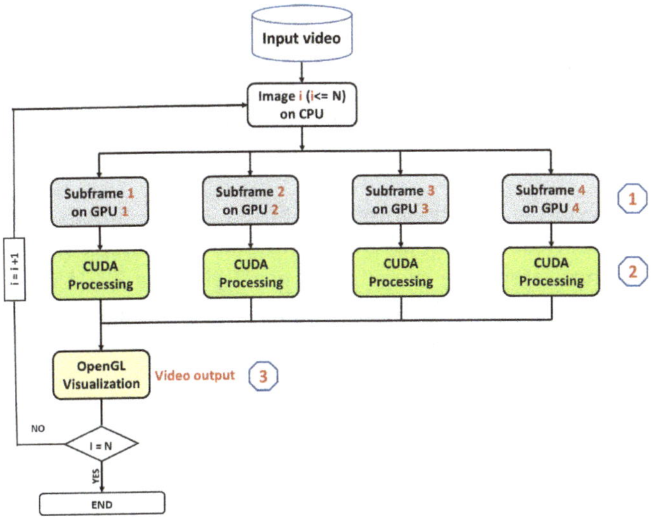

Figure 1. Multi-GPU based framework for real-time video processing (N : video frames number)

This framework could provide a unified way to implement advanced methods accordingly to each given scenario. These methods can exploit easily the primitive functions (described in Section 3.2). Otherwise, the framework user could provide the new required algorithms before integrating them into the processing primitives. Fig. 2 summarizes the method of exploitation of our framework. Notice that the selection of primitive functions depends mainly from the video processing application.

3.2. GPU-based primitive functions

In this section, we propose GPU (CUDA) based primitive functions that can be used within our framework as shown in Fig. 2. These primitives correspond to the functions of silhouette extraction, features (corners and contours) detection and tracking methods.

1. **GPU based silhouette extraction :** the computation of difference between frames presents a simple and efficient method for detecting the silhouettes of moving objects. We propose a GPU implementation of this method using three steps. First, we load the two first frames on GPU in order to compute the difference between them during the CUDA parallel processing step. Once the first image displayed, we replace it by the next

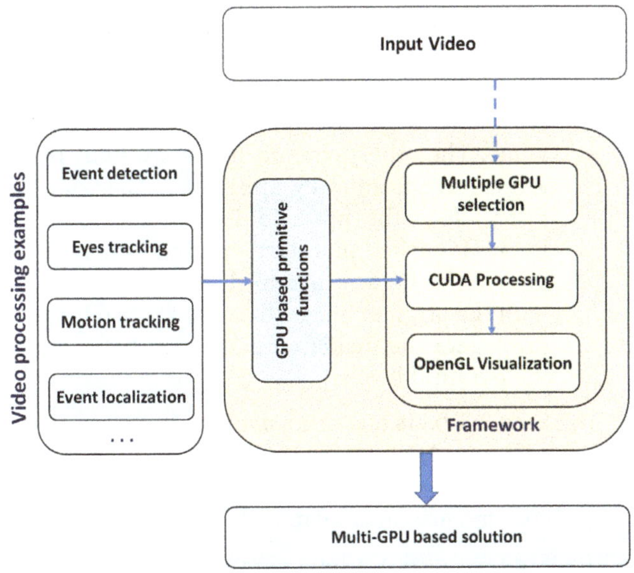

Figure 2. Exploitation process of our framework

video frame in order to apply the same treatment. Fig 7.(b) presents the obtained result of silhouette extraction. This figure shows two silhouettes extracted, that present two moving persons. In order to improve the quality of results, a threshold of 200 was used for noise elimination.

2. **GPU based features detection and tracking :** in this section, we propose the GPU implementation of both features detection and tracking methods. The first one enables to detect the good features to track, i.e. corners. To achieve this, we have exploited the Bouguet's corners extraction technique [14], which is based on the principle of Harris detector [15]. Moreover, we developed a CUDA version of Deriche-Canny edge detector. These two GPU implementations are detailed in [16].

The second step enables to track the features (corners) previously detected using the optical flow method, which presents a distribution of apparent velocities of movement of brightness pattern in an image. It enables to compute the spatial displacements of images pixels based on the assumption of constant light hypothesis which supposes that the properties of consecutive images are similar in a small region. For more detail about optical flow computation, we refer readers to [17] and [14]. In this work, we propose the GPU implementation of the Lucas-Kanade algorithm, which is well known for its high efficiency, accuracy and robustness. This algorithm disposes of six steps:

(a) **Step 1: Pyramid construction :** in the first step, the algorithm computes a pyramid representation of images I and J which represent two consecutive images from the video. The other pyramid levels are built in a recursive fashion by applying a Gaussian filter. Once the pyramid is constructed, a loop is launched that starts from the smallest image (the highest pyramid level) and ends with the original image (level 0). Its goal is to propagate the displacement vector between the pyramid levels.

(b) **Step 2: Pixels matching over levels :** for each pyramid level (described in the previous step), the new coordinates of pixels (or corners) are calculated.

(c) **Step 3: Local gradient computation :** in this step, the matrix of spatial gradient G is computed for each pixel (or corner) of the image I. This matrix of four elements (2×2) is calculated with the horizontal and vertical spatial derivatives. The computation of the gradient matrix takes into account the area (window) of pixels which are centered on the point to track.

(d) **Step 4: Iterative loop launch and temporal derivative computation :** a loop is launched and iterated until the difference between the two successive optical flow measures (calculated in the next step), or iterations, is higher than a defined threshold. Once the loop is launched, the computation of the temporal derivatives is performed using the image J (second image). This derivative is obtained by the subtraction of each pixel (or corner) of the image I (first image) and its corresponding corner in the image J (second image). This enables to estimate the displacement estimations which is then propagated between successive pyramid levels.

(e) **Step 5: Optical flow computation :** the optical flow measure $\bar{g}$ is calculated using the gradient matrix G and the sum of temporal derivatives presented by shift vector $\bar{b}$. The measure of optical flow is calculated by multiplying the inverse of the gradient matrix G by the shift vector $\bar{b}$.

(f) **Step 6: Result propagation and end of the pyramid loop :** the current results are propagated to the lower level. Once the algorithm reaches the lowest pyramid level (the original image), the pyramid loop (launched in the first step) is stopped. The vector $\bar{g}$ presents the final optical flow value of the analyzed corner.

Upon matching and tracking pixels (corners) between frames, the result is a set of vectors as shown in Equation (1):

$$\Omega = \{\omega_1 \dots \omega_n \mid \omega_i = (x_i, y_i, v_i, \alpha_i)\} \quad (1)$$

where:

- x_i, y_i are x a y coordinates of the feature i;
- v_i represents the velocity of the feature i;
- α_i denotes motion direction of the feature i.

We propose a GPU implementation of the Lucas-Kanade optical flow method by parallelizing its steps on GPU. These steps are executed in parallel using CUDA such that each GPU thread applies its instructions (among the six steps) on one pixel or corner. Therefore, the number of GPU threads is equal to the number of pixels or corners. Since the algorithm looks at the neighboring pixels, for a given pixel, the images, or pyramid levels are kept in the texture memory. This allows a faster access within the 2-dimensional spatial data. Other data, e.g. the arrays with computed displacements, are kept in the global memory, and are cached in the shared memory if needed.

4. Multi-GPU based motion detection using mobile camera :

The above-mentioned GPU implementations are exploited in an application that consists of real-time motion detection within moving camera. In this category, motion detection algorithms are generally based on background subtraction which presents a widely used technique in computer vision domain. Typically, a fixed background is given to the application and new frames are subtracted from this background to detect the motion. The difference will give the objects or motion when the frame is subtracted from the fixed background. This difference in resulting binary image is called foreground objects. However, some scenarios present a dynamic background which can change due to the movement of cameras. In this context, we propose an application for real-time background subtraction, which enables to detect automatically background and foreground using a moving camera. This application can be summarized in four steps:

1. **Corners detection :** the Harris corner detector [15] is applied to extract good features to track and examine for camera motion.

2. **Optical flow computation :** the Lukas-Kanade optical flow method [14] is applied to track the corners, detected previously.

(a) (b)

Figure 3. (a). Camera motion estimation (b). Motion detection

3. **Camera motion inhibition :** the camera motion is estimated by computing the dominant values of optical flow vectors. This enables to extract the common area between each two consecutive images and focus only on motions related to objects in the scene.

4. **Motion detection :** this step consists of detecting movements based on computing the difference between each two consecutive frames.

In order to achieve a real-time treatment of high definition videos, the method steps are ported on GPU. Their GPU implementation is described in section 3. Fig. 3.(a) shows a scene of camera motion. Dotted and dashed line presents the first image, dotted line presents the second frame and solid line shows the joint area of two frames. Once, the camera motion is estimated. The joint area between 2 consecutive frames is determined by cropping the incoming and outgoing areas as seen in the white area of Fig. 3.(a). Fig. 3.(b) shows the resulting image of background subtraction. White areas represent the difference around moving objects.

This method steps are integrated within our framework for providing an accelerated Multi-GPU based solution that offers an efficient and scalable exploitation of multiple GPUs.

5. Multi-GPU based event detection

Before presenting the GPU implementation, we start by describing our CPU based implementation of event detection method that consists of four steps: points of interest detection, points of interest tracking, clustering and event detection.

1. **Points of interest detection :** the Harris [15] corner detector is applied to extract good features, which will be tracked during the next step.

2. **Points on interest tracking :** the Lucas-Kanade [10] optical flow method is applied to track the previously detected corners. As a result, this method provides, for each corner, a related velocity and an angular direction of motion. The features (points on interest) that have a velocity close to zero are considered as static features. Noise features are isolated points that have a velocity or direction so different compared to their neighbors.

3. **Clustering :** this step consists of applying the K-means method to get clusters. The features which are used within the clustering process are the spatial coordinates (x; y), the velocity and the direction. The K-means method is a well-known geometric clustering algorithm. This classification allows to have in each class individuals moving with similar velocities and directions in the same region.

4. **Event detection :** this step consists of computing the distance between each cluster and its corresponding one in the next frame. If we have N clusters within each frame, the global distance between frames is represented by the sum of the N distances between clusters.

As compared to normal motion, the positions and sizes of the clusters (rectangles) of abnormal motion are significantly different between two consecutive frames. This is due to the sudden changes of the values of velocity and direction of optical flow fields. Fig. 4 presents the result of the steps listed above using a crowd video. For more detail about this implementation, we refer readers to [19].

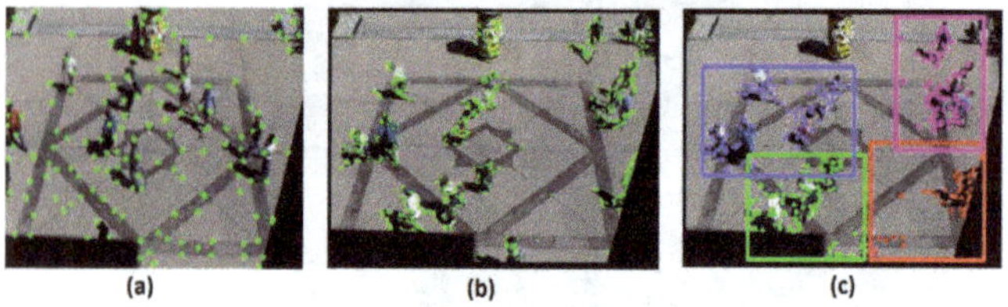

Figure 4. (a). Corners detection (b). Corners tracking (c). Clustering

For a better exploitation of graphic cards, we developed a GPU implementation of the above-mentioned event detection method by parallelizing its most intensive steps on graphic processing unit: corners detection and tracking. The related GPU versions are described in Section 3.2.2. Otherwise, these steps are integrated within our framework for providing an accelerated Multi-GPU based solution that offers an efficient and scalable exploitation of multiple GPUs.

To conduct the experiment, we used as data set videos from different outdoor places, comprising both normal and abnormal motions. The experimental results corresponding to the approach of event detection based on clustering optical flow fields, applied on one of the videos have been presented in Fig. 5. Abnormal motion includes a sudden situation when a group of people start running. From frame 1 to 550 the people's motion is normal. People tend to run since the frame number 551. More precisely, the assigned distance of frame 551 will be higher than any other assigned distances among frames. In Fig 5, the blue colored curve is the output of the proposed approach. The Gaussian like curve represents the abnormal motion when the group of people is trying to leave the place with very quick motion.

Fig. 6 presents another case of event detection, applied on a video that consists of four persons only. In frame 231, one person starts running and the assigned distance will be higher. Therefore, an abnormal event is detected in this frame. Moreover, the high values of optical flow vectors (compared to previous values) enabled to localize the persons which present an abnormal motion. The latter are presented by red optical flow vectors, as shown in Fig. 6.

6. Multi-GPU based event localization

Event localization allows to provide, in real-time, areas in video frames where motion behavior is surprising compared to the rest of motion in the same frame. We propose in this section an approach that takes into account the spatial occupation of the moving objects. It is based on the extraction of scene occupation model from video sequences. Once this model is acquired

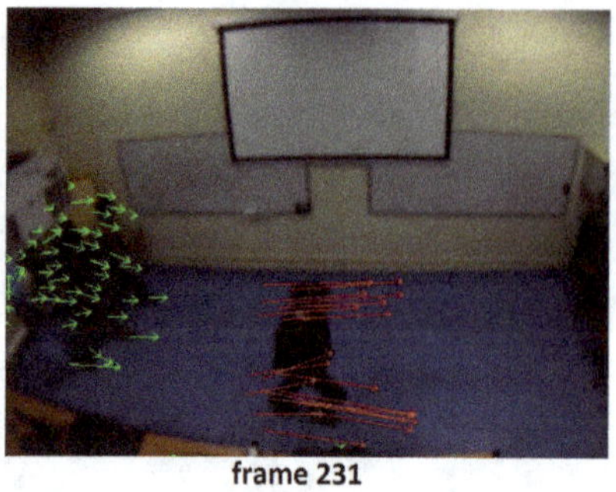

Figure 6. Event detection: video with several moving persons

(Fig. 7.(c)), motions close to this model are inhibited while only different motion (in other spatial locations) is highlighted. The event localization using scene occupation model follows three principal steps: frame difference, frame accumulation and motion inhibition.

1. **Frame difference :** we use the simplest motion detection technique which consists in the difference between every two consecutive frames in video. This method is very fast to implement. It needs no background modeling which can be a difficult task in some situations. Fig. 7.(b) shows the result given by this step.

2. **Frame accumulation :** the frame difference enables to detect moving individuals, so every frame represents individuals in movement. The video frames will be accumulated using a threshold accumulator in order to have as result a top-down model representing dominant regions. The threshold accumulator value depends of the computing time of model extraction. This extraction can be longer if one has a bigger value of this threshold. Fig. 7.(c) shows the model extracted after 1000 frames using a threshold of 100. Indeed, the white area presents the dominant regions on which the major individuals' movements have taken place.

3. **Motion inhibition :** in this step we apply a subtraction between the frames of moving objects (frame difference) and the model extracted. This subtraction allows to inhibit similar motions and focus only on abnormal motions.

On the one hand, we developed a GPU version of the above-mentioned event localization method. This parallel implementation is based on parallelizing these steps on GPU: frame difference, frame accumulation and motion inhibition. The GPU based frame difference is described in Section 3.2.1, while the GPU based frame accumulation consists on exploiting the result of frame difference step for each image of the video. Theses resulting frames will be accumulated (added),

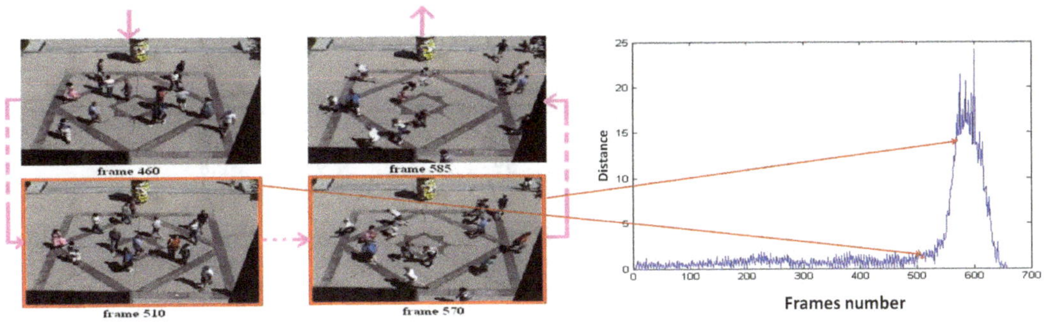

Figure 5. Event detection in crowd videos

Figure 7. (a). Input Video (b). Frame difference (c). Frame accumulation

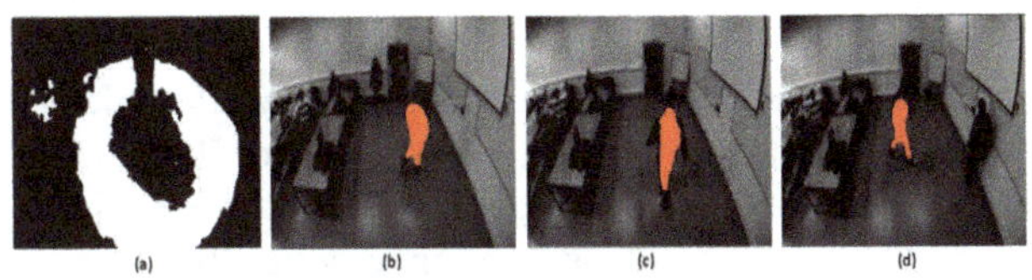

Figure 8. (a). Scene Occupation Model (b).(c).(d). Abnormal events localized

in a GPU buffer, using a threshold accumulator in order to have as result a top-down model representing dominant regions. The GPU implementation of this step consists for each frame of applying an addition and a comparison with threshold. The last step applies a subtraction on GPU between the frames of moving objects (frame difference) and the model extracted in previous step. This enables to inhibit similar motions and focus only on abnormal motions. On the other hand, these GPU based steps are included within our framework for providing an accelerated Multi-GPU based solution that offers an efficient and scalable exploitation of multiple GPUs. Experimentation are conducted with a video of 3800 frames representing normal and abnormal motions. The model extraction (accumulation) needed 1000 frames, the event localization can therefore begin at the same time. At the beginning, when the model is not finished all the moving objects will be salient, but after 1000 frames, some motion is no more interesting as it corresponds to the model. This approach is close to human perception when recurrent motions become annoying after a given time. Fig. 8 shows abnormal events detected due to the motion of a person out of the model representing dominant regions of movement at the frames number 1900, 2100 and 3150, this person is tracked using the red color.

The above-mentioned GPU implementations can be also exploited for accelerating an eye tracker application called "CVC Eye-Tracker" [22], presented by an open source eye tracking software. Indeed, this application consists mainly on selecting the 8 facial points (related to eyes, nose and frontal face), followed their tracking. These steps are well adapted for a Multi-GPU acceleration using our framework.

7. Performance

On the one hand, we can say that the quality of the above-mentioned applications of camera motion estimation, event detection and event localization remains identical since the procedure has not changed. Only the architecture and the implementation did. On the other hand, the exploitation of single or multiple GPUs enabled to accelerate the computation time. This acceleration allowed to obtain a real-time processing of high definition videos, even of 4K standards.

Table 1 presents a comparison between CPU, GPU and Multi-GPU performances of our approach of motion detection within mobile camera (Section 4). Notice that the exploitation of GPUs enabled a real-time processing of Full HD videos (1920×1080) and even 4K standards (3840×2160), with an acceleration ranging from 16 × to 47 ×.

Otherwise, Table 2 presents a comparison, in terms of the number of frames per second (fps), between sequential (CPU), parallel (GPU) and Multi-GPU implementations of event detection and localization methods respectively. Notice that use of GPUs enabled to achieve a real-time processing of high definition videos. This is due to the parallel treatment of the pixels within video frames, exploiting the large number of computing units in GPUs. We note also that we obtain reduced acceleration when processing low resolution

Resolution	2 CPU	1 GPU		2 GPU		4 GPU	
		fps	Acc (x)	fps	Acc (x)	fps	Acc (x)
512×512	5 fps	79	16.8 ×	89	17.8 ×	98	19.6 ×
1280×720	2,9 fps	51	17.6 ×	69	23.8 ×	95	32.8 ×
1920×1080	1,9 fps	45	23.7 ×	58	30.5 ×	84	44.2 ×
3840×2160	1,7 fps	35	20.6 ×	54	31.8 ×	80	47.1 ×

Table 1. GPU performances of motion detection using mobile camera

videos since we can't exploit sufficiently the high power of GPUs. As result, the use of multiple GPUs in this case is not so beneficial in terms of performance. Notice also that the number of exploited GPUs should take into account the treatment intensity. Indeed, Table 2 shows that using 2 or 4 GPUs can be significantly advantageous, in terms of performance, when treating Full HD or 4K video. This is due to an efficient exploitation of GPUs since the graphic cards can be fully exploited. In case of event localization, we present single GPU performances only since this method doesn't require a high intensive treatment. Thus, the use of more than one GPU doesn't improve performance.

A demonstration of GPU based features detection, features tracking, and event detection in crowd video is shown in this video sequence: https://www.youtube.com/watch?v=PwJRUTdQWg8.

In order to have a fair evaluation, we compare our GPU performance with the very recent and up to our best knowledge the fastest GPU based image processing library . *i.e.* OpenCV 2.4.9. The latter provides both CPU and GPU versions of algorithms. Table 3 presents a comparison between our GPU performances and those obtained with the GPU module of OpenCV. Notice that our implementations offers better performances thanks to the efficient exploitation of GPU (texture & shared) memories, the overlapping (streaming) of data transfers by CUDA executions and the fast OpenGL visualization.

Notice that the tests were run on the following hardware:

- CPU: Intel Core 2 Quad Q8200, 2.33GHz,
- GPU: 4 x NVIDIA GeForce GTX 580 with 1.5GB of RAM,
- RAM: 8GB,
- OS: 64-bit Linux.

8. Conclusion

We proposed in this paper a new framework for real-time video processing using multiple GPUs, and particularly in case of treating high definition videos such as Full HD formats or even 4K standards. This framework includes several GPU based primitive functions related to motion analysis and tracking methods, such as silhouette extraction, contours extraction, corners detection and tracking using optical flow estimation. Based on this framework, we developed three real-time video processing applications: motion detection using mobile camera, event detection and localization in multi-user scenarios. Performed tests show that our applications can turn in multi-user scenarios, and in real-time even when processing high definition videos such as Full HD or 4K standards. Moreover, the scalability of our results is achieved thanks to the efficient exploitation of multiple graphic cards. As future work, we plan to extend our framework in order to facilitate the implementation of new advanced monitoring and control systems exploiting parallel and heterogeneous platforms, with reduced energy consumption. We plan also to exploit the SDI capture cards[2] that allow for direct video stream capture into the GPU memory without any use of the CPU memory, which enables the software to decrease the amount of PCI-E bandwidth used for the video transmission.

References

[1] A. Fonseca, L. Mayron, D. Socek and O. Marques (2008) "Design and implementation of an optical flow-based autonomous video surveillance system", *Proc. IASTED* 1, p. 209-214.

[2] P. Bilinski, F. Bremond and M. B. Kaaniche (2009) "Multiple object tracking with occlusions using hog descriptors and multi resolution images", *3rd International Conference on Crime Detection and Prevention*, p. 1-6.

[3] F. Jiang, J. Yuan, S. A. Tsaftaris and A. K. Katsaggelos (2011) "Anomalous video event detection using spatiotemporal context", *Computer Vision and Image Understanding*, vol. 115, p. 323-333.

[4] M. Thida, H.-L. Eng, M. Dorothy and P. Remagnino (2011) "Learning video manifold for segmenting crowd events and abnormality detection", *10th Asian Conference on Computer Vision*, vol. 6492, p. 439-449.

[2]NVIDIA Quadro SDI Capture: http://www.nvidia.com/object/product quadro sdi capture us.html

Table 2. Performance of event detection and localization on GPUs: CPU Intel i5 and GPU GTX 580

Videos resolution	Event detection							Event localization		
	2 CPU	1 GPU		2 GPU		4 GPU		2 CPU	1 GPU	
	fps	fps	Acc (x)	fps	Acc (x)	fps	Acc (x)	fps	fps	Acc (x)
320×240	18	82	4.6 ×	74	04.1 ×	69	03.8 ×	20	61	3.1 ×
1280×720	7	44	6.3 ×	54	07.7 ×	66	09.4 ×	11	57	5.2 ×
1980×1080	4	32	8.0 ×	51	12.7 ×	65	16.3 ×	5	55	11 ×
3840×2160	1.6	21	13.1 ×	38	23.7 ×	63	39.4 ×	2	50	25 ×

Table 3. Performance of silhouette extraction, corners detection and tracking on GPU: CPU Intel i5 and GPU GTX 580

Videos resolution	Silhouette extraction		Edges & corners detection		Corners tracking	
	OpenCV GPU	GPU (ours)	OpenCV GPU	GPU (ours)	OpenCV GPU	GPU (ours)
1280×720	67 fps	90 fps	62 fps	79 fps	25.2 fps	30 fps
1920×1080	49 fps	62.7 fps	33 fps	39.9 fps	28.8 fps	38.8 fps
3840×2160	33.5 fps	40.1 fps	28.7 fps	34 fps	13.4 fps	20.1 fps

[5] J. Almeida, R. Minetto, T. A. Almeida, R. Torres and N. Leite (2009) "Robust Estimation of Camera Motion Using Optical Flow Models", *5th International Symposium, ISVC*, vol. 5875, p. 435-446.

[6] J. Marzat, Y. Dumortier and A. Ducrot (2009) "Real-time dense and accurate parallel optical flow using CUDA", *Proc. WSCG 1*, p. 105-111.

[7] Y. Mizukami and K. Tadamura (2007) "Optical flow computation on Compute Unified Device Architecture", *Proc. 14th International Conf. on Image Analysis and Processing*, 1, p. 179-184.

[8] Gwosdek. Pascal, Zimmer. Henning, Grewenig. Sven, Bruhn. AndrÃĺs and Weickert. Joachim (2012) "A Highly Efficient GPU Implementation for Variational Optic Flow Based on the Euler-Lagrange Framework", *Trends and Topics in Computer Vision*, vol. 6554, p. 372-383.

[9] S. Sinha, J.-M. Fram, M. Pollefeys and Y. Genc(2006) "GPU-based video feature tracking and matching", *EDGE, Workshop on Edge Computing Using New Commodity Architectures 1*.

[10] C. Tomasi and T. Kanade (1991) "Detection and tracking of point features", *Technical Report CMU-CS-91-132, Carnegie Mellon University, Pittsburgh*, p. 383-394.

[11] D. Lowe (2004) "Distinctive image features from scale-invariant key-points", *Int. J. Computer Vision (IJCV)*, 60(2), p. 91-110.

[12] J.M. Ready and C.N. Taylor (2007) "GPU acceleration of real-time feature based algorithms", *Proc. IEEE Workshop on Motion and Video Computing, WMVC'07*, vol. 1, p. 8-9.

[13] N. Sundaram, T. Brox and K. Keutzer (2010) "Dense point trajec-tories by GPU-accelerated large displacement optical flow", *In the 11th European Conference in Computer Vision, ECCV*.

[14] J. Y. Bouguet (2000) "Pyramidal Implementation of the Lucas Kanade Feature Tracker", *Intel Corporation Microprocessor Research Labs.*

[15] C. Harris and M. Stephens (1988) "A combined corner and edge detector", *in Proceedings of the 4th Alvey vision conference*, vol. 15, p. 147-151.

[16] S. A. Mahmoudi and P. Manneback (2012) "Efficient Exploitation of Heterogeneous Platforms for Images Features Extraction", *International Conference on Image Process-ing Theory, Tools and Applications, IPTA*, p. 91-96.

[17] B. D. Lucas and T. Kanade (1981) "An iterative image registration technique with an application to stereo vision", *Imaging Understanding Workshop*, p. 121.

[18] S.A. Mahmoudi, M. Kierzynka, P. Manneback and K. Kurowski(2014) "Real-time motion tracking using optical flow on multiple GPUs", *Bulletin of the Polish Academy of Sciences: Technical Sciences*, vol. 62, Issue. 1, p. 139-150.

[19] S. A. Mahmoudi, H. Sharif, N. Ihaddadene and C. Djeraba(2008) "Abnormal event detection in real-time video", *First International Workshop on Multimodal Interactions Analysis of Users in a Controlled Environment, ICMI.*

[20] Q. Y, G. Medioni (2008) " A GPU-based implementation of motion detection from a moving platform ", *IEEE Computer Society Conference on Computer Vision and Pattern Recognition Workshops CVPRW'08*, p. 1-6.

[21] P. Polatsek (2008), "Eye Blink Detection", *Research report. IIT.SRC, Bratislava, Slovakia*, p. 1-8.

[22] O. Ferhat, F. Vilariño and F. J. Sanchez, (2014) " cheap portable eye-tracker solution for common setups ", *Journal of Eye Movement Research*, vol. 7, p. 1-10.

11

Chronological states of viewer's intentions using hidden Markov models and features of eye movement

Minoru Nakayama*, Naoya Takahashi

Human System Science, Tokyo Institute of Technology
Ookayama, Meguro, Tokyo, Japan

Abstract

To determine the possibility of predicting viewer's internal states using the hidden Markov model, several features of eye movements were introduced to the model. Performance was measured using the data from a set of eye movement features recorded during recall tests which consisted of observations of three levels of task difficulty. The features were the temporal appearances of fixations and saccades, and combinations of 8 viewed directions during long and short eye movements. As a result, features of long eye movements, such as saccade information, contributed to prediction accuracy. Also, this prediction accuracy was regulated by the difficulty of the task.

Keywords: User intention, hidden Markov model, features of eye movements

1. Introduction

As eye movements reflect user's mental activity [1], the prediction of user intentions can be studied using eye movements [2]. Here, the definition of intention is the user's decision. In this approach, key features of eye movements are extracted, and discriminant analysis is employed to estimate intention. Recently, when these predictions were made using a combined set of features of eye movements and machine learning techniques, performance improved significantly [3, 4].

Since eye movements are measured as sequential data, time-domain analysis such as conventional scan-path analysis is conducted frequently. The sequential data of eye movements can also be employed to predict user intention [5]. As the scan-paths depend on the features of visual images, mathematical analytical techniques have been introduced to extract significant information [6]. In particular, the hidden Markov model (HMM) [7] is sometimes employed to detect viewer's intentions and to create a cognitive process model using eye movement [8]. For example, in the analysis of the intentions of drivers of cars, a number of implicit cognitive states were created for HMM using the sequential data of driver's eye movements [9]. Also, sentence relevance has been evaluated using eye movement data from an information retrieval task [10]. The HMM is a famous tool for speech recognition, handwriting recognition, etc. The attributes of handwritten data are similar to sequential data of eye movements. Some feature extraction techniques such as "sub-stroke" description [11] may be useful for eye movement analysis. The selection of combinations of features of eye movements is also important to improve prediction performance, while simple combinations of saccades and fixations can be used in an HMM model to predict viewer's intentions [12].

In regards to the above mentioned issues, this paper will examine the feasibility of presenting the viewer's internal states to predict his or her intentions using HMM and selected features of eye movements. Also, the possibility of detecting the process used for internal information processing using the temporal changes recorded in the experimental data is explored. In particular, the performance improvement should be

*Corresponding author. Email: nakayama@cradle.titech.ac.jp

confirmed by comparing it with previous results [12] when the selected features were introduced. Again, the possibility of predicting the viewer's temporal changes in internal states using HMM and sequential data should be determined, in addition to estimating the intention using the overall data [3, 13].

2. Method

To determine the feasibility of detecting chronological changes of intention, previous experimental data [3, 13] was analyzed and applied to the hidden Markov model.

2.1. Experimental task

The task was a recall test for contextual understanding and memorization. First, participants were asked to read a number of definition statements which described locational relationships between two objects (Figure 1a) [3, 13]. A set of definition statements containing a number of statements (K= 3, 5, or 7) was presented for 5.0 seconds for each statement, as shown in Figure 1a, and one of the 10 question statements in Figure 1b was presented for 10.0 seconds following that. The subject was asked to create a knowledge base from the definition statements, and then asked to evaluate whether a question statement was "True" or "False" using inference. Therefore, all statements used simple objects, and were carefully created to possess the same levels of comprehension. The question was a two alternative forced choice (2AFC) task where each question statement was "Yes (True)" or "No (False)" (Figure 1b).

Five sets of statements were created for each of the three levels of tasks. In total, there were 150 data responses divided into 15 task sets. The subjects were 6 male university students ranging from 23 to 33 years of age. They had normal visual acuity for this experiment.

The accuracy rates and mean reaction times of responses to question statements are summarized in Figure 2 [3]. Deviation in the reaction time suggests that the decision is sometimes not stable when thorough understanding does not occur. Therefore, the certainness of decision varies with the level of viewer's understanding, and this may influence their responses. When the response time was long, the certainty was low because participants could not decide quickly. Therefore, all answers were processed as incorrect responses when the reaction time was longer than 4.0 seconds. Since the mean reaction time for incorrect responses was less than 4.0 seconds, it may mean that the longer responses are not certain.

2.2. Eye-movement measuring

The task was displayed on a 20 inch LCD monitor positioned 60 cm from the subject. During the experiment, participant's eye movements regarding definition and question statements were continually observed, using a video-based eye tracker (EMR-8NL). Eye-movement was tracked on a 640 by 480 pixel screen at 60 Hz.

The tracking data was converted into visual angles, and eye movements were divided into saccades and fixations using a threshold of 40 degrees per second [14].

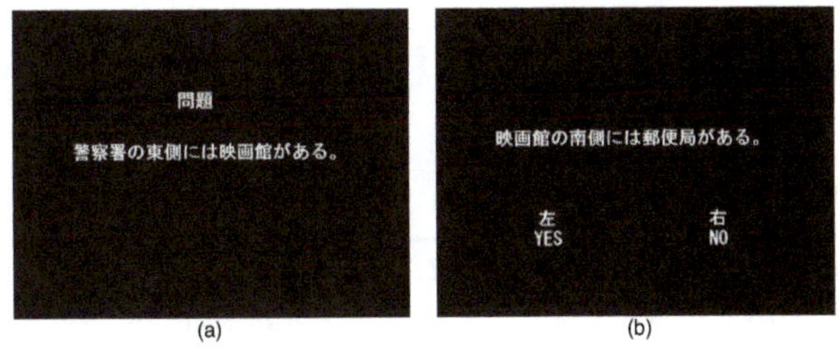

Figure 1. The screen on the left (a) shows a sample of a definition statement: "A theater is located on the east side of the police station." The screen on the right (b) shows a sample of a question statement: "There is a post office on the south side of the theater."

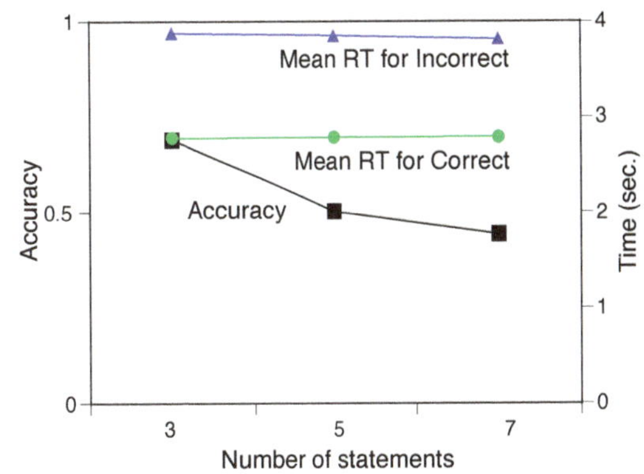

Figure 2. Accuracy rate and mean reaction times for responses

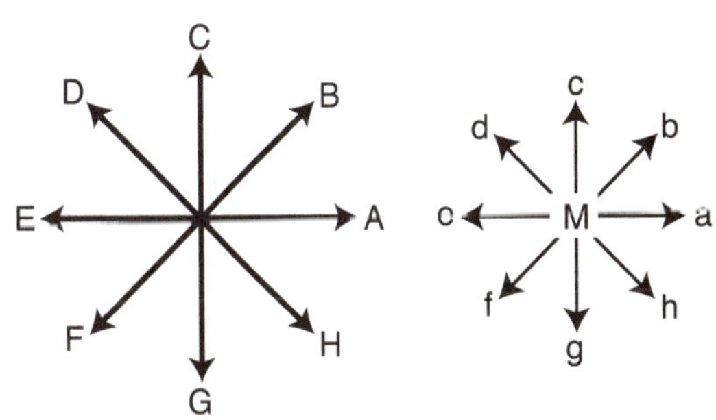

Figure 3. Eye movement categories: 'A'-'H' ('a'-'h') are directions of saccades (micro-saccades) and 'M' is a fixation.

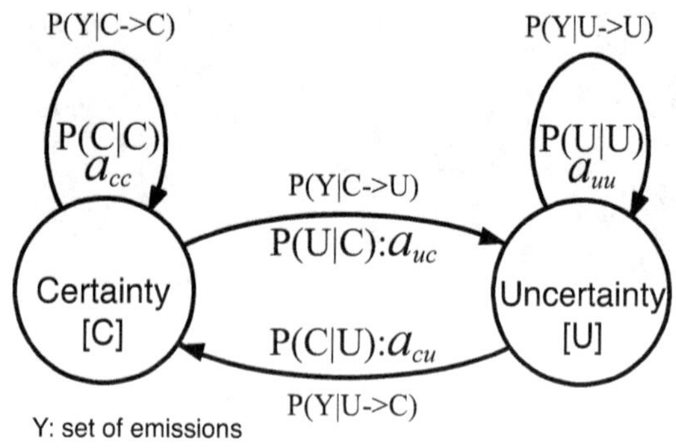

Figure 4. HMM diagram.

3. Hidden Markov modeling

According to the correct and incorrect responses in the task results, it is supposed that there are two levels of internal states used when making choices regarding the certainness mentioned above. These two internal states are defined as "Certainty" and "Uncertainty" for the prediction of response correctness. The probability of a correct response may be high when an internal state remains at "Certainty". The state of being certain starts from a state of "Uncertainty", and then moves between "Uncertainty" and "Certainty". During these state transitions, some emissions are observed as sets of eye movement features.

Using the series of observed eye movements as a set of time series data, the observation O and a set of eye movement features Y can be defined as follows:

$$\begin{aligned} O &= \{o_1, o_2, o_3, \cdots, o_t\} \\ o_t &\in Y \\ t &= T \times 60 - 1, T: sampling\ time(sec.) \\ & \quad 60: sampling frequency(Hz) \end{aligned} \quad (1)$$

A study of HMM-based handwriting recognition introduced features such as "sub-stroke" writing motion [11]. This trace analysis of handwriting was employed to extract features of eye movements [8]. A similar method of eye movement analysis has been conducted using EOG [15]. As mentioned above, the eye movements were classified into saccades and fixation, but were all scalar. Using the categories in Figure 3, the saccades and micro-saccades during fixation were categorized into one of 8 directions, as were "sub-stroke" classifications from the handwriting study. The directional categorization of eye movements was based on an idea from a previous study [3, 13]. The set of emissions Y can be noted using a combination of features, as in the following formula:

$$\begin{aligned} Y &\ni \{Y_{f-s}, Y_{f9-s8}, Y_{f9-s}, Y_{f-s9}\} \\ Y_{f-s} &= \{fix, sac\} \\ Y_{f9-s8} &= \{a, \ldots, h, M, A, \ldots, H, \} \\ Y_{f9-s} &= \{a, \ldots, h, M, sac\} \\ Y_{f-s9} &= \{fix, A, \ldots, H\} \end{aligned} \quad (2)$$

Figure 4 illustrates a hidden Markov model of two states of certainty of responses and a set of emissions during transitions. The two states indicate levels of high or low certainty for the two responses, known as "Certainty" and "Uncertainty".

As a result, the model of HMM λ can be defined as follows:

$$\begin{aligned} \lambda &= \{S, Y, A, B, \pi\} \\ S &= \{s_i | "C": certain \vee "U": uncertain\} \\ A &= \{a_{ij}, i, j = C, U | a_{cc}, a_{cu}, a_{uu}, a_{uc}\}, \sum_j a_{ij} = 1 \\ B &= \{b_{ij}(k), i, j = C, U, k \in Y\}, \sum_k b_{ij}(k) = 1 \\ \pi &= \{\pi_C = 0, \pi_U = 1\} \end{aligned} \quad (3)$$

A set of parameters, $\theta \equiv (A, B)$ is optimized using experimental data. The Baum-Welch algorithm, which uses a likelihood function, is employed as shown in (4) [7].

$$\log P(S|O, \theta) \quad (4)$$

The Forward algorithm provides a series of state transitions S, which maximize the likelihood function [7].

Additionally, the probability (Pr) of remaining in one state can be defined as c for "certain" and u for "uncertain", while $Pr(c) + Pr(u) = 1$. According to the features of the Markov transition, the transitional probability can be calculated as shown in (5). Then, the probability of certainty can be stated as a time series.

$$\lim_{m \to \infty} \begin{bmatrix} a_{cc} & a_{uc} \\ a_{cu} & a_{uu} \end{bmatrix}^m \times \begin{bmatrix} c^{ini.} \\ u^{ini.} \end{bmatrix} \to \begin{bmatrix} c \\ u \end{bmatrix} \quad (5)$$

4. Results

The parameters of HMM models were optimized using data sets as sequential data between the observation time and stimulus onset, and performance evaluations were conducted every 0.05 seconds after stimulus onset. The results were then summarized using the same format as in the previous study [12].

4.1. Prediction accuracy

The performance of predicting the states and responses was evaluated using the leave-one-out technique. As previously indicated, correct and incorrect responses were predicted, and the estimation accuracy was also evaluated during the time series. An example of prediction results for 3 statements using a feature set of directional saccades and micro-saccades (Y_{f9-s8}) is summarized in Table 1.

The temporal change in prediction accuracy using the 3 statement condition is summarized in Figure 5. Performance in three conditions (Y_{f-s}, Y_{f9-s8}, Y_{f-s8}) is compared in the figure. Most levels of accuracy for the models with Y_{f9-s8} and Y_{f-s8} are higher than for the ones with Y_{f-s}, as has been previously reported

Table 1. Prediction accuracy of 3 statements using set (Y_{f9-s8}) at 450 msec. (Accuracy=70.7%)

	Prediction	
	Certainty	Uncertainty
Correct	188	19
Incorrect	69	24

$\chi^2 = 13.1, df = 1, p < 0.01$

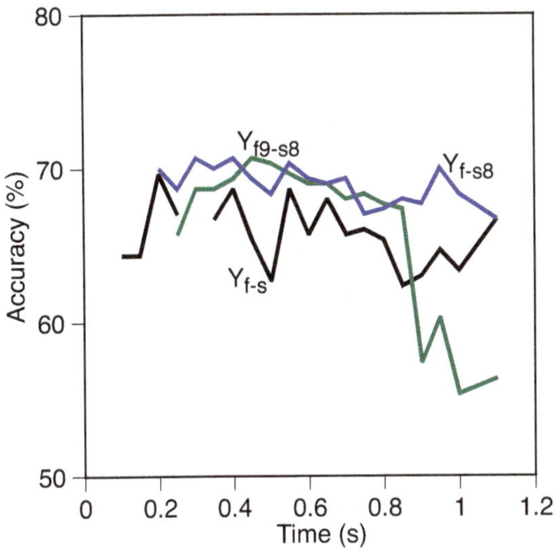

Figure 5. Prediction accuracy across the viewing time (K=3).

[12]. In particular, the levels of accuracy from 0.3 to 0.7 seconds are the highest. On the other hand, the level of accuracy for the models with Y_{f9-s8} suddenly decreases after 0.9 seconds, and the level of accuracy for the models with Y_{f9-s} remains at chance. These all suggest that the appropriate combination of features of eye movements provide accurate predictions of viewer's certainty during a chronological change of state, while the level of accuracy changes with the given data set of temporal eye movements. Since performance is high when detailed directional saccade information is employed, prediction performance is a significant source of information for the prediction of intention. This result coincides with previous analysis of the estimation of response correctness [13]. Additionally, the prediction can be made using shorter time series data, though this prediction performance is lower than the performance of previous analysis. This is a benefit of the proposed procedure.

However, performance remains at chance when the number of definition statements is 5 or 7 (K=5 and 7). Prediction performance using the original responses was significant if even the reaction time was longer than 4.0 seconds [12]. It may be possible that either the overall correct rates influence prediction accuracy, or there are other calculation-related issues. At least, chronological prediction is possible when the number of definition statements is 3 (K=3), since the mental workload in this condition is not high [13]. A detailed analysis of this will be a subject of our further study.

4.2. Probability of staying in "Certainty"

As mentioned in Equation (5), the probability can be calculated using the transition matrix, which is estimated using a set of features of eye movements in a time series. The probability of remaining in the "Certainty" state indicates the degree to which the content is understood, so that the viewer can make correct responses. The temporal changes in probability for the sets with Y_{f-s}, and Y_{f-s8} are calculated using the three levels of the number of definition statements (K=3,5,7), and summarized in Figure 6. Again, the number of definition statements determines the task difficulty. Figure 6 shows the probability relationship between the temporal time and the task difficulty. For example, a viewer has a strong inclination to make correct responses when question statements are presented, so the probability of remaining in the "Certainty" state exists. However the inclination sometimes decreases as the statement is understood, so the probability also decreases. The probabilities with Y_{f-s8} are higher than the ones with Y_{f-s} while the probabilities decrease with the number of definition statements. These results seem reasonable in view of the experimental conditions.

Most levels of probability are high during the early stage of reading, in the area of 0.1-0.3 seconds, and then decrease through the initial probability sets to 0 at stimulus onset. After 0.8 seconds, the levels of probability for the conditions (K=5, 7) are the lowest. As the results of temporal accuracy changes suggest, the features of eye movements during early stage of reading a statement affect the level of certainty of the response.

The level of probability of a set with Y_{f9-s8} is summarized in Figure 7 using the same procedure as with other sets. Using this method of calculation, the results present a different tendency. The levels of probability for a condition (K=3) remain at the same level, which is higher than 0.8 during the initial second following stimulus onset. The level of probability was high during a specific time when the number of definition statements increased (K=5, 7). As Figure 7 shows, the level of probability for condition (K=5) suddenly dropped after 0.6 seconds. Also, another condition (K=7) presents a high level of probability between 0.4 and 0.65 seconds. According to the results of transition simulations using this data set, viewers remain in the "Certainty" state for all conditions between 0.4 and 0.6 seconds. This result may show that there is a common period of certainness to be inclined to respond correctly when the knowledge

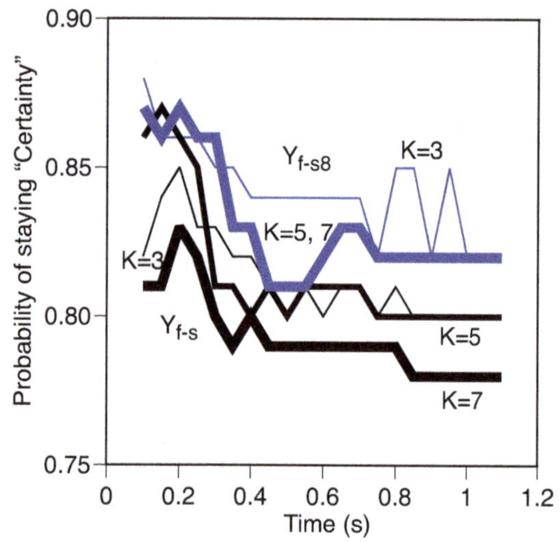

Figure 6. Probability changes of remaining in "Certainty" with Y_{f-s} and Y_{f-s8}.

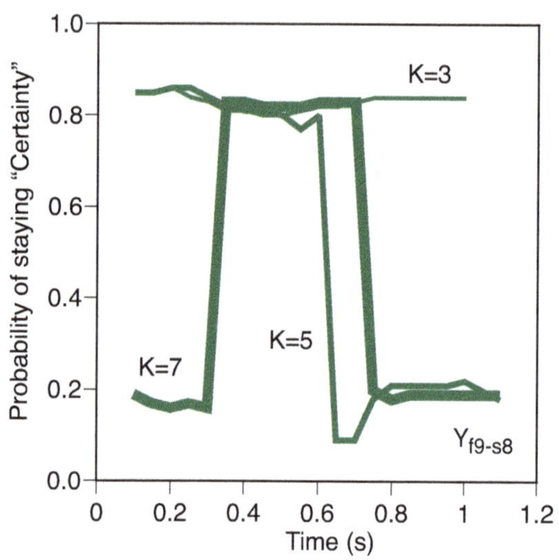

Figure 7. Probability changes of staying in "Certainty" with Y_{f9-s8}.

base is complicated (K=5,7) though it is almost always possible when the knowledge base is the simplest (K=3). Therefore, viewers may receive key information useful in a recall test during this short period of time.

This phenomenon is likely to be similar to previous studies regarding text comprehension and information processing, and should be discussed in more detail. However, it is not clear which exact type of processing is occurring during this period. A more detailed analysis should be conducted and will be a subject for our further study. Also, a model setting which includes model selection and tuning should be considered for use in future studies. Determination of the possibility that this procedure can be applied to other tasks will be a subject of our further study.

5. Conculusion

To obtain chronological transitions of internal states in order to determine viewer's intentions, the hidden Markov model (HMM) was applied to sequential data sets of eye movements. Regarding the features of handwriting motions for HMM, temporal eye movements such as saccades and fixation were categorized using eye movement directions, and then the parameters of HMMs were calculated using the data which was observed during a recall test. Levels of prediction accuracy regarding internal state transitions were measured using HMMs and sequential features of eye movements. Performance depended on details of features presented and sequential data in the early stage of viewing, and performance using those features improved rather than ones with simple combination of saccades and fixations. The probability of remaining in an internal state can be calculated using the sequential data, and the specific time period can be extracted as a stage of information processing using HMM models.

References

[1] Jacob, R.J.K. and Karn, K.S. (2003) Eye tracking in human–computer interaction and usability research: Ready to deliver the promises. In Hyona, Radach and Deubel [eds.] *The Mind's Eye: Cognitive and Applied Aspects of Eye Movement Research* (Oxford, UK: Elsevier Science BV).

[2] Goldberg, J.H. and Schryver, J.C. (1995) Eye-gaze determination of user intent at the computer interface. In Findlay, J., Walker, R. and Kentridge, R. [eds.] *Eye Movement Research, Volume 6: Mechanisms, Processes and Applications (Studies in Visual Information Processing)* (Elsevier), 491–502.

[3] Nakayama, M. and Hayashi, Y. (2010) Estimation of viewer's response for contextual understanding of tasks using features of eye-movements. In Hyrskykari, A. and Ji, Q. [eds.] *Proceedings of ACM Symposium on Eye-Tracking Research & Applications (ETRA2010)* (New York, USA: ACM): 53–56.

[4] Bednarik, R., Vrzakova, H. and Hradis, M. (2012) What do you want to do next: A novel approach for intent prediction in gaze-based interaction. In Mulligan, J.B. and Qvarfordt, P. [eds.] *Proceedings of ETRA 2012: ACM Symposium on Eye-Tracking Research & Applications* (New York, USA: ACM): 83–90.

[5] Goldberg, J.H. and Kotval, X.P. (1999) Computer interface evaluation using eye movements: methods and constructs. *Industrial Ergonomics* **24**: 631–645.

[6] Choi, Y.S., Mosely, A.D. and Stark, L.W. (1995) String editing analysis of human visual search. *Optometry and Vision Science* **72**(7): 439–451.

[7] Bishop, C.M. (2006) *Pattern Recognition and Machine Learning* (New York, USA: Springer Science+Business Media).

[8] Salvucci, D.D. and Anderson, J.R. (1998) *Tracing Eye Movement Protocols with Cognitive Process Models.* Tech. Rep. Paper 48, Carnegie Mellon University, Department Psychology.

[9] Liu, A. (1998) What the driver's eye tells the car's brain. In Underwood, G. [ed.] *Eye Guidance in Reading and Scene Perception* (Elsevier), 431–452.

[10] Puolamäki, Y., Salojärvi, J., Savia, E., Simola, J. and Kaski, S. (2005) Combining eye movements and collaborative filtering for proactive information retrieval. In Heikkil, A., Pietik, A. and Silven, O. [eds.] *Proceedings of ACM-SIGIR 2005*, ACM (New York, USA: ACM Press): 145–153.

[11] Nakai, M., Akira, N., Shimada, H. and Sagayama, S. (2001) Substroke approach to HMM-based on-line Kanji handwriting recognition. In *Proceedings of the Sixth International Conference on Document Analysis and Recognition (ICDAR 2001)*: 491–495.

[12] Takahashi, N. and Nakayama, M. (2013) Chronological prediction of certainty in recall tests using Markov models of eye movements. In *Proceedings of BIOTECHNO 2013: The Fifth International Conference on Biometrics, Biocomputational Systems and Biotechnologies* (IARIA): 55–60.

[13] Nakayama, M. and Hayashi, Y. (in Press) Prediction of recall accuracy in contextual understanding tasks using features of oculo-motors. *Universal Access in Information Society* .

[14] Ebisawa, Y. and Sugiura, M. (1998) Influences of target and fixation point conditions on characteristics of visually guided voluntary saccade. *The Journal of the Institute of Image Information and Television Engineers* **52**(11): 1730–1737.

[15] Bulling, A., Ward, J.A., Gellersen, H. and Tröster, G. (2011) Eye movement analysis for activity recognition using electrooculography. *IEEE Transasction on pattern analysis and machine intelligence* **33**: 741–753.

12

Context-aware Readings with Inbooki

Daniele Grassi[1], Giacomo Barigazzi[1,2], Anas Bouhtouch[1], Giacomo Cabri[2,*]

[1]DM Digital, Modena, Italy
[2]Università di Modena e Reggio Emilia, Modena, Italy

Abstract

E-books are more and more spreading among readers, who take advantage from the capability of reading them on different devices and of storing a whole library of books in a light and portable device. What is still missing is a technology that enables not only the adaptation to the exploited device, but also to the context in which the books are read. The experience of the users can be more immersive if the flow of the book content can be modified by the reading device depending on different aspects related to the moment and place of reading.

To overcome this limit, we have developed Inbooki, a system for writing and reading e-books that can take into consideration the context in which they are read and can adapt to it, making the experience of the readers more immersive in the story told by the book. The proposed model of the context take into consideration the information from the device sensors, the user profile, and the user choices.

Keywords: Context-dependency, E-book

1. Introduction

An *E-book* is "an electronic version of a printed book which can be read on a computer or a specifically designed handheld device" [4]. E-books can be seen from two points of view: from the one hand, they can be considered as an evolution of classic paper books; from the other hand, they represent a revolution in the way contents are delivered and read [6].

The e-book are spreading, they are likely to be more and more our companions and the trend seems very positive [5, 12]. In fact, today's e-books have several *advantages* over conventional books; for example: they can be read using different devices and technologies, letting people choose the preferred one; their visualization can be adapted to the devices' features or to readers' necessities; a single device can store a great number of books, relieving people from the need of carrying heavy books (or a heavy set of books).

Of course, they can present also some drawbacks or minor problems. Some studies point out that they cannot completely replace conventional books [7] in a very near future. Moreover, some challenges must be tackled: the studies carried out in [9] and in [13] point out some problems with the technological aspects of e-books, in particular related to the setup process, the processes of finding and downloading the e-books, the hardware, and the reading process itself.

Besides these minor problems, we point out an important aspect; current e-books mainly consist of an electronic version of their paper edition, and this leads to a main drawback: they do not take advantage of the possibilities offered by today's devices, which can be summarized as follows:

- the chance of gathering information received from *sensors* available in the reader devices (light, temperature, orientation, position, etc.);
- the capability of retrieving information from external *services* (weather, near places, news of the day, etc.);
- the access to the *user's profile and history* (age, gender, previous readings, etc.).

As a consequence, today's e-books cannot adapt their content to the context in which they are read, or to the specific user who is reading them.

*Corresponding author. Email: giacomo.cabri@unimore.it

Starting from the above considerations, we designed and built Inbooki, a system to develop and read e-books that take into account the *context* of the readers and adapt the content and the flow to it. So, they can be considered *context-aware* e-books. The contribution of this paper is the detailed description of Inbooki. Some similar approaches exist, but, as we explain the the next section, they are limited because either they are bound to a specific field or they exploit little context information; instead, our approach aims at being more general and at considering many sources of information. Inbooki enables writers to structure their book taking into consideration different aspects of the context where readers will actually read the e-book. The resulting e-books are called *immersive*-books (shortly *in-books*), meaning that the readers are "immersed" *in* the context where they read the e-books.

The paper is organized as follows. We first provide an overview of the related work in the field (Section 2). Then, we present the model of the context, the structure of an in-book and other details about in-books (Section 3). In Section 4 we explain the architecture of Inbooki along with a presentation of the different components. We will present an example of use of Inbooki (Section 5) and then we discuss about the possible applications of the proposed platform (Section 6). Finally, we discuss results and propose future directions in the conclusive Section 7.

2. Related Work

According to [2], existing systems related to our work can be classified as:

- Location-aware systems, which exploit only a part of the context;
- Context-aware systems, which fully exploit all the context information;
- Context-aware frameworks, which provide an abstract framework for specific context-aware applications.

As said, there are some systems that exploit information about the location (location-awareness) to display customized information to users [3, 8, 11]. Examples of location-aware systems are e-guides for tourists, as the Cyberguide project [1], which aims at building a mobile context-aware tour guide. The implemented system considered several context parameters to adapt the tour to the reader in a form similar to an e-book. However, this project is mostly focused on geographic information related to the tour, and it cannot be exploited for a general-purpose context-dependent book content.

A few examples of context-aware system are reported in the literature. For instance, Muñoz et al. [10] propose a context-aware system for hospitals that take into consideration several context parameters (not only the user's location) such as patient turnaround time, the location of other users, devices or artifacts, and the user's role. This is an interesting approach but strictly oriented to a specific field.

To summarize, the approaches we can find in the literature suffer of being very bound to a specific field or of exploiting little context information. Our aim is to propose a more general approach that considers a wider range of information available.

3. The Immersive Books

3.1. Context Model

The context of the reading is modeled around three sets of parameters: *environment* parameters, *profile* parameters, *reading* parameters.

Environment parameters are the set of external conditions surrounding the user while reading an in-book. Currently, the in-book structure supports GPS localization, weather, air temperature, season and time of the day. Weather and air temperature are determined using third-party web services and GPS location. *Profile* parameters contain the essential data about the user's profile: gender and age. The user sets these parameters during the registration process, and they can be modified in the reader apps. *Reading* parameters contain the "history" of the user's reading process for a given in-book: chapters read, reading conditions' status while reading each chapter, progress.

All these context parameters are used by the Inbooki mobile app[1] to determine which content to show the reader, as explained in the following.

3.2. Structure of an In-book

An *in-book* is an e-book composed of one or more graphs of chapters. Figure 1 reports examples of an in-book graph. Each chapter can have one or more parents and one or more children, and it is composed of an ordered series of blocks. Each block has a specific type ("title", "text", "image") and it can count on one or more variations, depending on the reading conditions. Variations can be grouped in 2 sets: *narrative* and *descriptive*.

Since an in-book is a graph, each chapter can have more than one child: narrative variations occur when the user completes the reading of a chapter and the Inbooki mobile app determines (using the user's reading conditions) which child chapter to show next among the possibilities. If the reading conditions meet the requirements of more than one chapter, a choice

[1] This is the app to read in-books, see 4.4

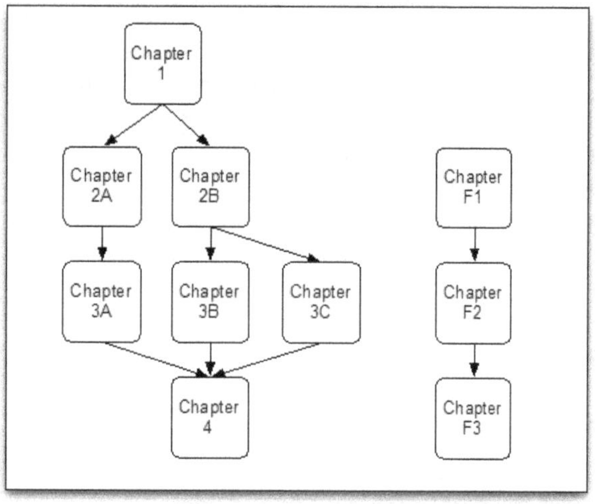

Figure 1. Example of in-book graphs

is presented to the user (a *fork*). If there is only one child chapter and its display conditions are not met, the user is forced to wait (or to take action) until they are met (e.g. the user can be required to go to a specific location, or he can be forced to wait until a specific time of the day comes). Descriptive variations occur inside a chapter's block. Each block can have one or more variations, each one of them having different display conditions. When the reader app has to render a chapter's text, it parses its whole content and determines which descriptive variations have their display conditions satisfied by the current user's reading conditions; if, for a given block, no variation has its display conditions met by the reading conditions, the whole block is not rendered. Figure 2 reports an example of structure of a chapter, writers can define their own structures.

3.3. Free Chapters

As described above, an in-book can be composed of more than one graph of chapters. This means that there can be chapters outside the main graph (being neither children nor parents of chapters belonging to the main graph). Each graph has one or more "root" chapters (not having a parent chapter): root chapters that do not belong to the main graph (called "free chapters") are listed as possible forks when the user finishes reading a chapter and their display conditions are met. For example, a free chapter can be set to become visible as a possible fork when the user is near a specific location at a specific time of the day.

3.4. Next Chapter Selection

When the user starts reading an in-book or finishes reading a previous chapter, the reader apps must determine how to continue the in-book rendering. The first step is to count how many children chapters have their visibility conditions fully satisfied by the user's reading conditions (environment parameters, profile parameters and reading parameters): if there are one or more results, they will be added to the possible following chapters set; if there are no results, the reader app tries to find the children chapters that have their profile and reading parameters satisfied: if there are one or more results, they are added to the possible following chapters set; if not, a virtual "book-end" chapter is added to the possible following chapters set. At this point the possible following chapters set consists of one or more chapters. A query is made to obtain free chapters that have their reading conditions satisfied, and the results are added to the possible following chapters set. The possible following chapters set is then analyzed: if it consists of two or more chapters, a fork is displayed to the user, who can choose which path to follow. At this point, only one chapter is chosen for rendering. If it is a virtual "book-end" chapter, the book ends. If it is a real chapter, this is further analyzed to determine whether its reading conditions are fully satisfied or not: if one or more parameters are not met yet, a "waiting screen" is shown to the user, who will have to wait in order to continue until the reading conditions are met; if the reading conditions are met, the chapter is rendered.

3.5. Chapter Rendering

Chapter rendering determines the text and the images that are shown to the user once they enter a chapter. The rendering process analyses each chapter's block, verifying that each variation's condition is met in the user's current reading conditions. Blocks that do not have at least one validated variation are not rendered. If a block has two or more variations validated, only the first one is rendered.

3.6. Environment Parameters Detection

The environment parameters can change while the user is reading an in-book. For this reason, the Inbooki mobile apps implement a timed polling cycle that records the parameters' states, which are used by the next chapter selection process to determine which chapters are eligible to be shown to the user when they finish a previous chapter. While time-related parameters are very fast to read, weather-related ones (current weather, current temperature) and GPS location present issues that need a caching system to obtain a smooth user experience. When the application starts, GPS location is usually approximate, and it takes some time for the device to determine the exact position; the Inbooki apps cache the last known location, and they query the device's GPS subsystem during the polling cycle of the environment parameters.

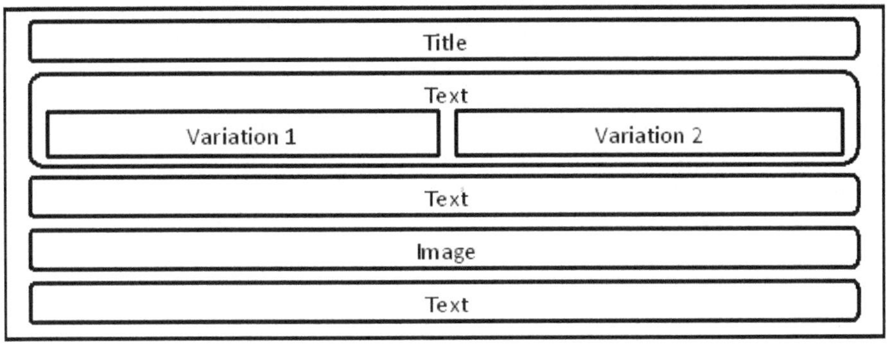

Figure 2. Example of chapter's structure

The same caching system is used for weather-related parameters: as the Internet services used for weather conditions retrieval can be unreachable from time to time (for example when the mobile phone signal or the internet connection are missing), the last weather conditions retrieved are cached for later use, and they are refreshed during the polling cycle of the environment parameters.

4. Inbooki's Architecture

The architecture powering Inbooki is composed of four subsystems:

- the *Inbooki Platform*, which provides API access to user accounts and to the marketplace;
- the *web interface*, which provides users with a website that allows them to register, apply for the Inbooki Author Program and browse the available in-books;
- the *Inbooki Editor*, a web application that allows authors to write in-books in a user-friendly environment;
- the *mobile apps* for iOS and Android devices, which allow the readers to enjoy the in-books using all the supported immersive and interactive features.

4.1. The Inbooki Platform

The platform is the *core* subsystem of Inbooki, providing a substrate which is used by the other subsystems to interact using a private API framework. There are two main API components: *AccountAPI*, which provides access to account utilities such as user registration, login, purchased books listing, account management, etc.; and *MarketAPI*, which provides an interface to the marketplace, including functions like in-book download, in-books information retrieval, marketplace search, etc. API security is ensured by a proprietary OAuth 2.0 implementation, written in PHP.

4.2. The Web Interface

This subsystem comprises a web interface to the marketplace and to account registration and management utilities. The main endpoint for the web interface is the Inbooki website (`http://www.inbooki.com`), which was developed using HTML, PHP and JavaScript using a Bootstrap template.

4.3. The Inbooki Editor

The Inbooki Editor plays an important role in the Inbooki architecture: it provides the authors with a user-friendly web application that allows them to write in-books in an easy-to-use yet complete environment. Using the editor, the authors can add *immersive* and *interactive* features to their in-books using a graphic interface. A built-in compiler gives the authors the possibility of creating deployable in-books that can be read on mobile devices running the Inbooki app for iOS or Android.

Using the editor authors can submit their in-books to the marketplace. The editor is written in PHP, HTML and JavaScript, using additional libraries as jQuery to enhance the user experience.

Snapshots of the web interface will be shown in Section 5.

4.4. Mobile Apps for iOS and Android Devices

An in-book contains several immersive and interactive features that are not supported by current e-books standards like ePub, mobi, etc. This is the reason why conventional e-book readers cannot display an in-book. iOS and Android Inbooki apps fill this gap, providing a reading experience that is extremely similar to the one provided by classic e-book readers (e.g. the user swipes to go to the next page), while introducing immersive and interactive features that allow the readers to fully enjoy the possibilities offered by the in-books.

The Inbooki apps give the reader the possibility of downloading the available in-books and automatically syncing the reading progress on all their devices (e.g. a given reader can use the same Inbooki account on an iPhone and on an Android device).

The immersive variations to the in-book's text–such as variations determined by weather conditions, time of the day, day of the year, user profile, etc.–are automatically processed during the reading, and each chapter is compiled on the fly using the conditions set by the author and determined by the reader's environment: the story itself, or single paragraphs, can thus change according to these conditions.

The Inbooki apps rely mainly on GPS localization to detect the reader's environment and to define the context: the data received is used to interact with external services to retrieve weather conditions and temperature, as well as to determine the "visibility" of geolocation-conditioned chapters (chapters that can be read only if the reader is within a certain distance from a specific geographical point).

The user's reading "history" is recorded and can be used by the authors while writing the in-book to determine story variations according to the path followed by each reader: for example, the story can change according to the path the reader chose in earlier chapters.

In-books also present interactive features: the author can design *forks* along the story where the reader is actually asked to make a decision. When the reader encounters a fork, the mobile app shows them the list of possible decisions and then calculates the following chapter according to this choice.

5. Example of Use and Impact

In this section we show how the Inbooki system can be exploited to write and to read in-books. We provide also an overview of possible applications.

5.1. Developing an In-book

To develop an in-book, the writer must first fill in the text of the chapters (see Figure 3)[2] as in a traditional editor.

The chapters have a customizable display order in the editor (see Figure 4) which doesn't necessarily correspond to the narrative order that is seen by the readers. For instance, the chapter "Scegli di confidarti" is a sibling of "Rispondi che non te lo ricordi": they both follow "Colazione in famiglia" (the reader is shown one of them depending on his context), but they continue in different narrative threads.

This leads to a non-linear chapters graph (see Figure 5); using the editor the writer can keep under control the structure of the in-book.

[2]The screenshots are from the Italian version of the software.

Furthermore, different conditions can be associated to each chapter. Figure 6 shows a dialog in which the following conditions can be defined:

- time of the day;
- weather conditions;
- age of the reader;
- external temperature;
- season;
- gender of the reader;
- location.

Advanced conditions can be specified as well.

Figure 7 details the capability of setting geolocalization conditions for a chapter: the writer can decide where in the world a given chapter can be read or, form the point of view of chapters, which ones will be triggered by a reader's given location.

5.2. Enjoying an In-book on Devices

We are aware that it is impossible to provide readers with the true experience of an in-book, since it is very dynamic, anyway in this section we sketch how the readers can enjoy the in-book using a mobile device (in this case, an Apple iPhone). As it can be seen from Figure 8, the reading of in-books is very similar to the reading of standard e-books.

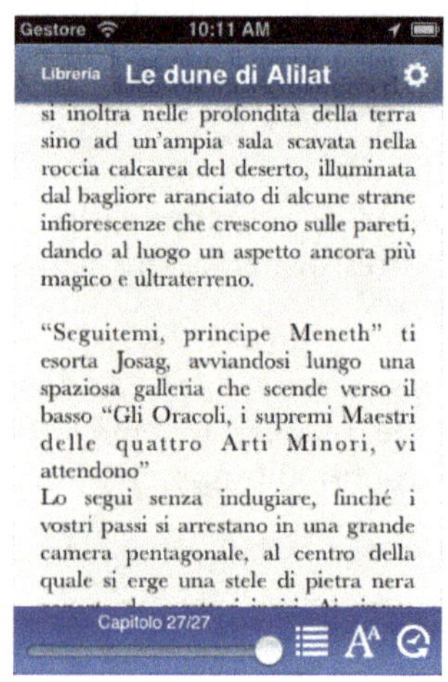

Figure 8. Reading interface (with menu)

One of the specific enhancements offered by in-books is the possibility to set some parameters before starting the reading; Figure 9 reports the reader's choice of gender, age, weather conditions, hour of the day, season, and temperature. The values of all these parameters can be defined in an automatic way: they can be retrieved from sensors, from (web) services, or from the user's profile.

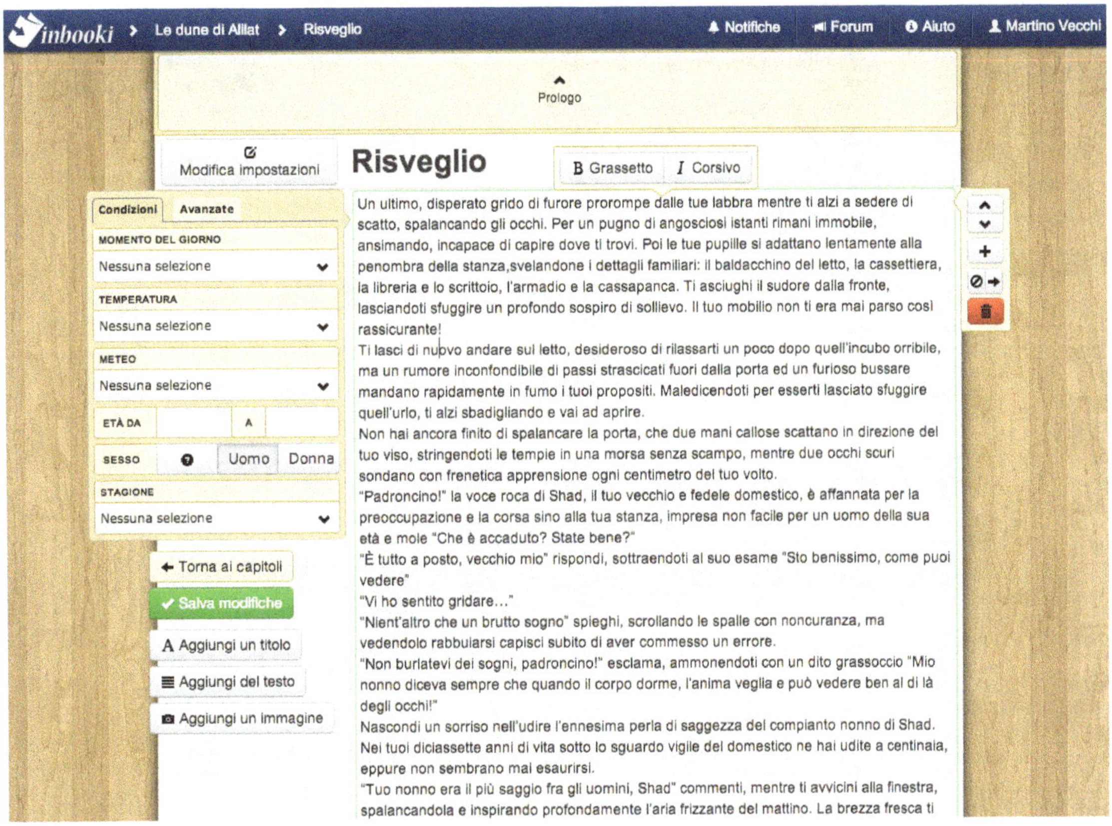

Figure 3. Inbooki chapter text

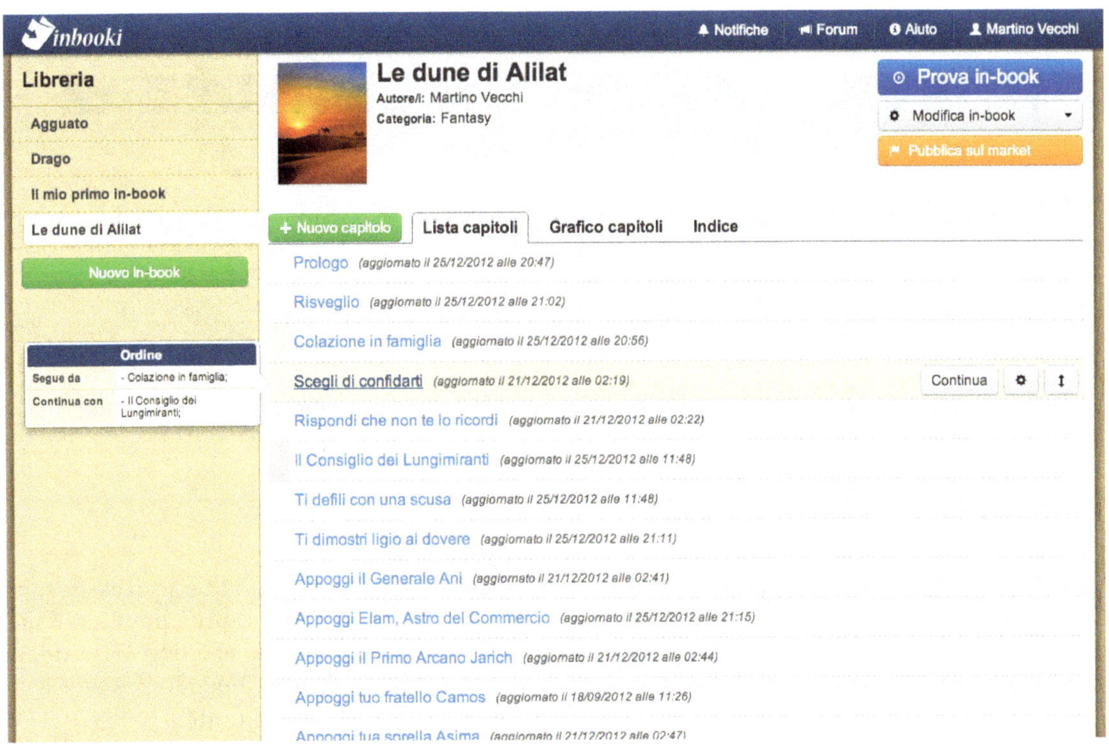

Figure 4. Inbooki chapters

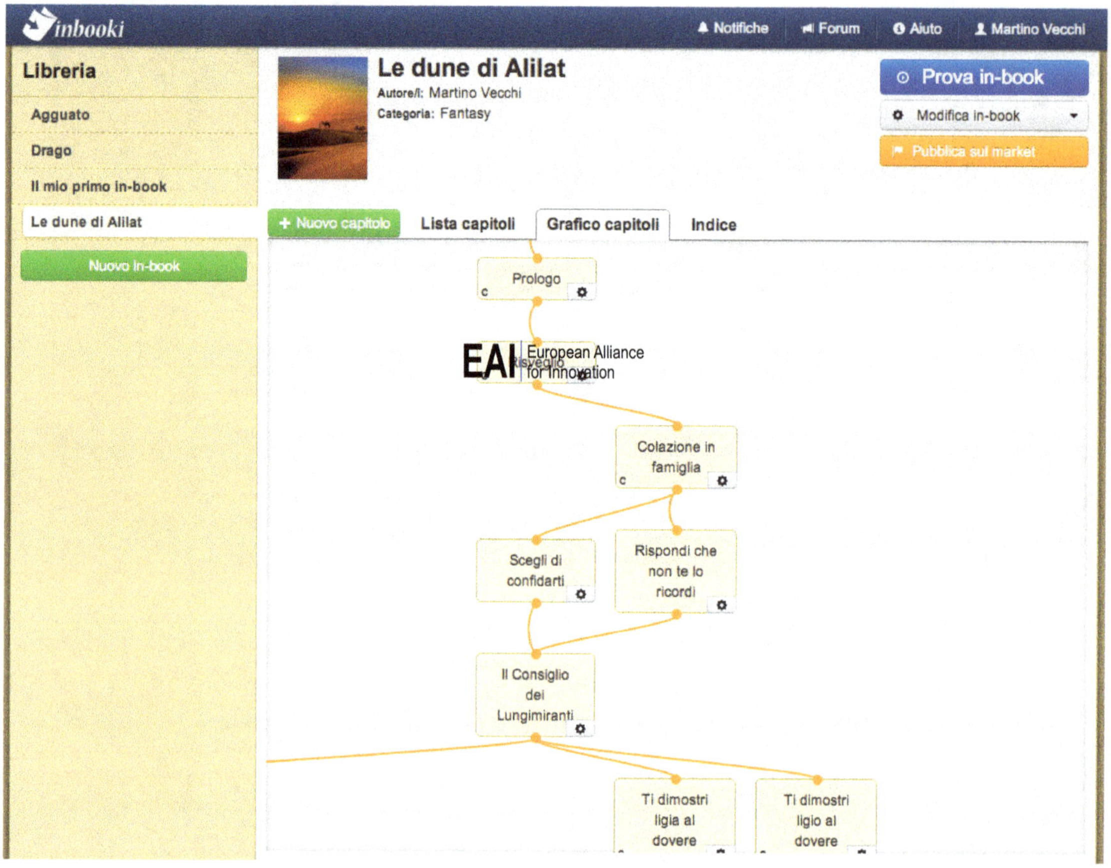

Figure 5. Inbooki graph

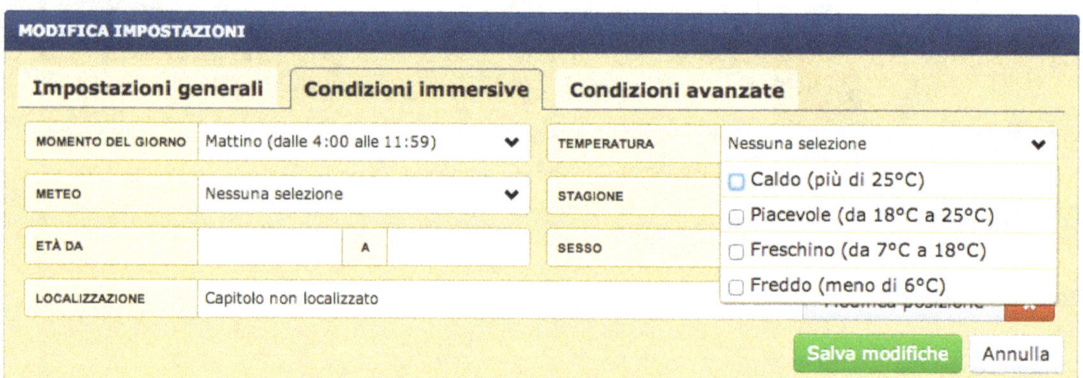

Figure 6. Inbooki chapter conditions

In Figure 10 we show a distinctive feature of in-books: facing a fork, the reader can choose which path to follow, which will lead to two different chapters, according to the structure of the in-book as defined by the writer.

6. Possible Applications of Inbooki

From a literary point of view, Inbooki introduces a new kind of e-book, in which stories can unfold in different directions depending on the user's choices and environment. An in-book could be seen as an enhanced version of the old gamebooks, quite popular in the '80s. Using the Inbooki Editor, the creation of an interactive and immersive e-book does not require any technical skills, and this represents an opportunity for authors to express their creativity without resorting to software developers to create their interactive e-books.

A practical example of use of Inbooki is found in the *guidebooks* field: currently, the creation of an interactive guidebook, able to use the reader's location

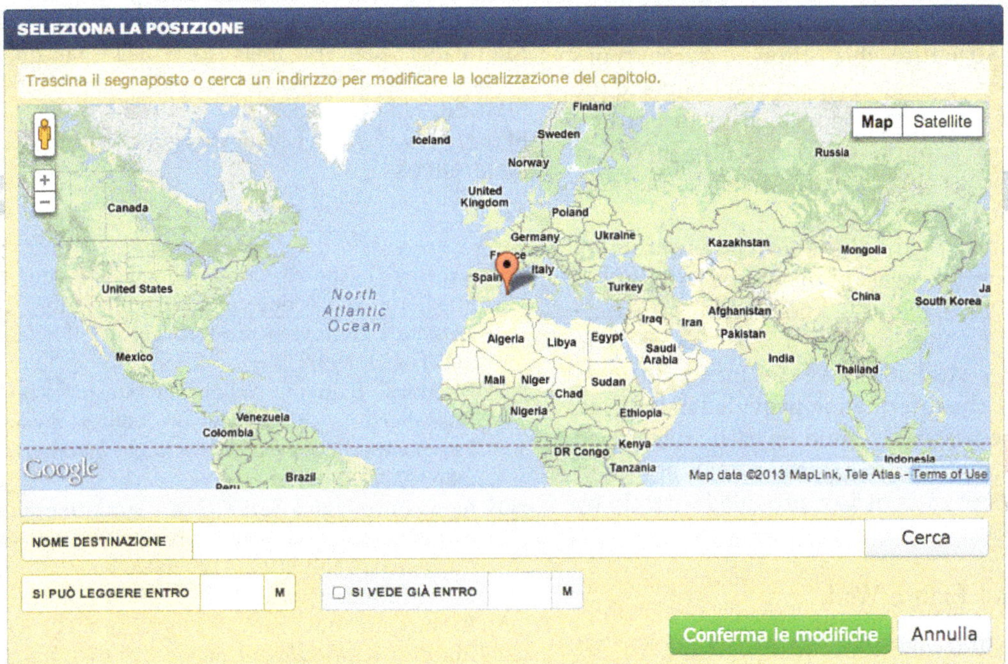

Figure 7. Inbooki localization

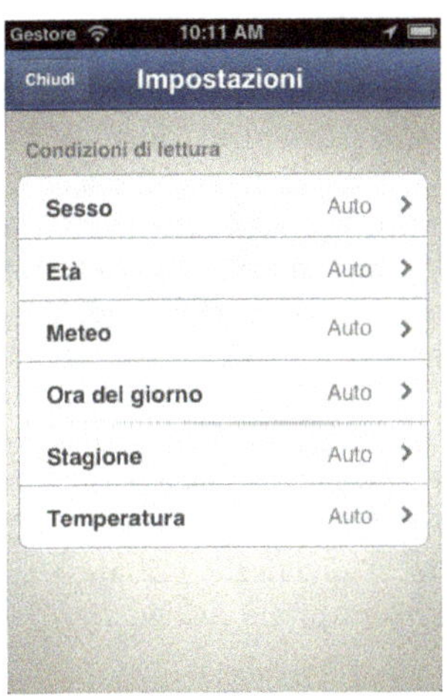

Figure 9. User's profile

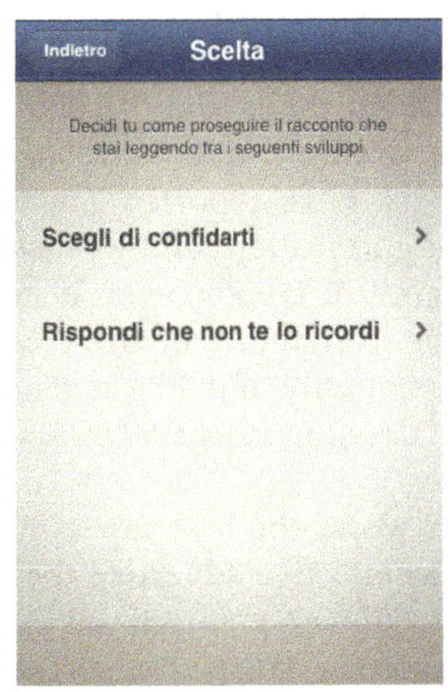

Figure 10. Reader's choice

to provide enhanced information, is restricted to who is willing and able to contract a software house and develop a dedicated mobile app. This means having to cope with the difficulties of deploying and distributing an app among different platforms (iOS, Android), and to depend on the software development team to provide the users with the updated content. Inbooki allows guidebook authors and editors to create their in-books using the Inbooki Editor and distribute them automatically to iOS and Android mobile devices, without needing any technical knowledge. The updating of the in-book content can be done using the Editor, and the release of new versions of the in-book is notified automatically to users, who can download it to their devices. In the guidebooks' field, an in-book can provide immersive and interactive e-books that can suggest to the user different tours and paths depending on the environment (e.g. the weather), on their personal interests (using forks) or their profile (gender and age). Using free chapters, specific content can be displayed depending on the location of the user and the environment conditions.

The concept of in-book can have a big impact on the e-book market, creating a niche for interactive and immersive stories and guidebooks. The supported features of the in-book can be easily further extended, thanks to its open structure which allows different types of blocks to compose a single chapter, and thanks to the flexible format of chapters' conditions. In the future, for example, video support could be introduced, or web-enhanced blocks could be defined to integrate web pages inside chapters.

Outside the literary field, Inbooki can be a useful tool to create simple context-aware applications without the need of any specific technical knowledge. A typical application could be the creation of business scenario tests, where MBA students have to make business decisions based on the information provided in the text. These business scenario tests could be easily developed by teachers using the Inbooki interface, making use of the "fork" features of in-books, or provided directly by authors or publishers.

7. Conclusion and Future Work

Inbooki is a system to develop adaptive e-books that take into consideration the context where they are read and which allow user interaction. These e-books are called "immersive books" or *in-books* because they consider the readers as "immersed" *in* the context in which they read the e-books themselves.

In this paper we have detailed different aspects related to Inbooki, from the model of context we have considered, to the architectural components of the system. Moreover, we have reported about the tools available for writers and readers, and we have proposed some uses of this kind of e-book, even if many others can be possible. The availability of these e-books will make people aware not only of their existence, but also of their usefulness, creating the basis of an interesting market.

Regarding future work, the first aspect on which to focus can be how to effective advertise the immersive books. From the technical point of view, today's e-readers and in particular smartphones are equipped with a broad range of sensors, whose contextual information can be taken into consideration by the authors to define the in-books' behavior. In a long-term work, interactivity can be enhanced enabling the readers to directly add content to the in-books, and encouraging the author-reader relationship adding specific communication tools to the Inbooki apps. Finally, also non-alphabetic language could be supported to make in-books available worldwide.

Acknowledgement. This work is supported by the project "Algorithms and Models for Building context-dependent Information delivery Tools" (AMBIT) sponsored by Fondazione Cassa di Risparmio di Modena (prot. Nr. 188.14.8C, SIME 2013.0660). Website:
http://www.agentgroup.unimore.it/ambit/

References

[1] Gregory D. Abowd, Christopher G. Atkeson, Jason Hong, Sue Long, Rob Kooper, and Mike Pinkerton. Cyberguide: A mobile context-aware tour guide. *Wireless Networks*, 3:421–433, 1997.

[2] Matthias Baldauf, Schahram Dustdar, and Florian Rosenberg. A survey on context-aware systems. *International Journal of Ad Hoc and Ubiquitous Computing*, 2:263–277, 2007.

[3] Jenna Burrell and Geri K Gay. E-graffiti: evaluating real-world use of a context-aware system. *Interacting with Computers*, 14(4):301 – 312, 2002.

[4] Oxford Dictionary. E-book, 2015.

[5] R. Duncan. Ebooks and beyond: The challenge for public libraries. *Australasian Public Libraries and Information Services*, 23(2):44, 2010.

[6] D. Ẑivković. The electronic book: evolution or revolution? *Bilgi dünyasi*, 9(1):1–19, 2008.

[7] Y.Y. Kang, M.J.J. Wang, and R. Lin. Usability evaluation of e-books. *Displays*, 30(2):49–52, 2009.

[8] Clemens Kerer, Schahram Dustdar, Mehdi Jazayeri, Danilo Gomes, Akos Szego, and Jose A. Burgos Caja. Presence-aware infrastructure using web services and rfid technologies. In Jacques Malenfant and BjarteM. Østvold, editors, *Object-Oriented Technology. ECOOP 2004 Workshop Reader*, volume 3344 of *Lecture Notes in Computer Science*. Springer Berlin Heidelberg, 2004.

[9] P. Lam, S.L. Lam, J. Lam, and C. McNaught. Usability and usefulness of ebooks on ppcs: How students' opinions vary over time. *Australasian Journal of Educational Technology*, 25(1):30–44, 2009.

[10] M.A. Munoz, M. Rodriguez, J. Favela, A.I. Martinez-Garcia, and V.M. Gonzalez. Context-aware mobile communication in hospitals. *Computer*, 36(9):38 – 46, sept. 2003.

[11] Nissanka B. Priyantha, Anit Chakraborty, and Hari Balakrishnan. The cricket location-support system. In *Proceedings of the 6th annual international conference on Mobile computing and networking*, MobiCom '00, pages 32–43, New York, NY, USA, 2000. ACM.

[12] A. Soules. The shifting landscape of e-books. *New Library World*, 110(1/2):7–21, 2009.

[13] W.D. Woody, D.B. Daniel, and C.A. Baker. E-books or textbooks: Students prefer textbooks. *Computers & Education*, 55(3):945–948, 2010.

13

Virtual Character Animations from Human Body Motion by Automatic Direct and Inverse Kinematics-based Mapping

Andrea Sanna, Fabrizio Lamberti, Gianluca Paravati, Gilles Carlevaris, Paolo Montuschi

Dipartimento di Automatica e Informatica, Politecnico di Torino, Torino 10129, Italy

Abstract

Motion capture systems provide an efficient and interactive solution for extracting information related to a human skeleton, which is often exploited to animate virtual characters. When the character cannot be assimilated to an anthropometric shape, the task to map motion capture data onto the armature to be animated could be extremely challenging. This paper presents two methodologies for the automatic mapping of a human skeleton onto virtual character armatures. Kinematics chains of the human skeleton are analyzed in order to map joints, bones and end-effectors onto an arbitrary shaped armatures. Both forward and inverse kinematics are considered. A prototype implementation has been developed by using the Microsoft Kinect as body tracking device. Results show that the proposed solution can already be used to animate truly different characters ranging from a Pixar-like lamp to different kinds of animals.

Keywords: virtual character animation, automatic armature mapping, motion capture, graph similarity, forward kinematics, inverse kinematics

1. Introduction

Several techniques are used to animate virtual characters. One of the most used is based on virtual skeletons (named rigs), which are related to meshes to be animated and deformed. The animator can move bones of the virtual skeleton in order to move mesh vertices. The so called keyframing method uses forward and inverse kinematics techniques to fix a set of key frames (poses), which will be automatically interpolated by the animation program [1][2]. The keyframing approach has been often outperformed by motion capture solutions [3]. Mocap solutions allow to directly gather the motion of the animator: motion data can been used both to animate interactively a virtual character and recorded and then applied to the character.

The keyframing approach allows animators to author animations that would be difficult or impossible to act out. On the other hand, complex actions usually need a lot of time (and skilled animators) to be produced by keyframing. Motion capture systems can gather complex motion data in a very accurate and smooth way, thus allowing animators to create very realistic (human) movements. Mocap systems are, in general, very expensive and data representing the animators' movements might be hard to be used for animating non-anthropometric characters (even if a solution to partially cope with this latter limitation has been proposed in [4]).

The goal of this paper is to automatically map the skeleton of the animator onto the armature of any virtual character, thus relaxing any possible constraints related to the armature topology to be animated. This task has been manually addressed in a number of previous works, such as [5] [6] and [7]. Manual association (mapping) can be a time consuming and difficult task, since only skilled animators are generally able to immediately identify the *best* match.

This work proposes and compares two different solutions that aim to find an efficient mapping of the human skeleton onto virtual character armatures. Exploiting graph similarity criteria, user preferences, motion constraints and information about the armature topology, the skeleton of the animator can be mapped onto the character armature bones both by a direct mapping and by a mapping of the end-effectors. The first solution (already presented in [8]) can be used when the character is animated by forward kinematics and the number of bones of the virtual character does not exceed the number of bones tracked by the mocap system. On the other hand, end-effectors are considered when the character is animated by Inverse Kinematics (IK); an end-effector is a bone on which the movement is applied, then the position of all the other bones

*Corresponding author Andrea Sanna. Email: andrea.sanna@polito.it

of the kinematics chain is updated automatically by an IK solver in order to place the end-effector in the selected position. Users can choose whether control a character by forward or inverse kinematics and then accept/reject/modify the association that has been automatically found.

The current implementation uses armatures defined in Blender [9] and the Microsoft Kinect sensor [10] as motion capture device. Nonetheless, the proposed methodology is general and it could be easily extended to any other tracking system. A mapping between the human skeleton tracked by the Kinect and the Blender armature allows the devised methods to translate the local movements of human skeleton bones into translations and rotations of the associated armature bones. In this way, a real-time markerless motion capture system for digital puppetry with generic characters is actually implemented.

The paper is organized as follows: Section 2 briefly reviews other approaches that have been designed to map a human skeleton onto a virtual character armature. Section 3 presents the first solution (which implements a direct mapping in forward kinematics) and shows how the similarity scores in the mapping matrix are computed. Section 4 proposes some applications of the matching algorithm to non-anthropometric virtual characters. On the other hand, Sections 5 and 6 present the solution based on the inverse kinematics and some examples, respectively. User feedbacks are presented in Section 7 where an assessment of both solutions is presented. In order to help reader to better understand automatic mapping algorithms, some mathematical steps, applied to a reference character, are also presented in the Appendix.

2. Background

Several works are known in the literature that propose solutions to use a performer as a sort of controller for animating virtual characters. Anthropometric limbs are interactively controlled by inverse kinematics in [11]; in this case, only sub-parts of the whole skeleton can be animated independently, thus this approach might be unable to cope with constraints that require the whole body to be animated. All body parts are considered in [12], where the main goal is to map the movements made by a performer onto an animated character by only considering constraints on the end-effectors, thus animating only by the inverse kinematics. An extension targeted to the control of non-anthropometric characters is proposed in [13]; an intermediate skeleton with less degrees of freedom is used and the remaining degrees of freedom are computed analytically in [13].

The above works basically relies upon specific representations of motion (e.g., through simplified skeletons). A comparable approach is also used in [14], where a data structure especially dedicated to motion adaptation is proposed. Moreover, in all the works considered, the focus is mainly on human-like animation.

Generic-shape characters are considered in a recent and impressive work [15]. The solution proposed in [15] allows animators to directly set a mapping between their own skeletons and the mesh of the object to be controlled. The character mesh is segmented from the background; then, the body of the animator is *embedded* to position vacated by the object. By a vocal command, limbs of the animator are attached to the parts of the mesh the body overlaps. These attachments serve as constraints for the deformation model that is inspired by the Embedded Deformation method proposed in [16].

Another solution proposing the animator as a direct controller of a virtual character has been presented in [17]. As [18], the solution proposed in [17] combines different feature mapping functions together with an intelligent blend scheme in order to generate highly realistic output motions. In particular, some features of the animator body are directly used by a direct mapping to animate the character, whereas another stage of motion classification allows the system to identify frames of pre-set clips of the virtual character to be blended with the direct mapping motion.

The approach proposed in [19] is tailored for non-humanoid characters. The motion of the animator is gathered and analyzed. The basic idea is to compare, by a statistical mapping function, a small set of corresponding key poses in order to define a mapping between the motion of the animator and the virtual character. Therefore, this approach is not completely automatic as a set of poses for the character has to be defined; moreover, it is not aimed to obtain a real-time puppetry animation.

This paper extends and improves the work presented in [8]. In [8], only an automatic mapping for armature controlled by forward kinematics was presented: this part is reviewed in Sections 3 and 4; on the other hand, the solution presented in [8] is not able to cope with armatures composed by a number of bones larger than the animator's one. The present manuscript tackles this issue and presents a second automatic mapping algorithm, which is based on the animation of end-effectors by inverse kinematics. Moreover, both automatic mapping techniques are evaluated in this paper; user tests have been performed in order to gather subjective and objective user feedbacks.

3. The first solution: a direct skeleton mapping

The first technique to map the skeleton of the animator onto an armature of a virtual character is shown in Figure 1. Basically, different blocks contribute to

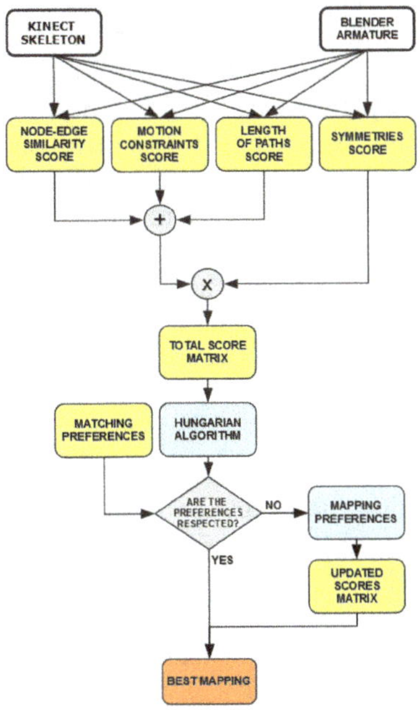

Figure 1. The first mapping algorithm described by a flow-chart.

compute a *score matrix*; coefficients of this matrix are used to automatically map the human skeleton onto the Blender armature. A set of parameters/constraints are considered:

- topology details (e.g., node-edge similarity);
- motion constraints (related both to the animator and the character to be animated);
- length of kinematic chains;
- symmetries of the armature.

Moreover, user preferences can be exploited to *force* some mappings, thus possibly overriding the above mentioned criteria (e.g., arms might be favorite with respect legs, and so on).

3.1. Graph representation

The first step of the mapping algorithm considers both the human and the virtual character skeletons. The Microsoft Kinect used as mocap device in this work provides tracking data containing information related to the center of mass and the position of each of the twenty joints of the captured skeleton. Moreover, a status information is associated with each joint: it indicates whether the joint position is being tracked or inferred (which happens when the Microsoft Kinect cannot see this point and tries to accurately *guess* it on the basis of information from previous frames and neighboring joints). Joints tracked for the skeleton are split into three main sections:

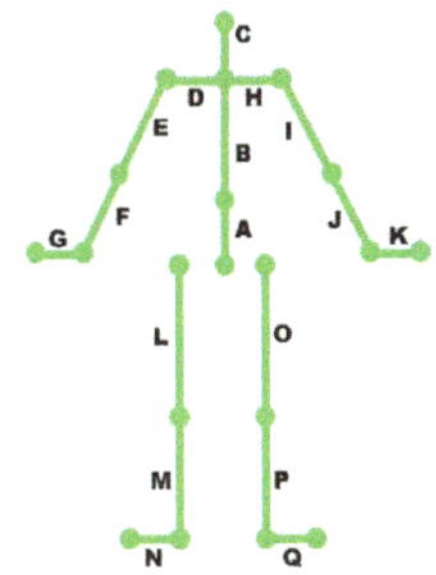

Figure 2. The skeleton extracted by the Kinect application: 20 joints are tracked.

1. the central area, containing the head, the neck, the spine and the hip center;
2. the arms, containing for each arm the shoulder, the elbow, the wrist and the hand;
3. the legs, containing for each leg the hip, the knee, the ankle and the foot.

The skeleton extracted by the Kinect application is shown in Figure 2. The Kinect application is connected by a socket to the Blender Game Engine (BGE): a Python script receives joint data and translates them into valid transformations for controlling the armature of the virtual character. A bone-to-bone mapping is performed; in other words, the position/rotation of a bone tracked by the Kinect is used to set up the position/rotation of a well defined bone of the character armature (for more details about the software architecture see [5]).

Blender armature is also analyzed: the starting point is the root bone. This step describes the armature as a sort of graph where bones are associated to nodes/vertices and relations between bones are considered as arcs/edges. The graph can be represented by an adjacency matrix [20]; given a graph G_A, $G_A = G(V_A, E_A)$ where V_A are the vertices and E_A are the edges, if the cardinality of V_A is n_a, then the adjacency matrix A of this graph is a $n_a \times n_a$ matrix in which entry $[A]_{ij}$ is equal to 1 if and only if $(i, j) \in E_A$, 0 otherwise. The adjacency matrix of an undirected graph will always be symmetric. Another useful graph representation is obtained by means of pair of matrices called edge-source matrix A_s and edge-terminus matrix A_t. This representation allows self-loops to be considered in the graph [20]. Let $s_A(i)$ denote the source of edge i, and let $t_A(i)$ denote the terminus of edge i. Then A_s and A_t can be defined as follows:

$$[A_s]_{ij} = \begin{cases} 1 & \text{if } s_A(j) = i \\ 0 & \text{else} \end{cases}$$

$$[A_t]_{ij} = \begin{cases} 1 & \text{if } t_A(j) = i \\ 0 & \text{else} \end{cases}$$

The graph representation given by A_s and A_t has the following properties:

- the adjacency matrix A is equal to $A_s A_t^T$;
- $A_s A_s^T$ is equal to a diagonal matrix D_{A_s} with the out-degree (i.e., the number of outgoing edges) of node i in the i-th diagonal position;
- $A_t A_t^T$ is equal to a diagonal matrix D_{A_t} with the in-degree (i.e., the number of incoming edges) of each node in the corresponding diagonal entry.

3.2. Node-edge similarity scores

Describing armatures as graphs allows the automatic procedure to assign a similarity score between pairs of nodes belonging to two different graphs (e.g, armatures); in this way, a match between two bone configurations (i.e., the skeleton extracted by the Kinect application and the Blender armature) is possible from a mathematical point of view. The applied similarity criterium considers both node and edge similarity as proposed in [20].

Given two graphs G_A and G_B, a simple way to give a definition of an edge score is: an edge in G_B is like an edge in G_A if their source and terminal nodes are similar, respectively. A is the adjacency matrix of G_A and B the adjacency matrix of G_B. D_{A_s}, D_{A_t} and D_{B_s}, D_{B_t} are the diagonal matrices containing the out-degree and the in-degree values of every node in G_A and G_B, respectively.

By iterating an certain number of times (usually, a satisfactory convergence is obtained with 11 iterations [20]) equation (1), a $n \times m$ scores matrix X is obtained, where n is the total number of bones in the Blender armature and m is the number of bones in the Kinect skeleton.

$$x_k \longleftarrow (A \otimes B + A^T \otimes B^T + D_{A_s} \otimes D_{B_s} + D_{A_t} \otimes D_{B_t})x_{k-1}. \quad (1)$$

The symbol $\otimes$ represents the Kronecker's matrix product, k the k-th iteration and x_k a column vector obtained by concatenating the columns of the scores matrix X. The iteration method presented well recognizes nodes that are very similar and provides good results if one of the two graphs is a subgraph of the other one.

A topological similarity is not sufficient to denote how two skeletons/armatures are similar, thus other parameters have to be taken into account.

3.3. Motion constraints scores

Bone motion constraints are another important parameter to be considered in the score matrix generation: the mapping technique should select pairs of bones that exhibit similar degrees of freedom. For example, it is easier to control a virtual segment exhibiting a high degree of freedom by means of a hand rather than with the spine. Constraint similarity is taken into account by assigning a *penalty* proportional to the different degree of freedom of each pair of bones; moreover, if two bones have completely different movement types, their score is set in the matrix X in such a way that they cannot be matched. Motion constraints scores are added to node-edge similarity scores (see Figure 1).

3.4. Length of paths scores

Length of kinematics chains are another parameter to be taken into account: sub-chains of similar length of the human skeleton and of the virtual armature character should have a greater mapping score than not similar long paths. A sort of *bonus* is added (see Figure 1) to the score of all those bone pairs that share the same position in a chain starting from the bone root. The bonus is added to enhance the probability of mapping a bone of the virtual character, place in a well given position of the chain, onto a bone placed at the same (or similar) position in the human skeleton. Updating the matrix of scores according to this criterion brings another advantage: the probability of mapping sequential bones in a chain by respecting the natural rank order is implicitly increased.

3.5. Symmetries scores

Armatures can exhibit some kind of symmetry; for instance, parts of the body of animals (e.g., legs) can be subdivided into small groups. On the other hand, a main symmetry in the human skeleton can be obtained by splitting left parts from right parts. This kind of symmetry can be present also in Blender armatures where bones are labeled by the suffix ".L" or ".R" to denote left and right parts with respect to the spine, respectively. By searching for the bones containing ".L" or ".R" in their names, it is possible to force a mapping of these bones onto the corresponding parts of the human body. The skeleton tracked by the Kinect application is already split in five groups: body, left arm, right arm, left leg and right leg. Depending on the bone type in Blender, the search space is reduced to a few groups that compose the Kinect skeleton. In this way, the left part of an armature will be always mapped onto the left arm or the left leg of the animator and the right part of the armature onto the right arm or leg. Scores can assume the following values: 0 (to avoid mapping a bone in the left part onto a bone in the right part and vice versa), 1 (to leave unchanged the score of bones not related to symmetries, like the spine bones) and 2 (to force left part bones to be mapped onto left part bones and vice versa). Symmetries scores multiply the previously obtained scores recorded in the matrix.

3.6. Evaluation component: Hungarian algorithm

In order to identify the best matching between two graphs, the node pairs with the highest scores in the matrix, according to a certain evaluation criterion, have to be found. This problem is known as the maximum weight bipartite graph matching problem. A common algorithm for identifying such a maximum weight is the Hungarian algorithm, described in [21]. In the proposed technique, the Hungarian algorithm is applied to the scores matrix to obtain a matching between the nodes of the graph representing the Blender armature and the nodes of the graph representing the Kinect skeleton, which maximizes the sum of squared matched scores. Thus, for each bone in the Blender armature, the Kinect bone it will be mapped onto is obtained.

3.7. Preferences

Preliminary tests that considered all the above mentioned criteria in order to generate the score matrix outlined how the resulting mapping can propose users to use some parts of the Kinect skeleton (e.g., the shoulders) which are very difficult to control for animating characters. This issue has been tackled by adding user's preferences to the overall mapping strategy.

If some armature bones are mapped onto the arms, forearms or hands are preferred to the shoulders. Of course, any other preferences could be coded in the matrix. Furthermore, the user can choose to avoid the mapping of the possible unused parts belonging to a mapped arm. The same approach is applied for the legs. After the application of the preferences, the Hungarian algorithm is used again to assign all the other unused bones.

4. Experimental results: forward kinematics

Three non-anthropometric characters (a Pixar-like lamp, a fish and a dog) have been used to the the first mapping technique. Blender armatures have been mapped onto the human skeleton extracted by the Kinect application and results are shown in Figures 3, 4 and 5, respectively. The kinematics chain of the lamp is mapped part onto two bones of the left arm (bones 1 and 2) and part onto three bones of the right arm (bones 3, 4 and 5). Arms are selected here, since their preferences score is the highest one (see Section 3.7). A completely different mapping has been obtained for the fish. In this case, it can be noticed how the symmetry (see Section 3.5) related to the fins is taken into account for mapping bones 4 and 5 on the right arm and bones 6 and 7 on the left arm. Since bones

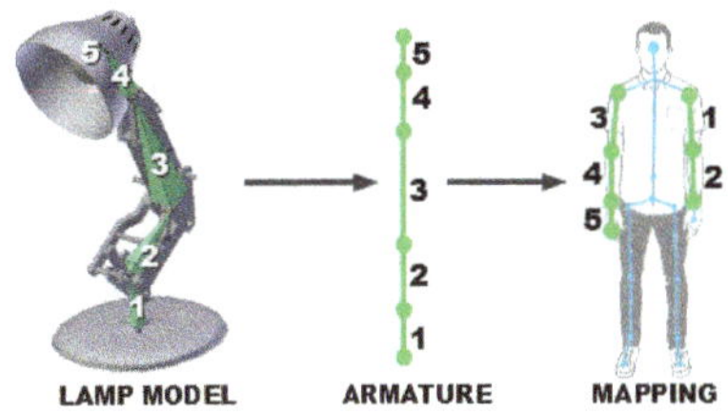

Figure 3. Mapping of the human skeleton onto the armature used to animate a Pixar-like lamp (a video is available at `http://youtu.be/iEuT7pzdMqw`).

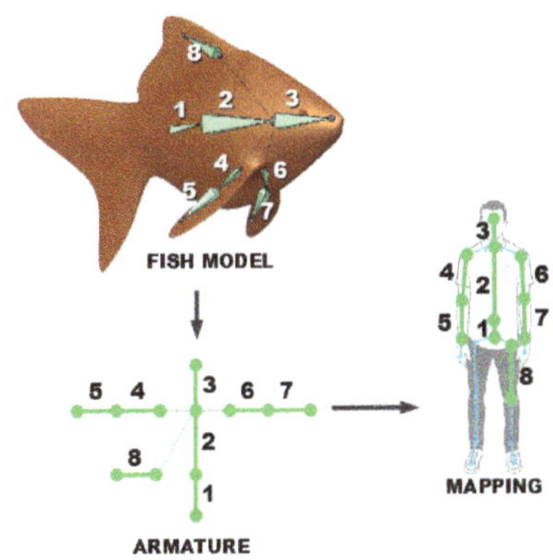

Figure 4. Mapping of the human skeleton onto the armature used to animate a fish (a video is available at `http://youtu.be/v1mMtAgFhzQ`).

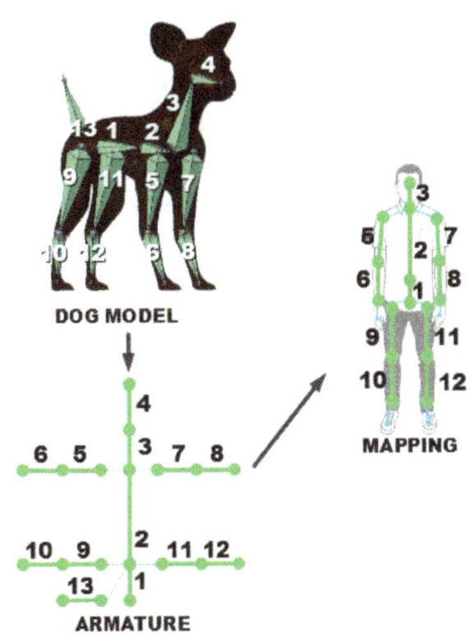

Figure 5. Mapping of the human skeleton onto the armature used to animate a dog (a video is available at `http://youtu.be/cQIKeOAwN24`).

1, 2 and 3 are topologically similar to the spine of the human skeleton, they are mapped onto it. The computation of mapping scores is detailed, for this character, in the Appendix. The dog armature is the most complex one among the three considered. Again, because of topological similarity, the spine of the dog is

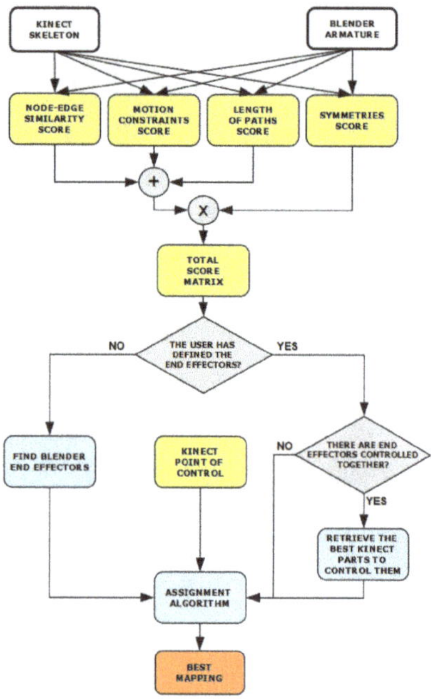

Figure 6. Flow-chart of the second mapping algorithm.

mapped onto the spine of the human skeleton (bones 1, 2 and 3). The symmetry is used to map the left parts of the skeleton onto the left parts of the armature and vice versa. Moreover, still because of topological similarity, the arms are mapped onto the front paws (the tail makes the similarity score for the back paws, with respect the skeleton arms, lower than the front paws). In this case, the user can choose to obtain a mapping also for the head and the tail or, as shown in the Figure 5, to leave these two bones unmapped as the five main kinematics chains of the skeleton have been already (even if not completely) used.

5. The second solution: a direct mapping of the end-effectors

The solution shown in Section 3 presents a main drawback: it is impossible to cope with armatures including a number of bones larger than the one provided by the tracking device. Unmapped bones cannot be animated (e.g., the dog's head of the previous example: the bone N. 4 is unmapped and the dog's head cannot be animated). In this case, forward kinematics cannot be used for the mapping, unless the target armature is reduced by collapsing pairs of bones. In order to control a chain of bones by the inverse kinematics it is necessary to consider the end-effectors (EEs) and update, by a IK solver, positions and orientations of all the other bones of the chain. Multiple solutions are generally possible and constraints are used to "help" the solver in order to find the best solution. Therefore, before to start the mapping step, all possible end effectors for an armature model have to be found. The armature is explored by starting from the root bone; following all possible chains it is possible to obtain all the paths forming the whole armature. Each of these paths should be animated in order to provide a realistic motion, thus one end effector has to be identified for all of the paths. The best choice is to take the terminal bone of a chain as the end effector, as moving the terminal all the other bones of the chain will be animated.

When the end-effectors have been identified, a suitable mapping for a human animator has to be found. It has been assumed that the skeleton provided by the Microsoft Kinect could have a maximum of five control points (CPs) able to drive kinematics chains: the head, the two hands and the two feet. At the next step, the best association between end-effectors of the character and control points of the animator has to be determined. Through the use of similarity scores calculated by the graph similarity theory, it is possible to evaluate each combination EE-CP, thus identifying the highest score. The similarity score is not only the similarity between two bones, but it also depends on all the other similarity scores of the bones connected to them (as illustrated in the coupled node-edge scoring of the graphs similarity theory). It results to be very suitable for character animation, as the algorithm aims to maximize the similarity of the human part (corresponding to a control point) with a chain of bones (corresponding to an end-effector). When the best scored combination is found, the corresponding control point is labeled as "no longer available" and the algorithm iteratively considers the next best pair EE-CP. When the control points are terminated, two events can occur: 1. there are not other EEs to be mapped, that is, the mapping is complete. 2. other EEs are not mapped and the 5 control points have been already assigned. The event 2 can be managed only by grouping two (or more) EEs under the influence/control of a *master* bone. In this case, the algorithm has to be able to retrieve bones controlling multiple EEs; this can happen when an animator wants to force the animation of multiple model parts with a single EE (see for instance Figure 8 where both wings are controlled by a master bone). In this case, the score of a master bone is determined by averaging scores of the EEs it controls. The complete flow chart of the second mapping algorithm is shown in Figure 6.

6. Experimental results: inverse kinematics

This section shows the application of the second mapping algorithm to three non-anthropomorphic characters (a seagull, a snake and an ostrich). Moreover, for the seagull model, both the application without master bones and with a master bone controlling both

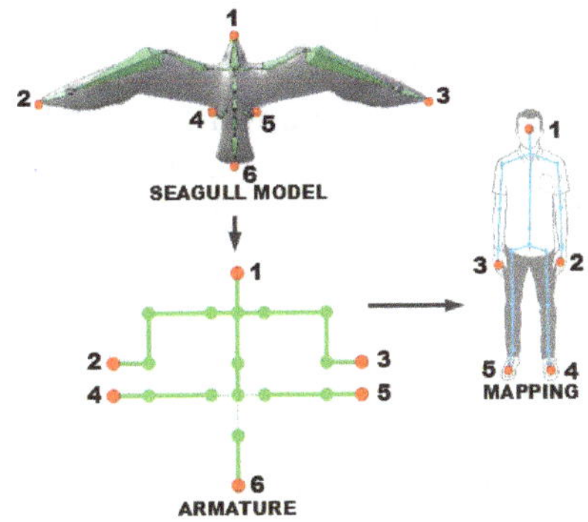

Figure 7. Mapping of the human skeleton onto the armature used to animate a seagull model(a video is available at http://youtu.be/KL55NuRKQAk).

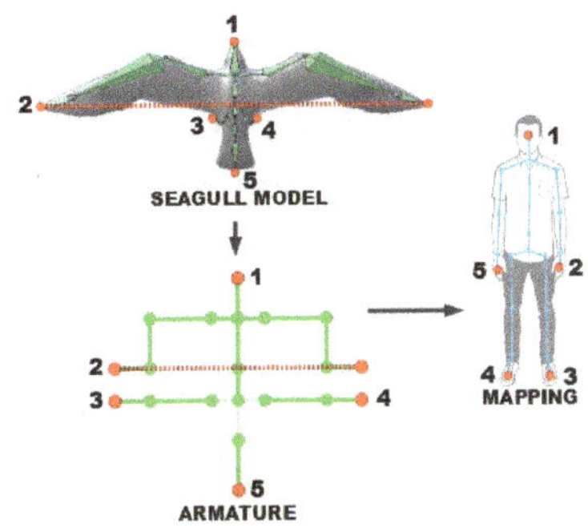

Figure 8. Mapping of the human skeleton onto the armature used to animate a seagull model with master bone controlling two end-effectors (a video is available at http://youtu.be/qaqP3za2dJM).

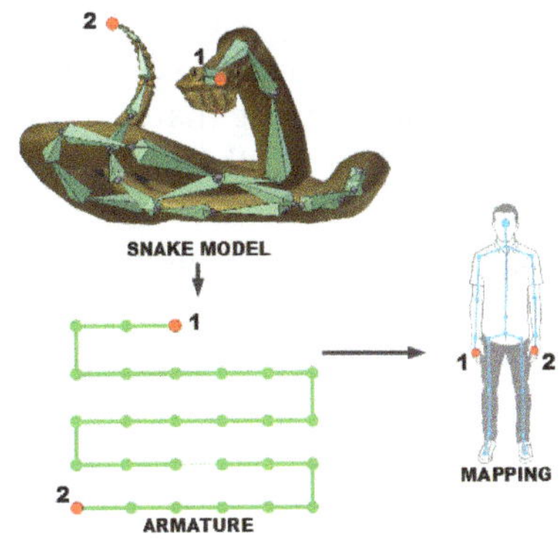

Figure 9. Mapping of the human skeleton onto the armature used to animate a snake model (a video is available at http://youtu.be/1ROzOmNgt8g).

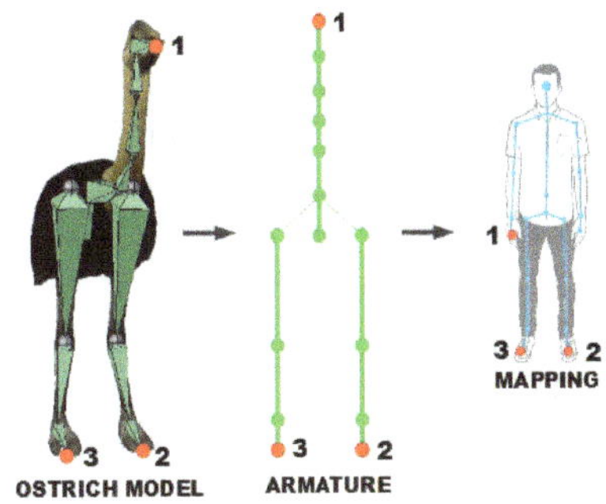

Figure 10. Mapping of the human skeleton onto the armature used to animate an ostrich model (a video is available at http://youtu.be/sM_ZiHVHFtY).

the EEs of the wings have been tested. Results are shown in Figure 7, 8, 9 and 10, respectively. In the seagull armature (see Figure 7) there are six different chains of bones and six corresponding EEs. This armature is very similar to the human one, apart the presence of an extra segment (the tail). The proposed system maps the head (end-effector 1) onto the human head, the two wings (EE 2 and 3) onto the two arms, the paws (EEs 4 and 5) onto the legs and the tail (EE 6) is unmapped. A different mapping is obtained considering the seagull model where a master bone to control both wings has been inserted (Figure 8). In this case the five end effectors are mapped onto each animator control point. The head of the seagull (EE 1) is assigned to the animator head, the master bone controlling the wings is mapped onto one hand, the two paws (EEs 3,4) are matched with the two legs and the tail (EE 5) is animated with the other hand. A completely different mapping is obtained for the snake model. The armature is formed by two chains of bones: one for the body and the head and one for the tail. The resulting mapping is as follows: the snake head (EE 1) is associated with the right hand and the snake tail (EE 2) is mapped onto the left hand.

The last character is an ostrich model. Its armature configuration is not complex and it includes two chains of bones for the legs and a long chain composed by many bones for the body, the neck and the head. The two paws (EEs 2 and 3) are mapped onto the two animator's legs, whereas the body-neck-head part (EE 3) is controlled by the right hand of the animator.

7. User feedbacks

In order to validate proposed automatic mapping algorithms, 10 students of the Computer Animation course of the Master of Science degree in Computer Science at Politecnico di Torino have performed some tests. The main goal was to compare the proposed mapping strategies with manually selected mappings, thus focusing the analysis on how good and useful are the proposed hints and how automatic mappings can help the user to efficiently animate virtual characters.

Two different models have been chosen for the tests. The lamp model has been selected to be animated by forward kinematics, whereas the ostrich model to be animated by inverse kinematics. The two models have been chosen for their particular armature configurations and consequently for the large number of mappings that could be associated to each model; this allows students to find the mapping that better fits their preferences. Students had to set their own mappings between the Kinect skeleton and the armature of virtual characters, then they were asked to animate models by using the chosen mapping and subsequently to retry the animation by using the automatic mapping. Students had to animate the lamp and the ostrich reproducing some very simple motions presented by two reference videos.

After the animation step, students evaluated some characteristics of mappings such as: intuitiveness, usability, difficulty to be retrieved and so on. Moreover, the time necessary to manually define a mapping has been also measured. Tests have shown as an average time of 48s has been measured for the lamp and 33s for the ostrich. Taking into account that these models are composed by a limited number of bones, the initial step of mapping definition might be much more time consuming for complex model armatures. Furthermore, a careful mental effort and a good model analysis were required to students in order to reach a satisfactory mapping.

For the lamp model test, for which students were asked to find a direct mapping for each bone belonging to the armature, exactly half of students have preferred the automatic mapping (the technique proposed in Section 3), whereas the other half have preferred their personal mapping in terms of ease of use. Then, it was asked students to evaluate the intuitiveness of mappings providing a grade in a range between 1(lowest rate) to 7(highest rate). It was found that the students have better evaluated the automatic mapping (an average rate of 4,7) respect their mapping (an average rate of 4,2). For the ostrich test, to be animated by inverse kinematics, students were asked to find a mapping for the three possible end-effectors (head, left paw, right paw). In this case, only two students have preferred their mapping, whereas the others have preferred the automatic one (the technique presented in Section 5) or have given an equal preference. Also the intuitiveness of the suggested mapping was better evaluated than the one chosen by the students (5,6 for the automatic mapping versus 5,3 for the manual mapping).

Another interesting factor to be analyzed is how many manual mappings were similar (or exactly the same as) the suggested one. Considering the two tests aggregated, about the 22% of students selected a mapping completely identical to the one automatically provided, whereas the 89% have identically defined the mapping of at least one bone (or end-effector). Generally, the 56% of the bones to be mapped have received the same point of control than the one automatically provided. In conclusion, it was studied the user satisfaction and the personal opinion about the automatic mapping. Approximately all the students have said that the automatic mapping could be useful for animating a model and about the 55% of them has said that would have defined a different mapping after the usage of the automatic one.

8. Conclusion and future works

This paper presents two automatic procedures to map the human skeleton captured by a tracking system (the Microsoft Kinect, in the proposed implementation) onto the armature of a virtual character to be animated. Proposed solutions are able to efficiently tackle issues related to the control of non-anthropometric characters by taking into account topological similarity scores as well as other parameters such as motion constraints, kinematics chains length, user preferences and so forth. Both skilled animators and, above all, an audience with little or no experience in computer animation could take advantage of the devised approach. The step of mapping the animator's skeleton onto the armature of the character to be animated is efficiently automated, thus reducing time losses and frustrating attempts.

User feedbacks have been also gathered and they show how proposed solutions well fit user expectations, thus they can be considered as a valuable alternative to manual mappings, which require both time to be correctly defined and skills of an expert animator.

Several other open problems need to be further investigated. At the moment, all approaches known in the literature try to migrate the human motion toward a corresponding/similar motion of the virtual character; on the other hand, other forms of mapping could be considered in order to move toward a context awareness animation.

Appendix A. Mapping process for the fish character

This Appendix provides the details of the mapping process for the fish character. Figure A.1 shows the two matrices produced by the node-edge similarity and the motion constraints blocks, respectively. Matrices have $n = 7$ rows (corresponding to the number of bones in the Blender armature for the fish) and $m = 17$ columns (corresponding to the number of bones in the human skeleton). A value in position i, j of the node-edge similarity scores matrix represents the similarity between bone i in the Blender armature and bone j in the Kinect skeleton. The motion constraints score matrix penalizes bones by a value that is proportional to the difference between their degrees of freedom; for instance, bone 2 in the fish armature cannot have the same type of rotational movement of bone 1 in the Kinect skeleton. Hence, the mapping of this pair of bones is penalized by a factor -0.067. The left matrix of Figure A.2 represents the length of paths scores, which

$$
\begin{vmatrix}
0.086 & 0 & 0 & 0 & 0 & 0 & 0 & 0 & 0 & 0 & 0 & 0 & 0 & 0 & 0 & 0 & 0 \\
0.204 & 0.938 & 0 & 0 & 0 & 0 & 0 & 0 & 0 & 0 & 0 & 0.009 & 0 & 0 & 0.009 & 0 & 0 \\
0 & 0.017 & 0.072 & 0.072 & 0 & 0 & 0 & 0.072 & 0 & 0 & 0 & 0.017 & 0 & 0 & 0.017 & 0 & 0 \\
0 & 0.023 & 0.072 & 0.078 & 0 & 0 & 0 & 0.078 & 0 & 0 & 0 & 0.019 & 0 & 0 & 0.019 & 0 & 0 \\
0 & 0 & 0.002 & 0.002 & 0.006 & 0 & 0 & 0.002 & 0.006 & 0 & 0 & 0 & 0.001 & 0 & 0 & 0.001 & 0 \\
0 & 0.023 & 0.072 & 0.078 & 0 & 0 & 0 & 0.078 & 0 & 0 & 0 & 0.019 & 0 & 0 & 0.019 & 0 & 0 \\
0 & 0 & 0.002 & 0.002 & 0.006 & 0 & 0 & 0.002 & 0.006 & 0 & 0 & 0 & 0.001 & 0 & 0 & 0.001 & 0 \\
0 & 0.017 & 0.072 & 0.072 & 0 & 0 & 0 & 0.072 & 0 & 0 & 0 & 0.017 & 0 & 0 & 0.017 & 0 & 0
\end{vmatrix}
\quad
\begin{vmatrix}
0 & 0 & 0 & 0 & 0 & 0 & 0 & 0 & 0 & 0 & 0 & 0 & 0 & 0 & 0 & 0 & 0 \\
-0.067 & 0 & 0 & 0 & 0 & 0 & 0 & 0 & 0 & 0 & 0 & 0 & 0 & 0 & 0 & 0 & 0 \\
0 & 0 & 0 & 0 & 0 & 0 & 0 & 0 & 0 & 0 & 0 & 0 & 0 & 0 & 0 & 0 & 0 \\
0 & 0 & 0 & 0 & 0 & 0 & 0 & 0 & 0 & 0 & 0 & 0 & 0 & 0 & 0 & 0 & 0 \\
0 & 0 & 0 & 0 & 0 & 0 & 0 & 0 & 0 & 0 & 0 & 0 & 0 & 0 & 0 & 0 & 0 \\
0 & 0 & 0 & 0 & 0 & 0 & 0 & 0 & 0 & 0 & 0 & 0 & 0 & 0 & 0 & 0 & 0 \\
0 & 0 & 0 & 0 & 0 & 0 & 0 & 0 & 0 & 0 & 0 & 0 & 0 & 0 & 0 & 0 & 0 \\
0 & 0 & 0 & 0 & 0 & 0 & 0 & 0 & 0 & 0 & 0 & 0 & 0 & 0 & 0 & 0 & 0
\end{vmatrix}
$$

Figure A.1. The left matrix shows the node-edge similarity scores, whereas the right matrix holds the motion constraints scores.

$$
\begin{vmatrix}
0 & 0 & 0 & 0 & 0 & 0 & 0 & 0 & 0 & 0 & 0 & 0 & 0 & 0 & 0 & 0 & 0 \\
0 & 1 & 0 & 0 & 0 & 0 & 0 & 0 & 0 & 0 & 0 & 0.01 & 0 & 0 & 0.01 & 0 & 0 \\
0 & 0.017 & 0.22 & 0.22 & 0 & 0 & 0 & 0.22 & 0 & 0 & 0 & 0.017 & 0.002 & 0 & 0.017 & 0.002 & 0 \\
0 & 0.023 & 0.22 & 0.24 & 0 & 0 & 0 & 0.24 & 0 & 0 & 0 & 0.018 & 0.002 & 0 & 0.018 & 0.002 & 0 \\
0 & 0 & 0.005 & 0.005 & 0.034 & 0 & 0 & 0.005 & 0.034 & 0 & 0 & 0.003 & 0 & 0 & 0.003 & 0 & 0 \\
0 & 0.023 & 0.22 & 0.24 & 0 & 0 & 0 & 0.24 & 0 & 0 & 0 & 0.018 & 0.002 & 0 & 0.018 & 0.002 & 0 \\
0 & 0 & 0.005 & 0.005 & 0.034 & 0 & 0 & 0.005 & 0.034 & 0 & 0 & 0.003 & 0 & 0 & 0.003 & 0 & 0 \\
0 & 0.017 & 0.22 & 0.24 & 0 & 0 & 0 & 0.24 & 0 & 0 & 0 & 0.018 & 0.002 & 0 & 0.018 & 0.002 & 0
\end{vmatrix}
\quad
\begin{vmatrix}
2 & 2 & 2 & 1 & 1 & 1 & 1 & 1 & 1 & 1 & 1 & 1 & 1 & 1 & 1 & 1 & 1 \\
2 & 2 & 2 & 1 & 1 & 1 & 1 & 1 & 1 & 1 & 1 & 1 & 1 & 1 & 1 & 1 & 1 \\
2 & 2 & 2 & 1 & 1 & 1 & 1 & 1 & 1 & 1 & 1 & 1 & 1 & 1 & 1 & 1 & 1 \\
1 & 1 & 1 & 2 & 2 & 2 & 2 & 0 & 0 & 0 & 0 & 2 & 2 & 2 & 0 & 0 & 0 \\
1 & 1 & 1 & 2 & 2 & 2 & 2 & 0 & 0 & 0 & 0 & 2 & 2 & 2 & 0 & 0 & 0 \\
1 & 1 & 1 & 0 & 0 & 0 & 0 & 2 & 2 & 2 & 2 & 0 & 0 & 0 & 2 & 2 & 2 \\
1 & 1 & 1 & 0 & 0 & 0 & 0 & 2 & 2 & 2 & 2 & 0 & 0 & 0 & 2 & 2 & 2 \\
2 & 2 & 2 & 1 & 1 & 1 & 1 & 1 & 1 & 1 & 1 & 1 & 1 & 1 & 1 & 1 & 1
\end{vmatrix}
$$

Figure A.2. The left matrix contains the symmetries scores, whereas the right matrix hosts the length of paths scores.

$$
\begin{vmatrix}
0.043 & 0 & 0 & 0 & 0 & 0 & 0 & 0 & 0 & 0 & 0 & 0 & 0 & 0 & 0 & 0 & 0 \\
0.035 & 0.94 & 0 & 0 & 0 & 0 & 0 & 0 & 0 & 0 & 0 & 0.005 & 0 & 0 & 0.005 & 0 & 0 \\
0 & 0.017 & 0.143 & 0.072 & 0 & 0 & 0 & 0.072 & 0 & 0 & 0 & 0.008 & 0 & 0 & 0.008 & 0 & 0 \\
0 & 0.013 & 0.07 & 0.015 & 0 & 0 & 0 & 0 & 0 & 0 & 0 & 0.018 & 0.001 & 0 & 0 & 0 & 0 \\
0 & 0 & 0.001 & 0.004 & 0.024 & 0 & 0 & 0 & 0 & 0 & 0 & 0 & 0.002 & 0 & 0 & 0 & 0 \\
0 & 0.011 & 0.067 & 0 & 0 & 0 & 0 & 0.015 & 0 & 0 & 0 & 0 & 0 & 0 & 0.018 & 0.001 & 0 \\
0 & 0 & 0.002 & 0 & 0 & 0 & 0 & 0.004 & 0.02 & 0 & 0 & 0 & 0 & 0 & 0 & 0.002 & 0 \\
0 & 0.016 & 0.138 & 0.069 & 0 & 0 & 0 & 0.069 & 0 & 0 & 0 & 0.008 & 0 & 0 & 0.008 & 0 & 0
\end{vmatrix}
\quad
\begin{vmatrix}
0.043 & 0 & 0 & -1 & -1 & -1 & -1 & -1 & -1 & -1 & -1 & 0 & 0 & 0 & 0 & 0 & 0 \\
0.035 & 0.94 & 0 & -1 & -1 & -1 & -1 & -1 & -1 & -1 & -1 & 0.005 & 0 & 0 & 0.005 & 0 & 0 \\
0 & 0.017 & 0.143 & -1 & -1 & -1 & -1 & -1 & -1 & -1 & -1 & 0.008 & 0 & 0 & 0.008 & 0 & 0 \\
-1 & -1 & -1 & -1 & -1 & -1 & -1 & -1 & -1 & -1 & -1 & -1 & -1 & -1 & -1 & -1 & -1 \\
-1 & -1 & -1 & -1 & -1 & -1 & -1 & -1 & -1 & -1 & -1 & -1 & -1 & -1 & -1 & -1 & -1 \\
-1 & -1 & -1 & -1 & -1 & -1 & -1 & -1 & -1 & -1 & -1 & -1 & -1 & -1 & -1 & -1 & -1 \\
-1 & -1 & -1 & -1 & -1 & -1 & -1 & -1 & -1 & -1 & -1 & -1 & -1 & -1 & -1 & -1 & -1 \\
0 & 0.016 & 0.138 & -1 & -1 & -1 & -1 & -1 & -1 & -1 & -1 & 0.008 & 0 & 0 & 0.008 & 0 & 0
\end{vmatrix}
$$

Figure A.3. The resulting scores matrix (left) and the updated scores matrix (right) after the application of preferences.

add a bonus to bone pairs that share the same position in a chain. The three matrices above are added together and then multiplied by the symmetry scores matrix (see the right matrix of Figure A.2). The symmetries scores matrix avoids (values 0) the matching of bones placed in armature parts with a different symmetry and amplifies the values of bones in the correct parts (values 2). The highlighted values in the final scores matrix are those selected by the Hungarian algorithm to produce the mapping (see the left matrix of Figure A.3). In the last step, bones are matched by also considering preferences. The remaining bones are mapped by applying the Hungarian Algorithm to an updated matrix. In fact, after the evaluation of preferences, the final scores matrix is updated by setting positions corresponding to mapped bones to a negative value. Highlighted values are those selected at the second application of the Hungarian Algorithm. The right matrix of Figure A.3 shows the updated scores matrix.

References

[1] Burtnyk, N. and Wein, M. (1971) *Computer generated key frame animation* Journal of the Society of Motion Picture and Television Engineers vol. 8, no. 3, pp. 149-153.

[2] Burtnyk, N. and Wein, M. (1976) *Interactive skeleton techniques for enhancing motion dynamics in key frame animation* Communication of the ACM vol. 19, no. 10, pp. 564-569.

[3] Menache, A. (2000) *Understanding motion capture for computer animation and video games* Morgan Kaufmann, New York.

[4] Gleicher, M. (1998) *Retargeting motion to new characters* In Proceedings of the ACM Siggraph'98, pp. 33-42.

[5] Sanna, A. and Lamberti, F. and Paravati, G. and Domingues Rocha, F. (2013) *A Kinect-based Interface to Animate Virtual Characters* International Journal of Multimodal User Interfaces. vol. 7, pp. 269ï£¡279, DOI 10.1007/s12193-012-0113-9.

[6] The Bloop project, http://dm.tzi.de/research/hci/bloop

[7] The Brekelmans Jasper web site, http://www.brekel.com

[8] Sanna, A. and Lamberti, F. and Paravati, G. and Carlevaris, G. and Montuschi, P. (2013) *Automatically mapping human skeletons onto virtual character armatures* Proceedings of the 5th International Conference on Intelligent Technologies for Interactive Entertainment, 2013, pp. 80-89.

[9] The Blender project, http://www.blender.org

[10] The Kinect web site, http://www.xbox.com/kinect/

[11] Tak, S. and Young Song, O. and Ko, H.S. (2002) *Spacetime sweeping: An interactive dynamic constraints solver* In Proceedings of the Computer Animation, IEEE Computer Society, pp. 261.

[12] Shin, H.J. and Lee, J. and Shin S.Y. and Gleicher M. (2001) *Computer puppetry: An importance-based approach* ACM Trans. Graph. 20, 2:67ï£¡94.

[13] Monzani, J.S. and Baerlocher, P. and Boulic, R. and Thalmann, D. (2000) *Using an intermediate skeleton and inverse kinematics for motion retargeting* Computer Graphics Forum 19, 3.

[14] Kulpa, R. and Multon, F. and Arnaldi, B. (2005) *Morphology-independent representation of motions for interactive human-like animation* In Computer Graphics Forum, vol. 24, pp. 343ï£¡352.

[15] Chen, J. and Izadi, S. and Fitzgibbon, A. (2012) *KinÊtre: animating the world with the human body*

ACM SIGGRAPH 2012 Talks, pp. 39:1-39:1, DOI 10.1145/2343045.2343098.

[16] SUMNER, R.W. and SCHMID, J. and PAULTY, M. (2007) *Embedded deformation for shape manipulation* In Proceedings of the ACM Siggraph'07.

[17] SEOL, Y. and O'SULLIVAN, C. and LEE, J. (2013) *Creature Features: Online Motion Puppetry for Non-human Characters* Proceedings of the 12th ACM SIGGRAPH/Eurographics Symposium on Computer Animation, 2013, pp. 213-221.

[18] ISHIGAKI, S. and WHITE, T. and ZORDAN, V.B. and LIU, C.K. (2009) *Performance-based Control Interface for Character Animation* ACM Trans. on Graph. vol. 28, no. 3.

[19] YAMANE, K. and ARIKI, Y. and HODGINS, J. (2010) *Animating Non-humanoid Characters with Human Motion Data* Proceedings of the 2010 ACM SIGGRAPH/Eurographics Symposium on Computer Animation, pp. 169-178.

[20] ZAGER, L. (2005) *Graph Similarity and Matching* Department of Electrical Engineering and Computer Science, Massachussets Institute of Technology.

[21] ANDRÀS, F. (2004) *On Khun's Hungarian method - a tribute from Hungary. Egervàry research Group on Combinatorial Optimization* Technical report, TR-2004-14. http://www.cs.elte.hu/egres/tr/egres-04-14.pdf

14

A 802.22b protocol for inter/intra WRAN communication

Carla Passiatore[1], Pietro Camarda[1,*]

[1]DEI, Politecnico di Bari, Bari, Italy

Abstract

In this paper we focus on cognitive 802.22b Wireless Regional Area Networks (WRANs). In particular a protocol, named *inter intra-resource sharing algorithm* (2I-RSA), is presented to address the problem of self coexistence for WRANs and secondary users of the WRAN. The purpose of the proposed resource sharing mechanism is to allow P2P communications avoiding interference among users of the same network or of neighbouring cells, optimizing, at the same time, fairness and network capacity.

Keywords: cognitive networks, 802.22b, fair spectrum allocation,Intra-cell communication, Inter-cell communication, TV white space

1. Introduction

In the last years we are observing the continuous and growing request of various services and smartphone wireless applications, corresponding to a continuous and growing request of spectrum availability. At this problem a inefficient resource usage is combined. The spectrum inefficient exploiting is related to a static channel assignment policy. Proof of this is the analysis directed by the Federal Communications Commission (FCC) which has demonstrated the usage of the spectrum [1]. The point of issue is that the band is not exploited with time continuity but there are temporal intervals where the users are idle and, consequently the transmission channel is unused. As an example, below some spectrum occupancy measurements are reported, [1]: in New York City the maximum total spectrum occupancy is only 13.1% from 30 MHz to 3 GHz; in Washington the band occupancy is less than 35% for the radio spectrum below 3 GHz.

Cognitive Network (CN) paradigm focuses on the idea to increase the spectrum utilization by allowing unlicensed users to exploit licensed spectrum during the inactivity period of licensed users, without causing harmful interferences each other. A CN is composed by *primary* and *secondary users* well as *spectrum holes*. A spectrum hole, also called white space, is a region of space-time-frequency temporally unused for communication by licensed users. Primary Users (PUs) are the licensed consumers which have the right to access a channel whenever they need. While Secondary Users (SUs) are unlicensed users who have the capability to sense the spectrum, and transmit on the temporally unallocated band, (i.e. spectrum holes), with the condition of not cause any harmful interference to the PUs. In this context, one of the major challenge is to share the licensed spectrum among SUs without interfering with PUs, and the same SUs. Moreover, one of the most important challenges for enabling efficient cognitive network systems is the coexistence problem [2]. Coexistence can be between SUs of the same network, called self-coexistence, or between secondary networks. The self-coexistence problem is emerging for network overlapped in the same geographical area. In this paper we propose a protocol to address self-coexistence problems, and coexistence among different cognitive networks. This paper is an extension of the previous work [3], where only the CPE coexistence problem in a WRAN was studied.

IEEE 802.22 is a standard for Wireless Regional Area Network (WRAN) which exploits cognitive radio techniques using in opportunistic way the white spaces in the TV frequency spectrum. In particular,

*Corresponding author. Email: carla.passiatore@poliba.it

when the TV bands are underutilized, WRAN service providers can exploit these resources. IEEE 802.22 standard adopts a centralized topology with a Base Station (BS) and Customer Premise Equipments(CPEs) which are the SUs of the network. According to the centralized structure all the messages have to transit through the BS. This uselessly increases the data traffic in the network if the two nodes that attempt to communicate are within the same WRAN. Furthermore, this model leads to a limited network capacity, because in downlink transmission a channel can be exploited by only one CPE per time slot. At the aim to overcome also this limitation a new working group, namely 802.22b, was approved by the IEEE 802 committee. According to the 802.22b standard the peer to peer (P2P) communication is made available, in other words direct CPE to CPE is allowed if the users are managed by the same BS. In this way, inside a WRAN the same channel can be used simultaneously by different couples of CPEs, provided that the transmissions are not interfering each other and with PU activities.

An issue of this topic is the channel allocation problem which influences the network performance significantly in different features, as fairness, network capacity and interference. The resource sharing problem happens at two levels, among the CPEs of the same WRAN, i.e. *intra-network*, and among neighbouring WRANs, *inter-network*. In this paper we present a two level algorithm to improve intra and inter network resource sharing, exploiting the frequency reuse and maximizing the total throughput. The proposed protocol holds all these characteristics, providing a spatial reuse mechanism, for inter and intra resource sharing, named Inter and Intra Resource Sharing Algorithm (2I-RSA) . The main idea of the presented method is based on the Interference Map (IM), which is a picture of the possible interference among CPEs and WRANs. This is a method already introduced in literature but we exploit IM and channel reuse in such a way to maximize fairness and network capacity. In literature, other works were proposed which focus on channel management for cognitive P2P networks ([4],[5],[6] or [7]), but in these papers the number of available channels is variable while the bandwidth is always the same, what's more these works focuses or on the inter networks resource sharing or intra networks. In our work we have considered also the bandwidth, which, according to the 802.22 standard, can be equal to 6,7 or 8 MHz. Channels with different bandwidth have different weights in the network, allowing to transmit, in the same time, more or less data. Moreover, we have proved, computing the Jain's index [8], that our method compared with the proposed one in [5] is more fair. In [7] is proposed a resource sharing algorithm for P2P network which tries a trade off between fairness and network capacity, but the proposed solution is static which is not suitable in a cognitive environment. While our protocol adapts to a dynamic environment, which is the feature of a cognitive network.

The rest of the paper is organized as follows. In Section 2 we show a brief overview of the latest update on IEEE 802.22 working group activities. In Section 3 we give a detailed description of the proposed resource sharing algorithm In Section 4, the performance is investigated through a simulation tool, and the results are compared with those obtained in [5]. Section 5 draws the conclusions.

2. 802.22b Overview

In this Section an overview of the 802.22b working group is presented. The new IEEE 802.22b task group has been formed by IEEE 802 working group to extend and improve IEEE 802.22 applications in TV white space. In particular 802.22b architecture proposes an enhancement for broadband services and monitoring applications.

After the switch from analog to digital broadcasts many frequency bands are unexploited, this trend is in opposition with the continuous growing request of spectrum availability. To compensate for the lack of bandwidth, cognitive radio networks have been developed. IEEE 802.22 is a standard which allows to exploit unused TV frequency spectrum, in the specific TV channels from 54 to 862 MHz with a bandwidth of 6, 7 or 8 MHz. IEEE 802.22 WRAN has a centralized topology composed by one BS which manages CPEs (which are the SUs of the network) and the medium accesses control. The BS is responsible of spectrum sensing, channel distribution among CPEs, and routing messages to the CPEs. On the one hand the present of the BS, in the centralized structure, makes the implementation of the resource management very simple. On the other hand this structure has a negative impact on the network capacity; because every communication transits through the BS, both if the messages are addressed from CPE to the BS either if the communication is between CPEs of the same 802.22 network. Consequently, in downlink transmission, a channel can be exploited by a CPE at a time slot. Moreover, the coverage area of a WRAN is on the order of some kilometres [6], so the number of inter communications is very high.

In the light of what is said, the availability of WRAN peer to peer communication is necessary to improve the network performance, such as capacity, and energy consumption. So IEEE 802.22b working group was introduced, at the aim to implement direct communication between CPEs of the same network.

The purpose of this amendment is to enhance the MAC and PHY layer to accommodate broadband extensions and monitoring use cases for IEEE 802.22

devices operating is VHF/UHF TV broadcast bands, such as real-time, emergency broadband services or remote medical services, [9]. None of these use cases are implemented by the IEEE Std. 802.22-2011.

In the following the main features of 802.22b standard are listed, [10] [9] [11]:

- working in TV white spaces
- working in Very High Frequency (VHF)/Ultra High Frequency (UHF) TV broadcast
- enabling monitoring applications
- enabling peer-to-peer connection
- individual spectrum sensing technologies
- supporting very large number of CPEs, more than 512 devices in a network.
- supporting high reliability and QoS
- supporting real-time monitoring system with low latency, for channel sensing
- exploiting interface with various sensors, also to improve sensing ability
- supporting high data rate
- managing mechanisms to enable coexistence with other 802 systems.

CPE to CPE communication is an essential feature and key function of the smart grid, broadband services, monitoring application, or other applications based on IEEE 802.22b networks.

3. 2I-RSA

In this Section 2I-RSA is explained.

The features of the assignment policy are:

- reuse of the spectral frequencies;
- resource sharing with regards to the spatial diversity, i.e. the availability of some channels only for some users;
- resource sharing according to the variance of the channel availability;
- fairness, i.e. channel distribution according to the CPE spectrum demands.

Before presenting the proposed algorithm some definitions are given. Two CPEs are defined *overlapped* if the communication (transmitting or receiving) of the first CPE could interfere with the transmission of the second one and vice versa. Otherwise the CPEs could exploit the same channel without causing interference each other. The BS has to coordinate the transmissions in its WRAN in such a way to avoid harmful interference. Nevertheless this approach is not sufficient to totally avoid collisions, because there is the risk of interference among CPEs that belong to different, but neighbouring, WRANs. We say that two neighbouring WRANs are overlapped if the transmissions of CPEs , that are in the two different cells, exploiting the same channel at the same time, interfere each other.

The algorithm is composed by two steps, the first one is the *inter network-RSA*, and the second one named *intra network-RSA*. The scenario is composed by multiple cognitive 802.22 networks, each of which composed by multiple CPEs and one BS. The BS is responsible of the device channel access of its network. The scope of the proposed protocol is to create a two level resource sharing map for interfering users. The channel assignment is used to allocate the set of idle channels to the overlapped WRAN cells in the first level, and to the overlapped CPEs in the second level.

The WRAN could be divided into two parts, the inner and the outer zone [12]. Considering that the typical 802.22 cell radius is in the range 30-100 Km [13], and that the range for a CPE is 33 Km, CPEs in the inner zone do not have the possibility to interfere with user of another 802.22 network, the same cannot be said of users in the outer zone. Referring to the figure1, two overlapped WRANs are represented. The figure shows as the communication between two outer zone users of the WRAN-A interferes with the transmission of the outer CPEs of the neighboured WRAN-B.

The algorithm is divided into two phases, after the first phase a Channel Access Map (CAM) is computed (which is shared among WRANs), while during the second one each BS evaluates a channel access map for the CPEs of its WRAN. The first part of the algorithm is exploited to determine the resources which are allowed to the the outer zone CPEs. While the intra-RSA is implemented to elaborate a channel access map in each network.

In figure 2 a 2I-RSA summarizing scheme is showed. The roles of the first step are represented with blue lines, which are I) the CPEs-BS and BSs-Coordinator communications, II) the data elaboration at the coordinator to compute the WRAN CAM, and III) the transmission of this to the BSs. While in the second step the operations are illustrated with green lines, in particular they are the procedures made at the single BSs to compute the CPE CAM , and the broadcasting of this in the cell.

The figure 2 is used to introduce the algorithm, while the 2I-RSA specific parts are well described in the follow subsections.

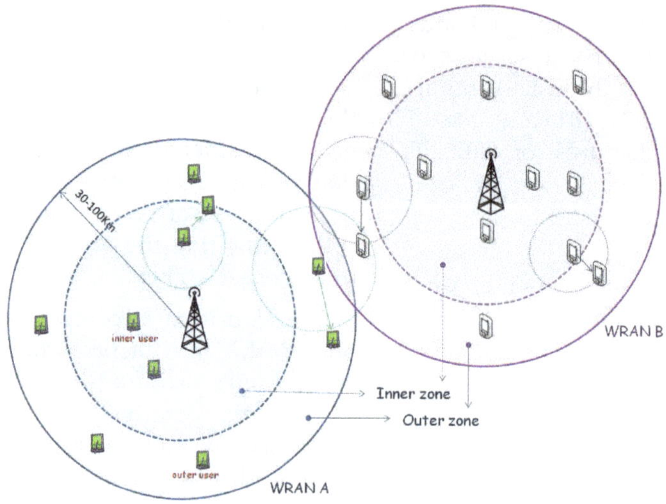

Figure 1. Interference among CPEs of overlapped WRANs.

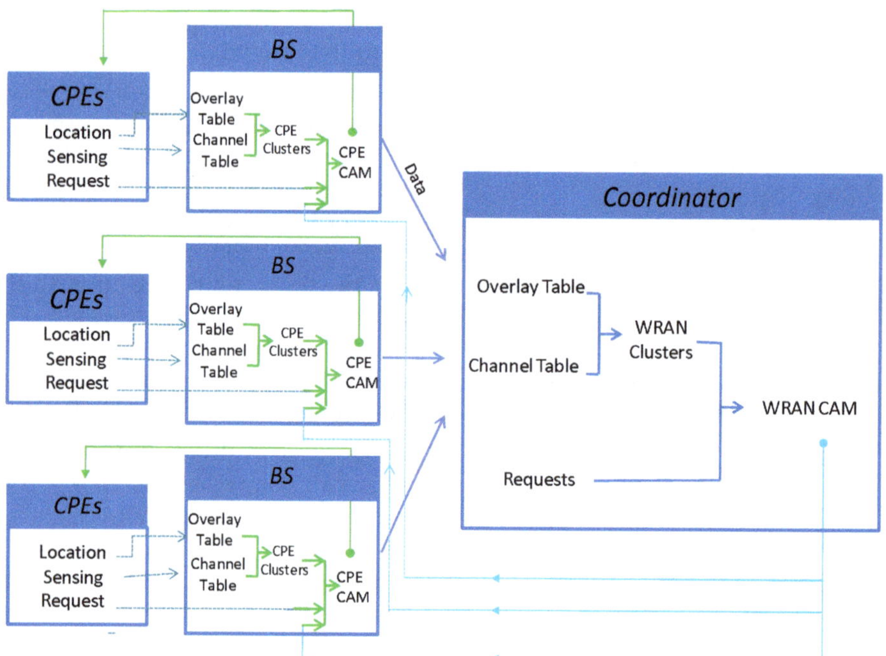

Figure 2. 2I-RSA Scheme.

3.1. Inter Network-RSA

As in [14] we refer to a scenario composed by multiple overlapped 802.22 networks, which form a community made by a coordinator and other memberships, namely WRANs. The coordinator, which is the community leader, is elected among the BSs. In literature different methods are well known for electing a leader, as [15] or [16]. In [14] the coordinator manages the membership access to the channels, in such a way to avoid harmful interference among the WRANs. Each BS is able to transmit information to the coordinator which exploits these data for resource sharing optimization. For the inter-network communication the 802.22 standard specifies a protocol, Coexistence Beacon Protocol (CBP) [17], a self-coexistence mechanism based on beacon transmissions among the coexisting WRAN cells.

The information transmitted by the BS to the coordinator is previously collected inside the WRAN with the CPE collaboration.

The main activities of the CPEs are three: sensing, communication with the BS, transmission according to the BS decisions. In a IEEE 802.22 network the sensing operation must be done periodically, with a period no larger then two second [18]. In particular each CPE communicates to the BS: I) its geographic position II) resource request, III) available channels, i.e. their sensing results. The concept of request will be clarified in the following. Note that the amount of information exchanged among CPEs and the BS is exiguous; each device has to communicate only the channel availability, the CPE in interfering area and its requests. The BS collects this information, computing the total request of its network, which is the sum of all the requests of the SUs of its cell, and WRAN's available channels. The BS exploits these data to avoid interference among users of the neighbouring WRANs. According to these data the inter-network RSA is implemented to define the frequency bands which are allowed for the transmission of the CPEs in the outer zone.

After the sensing period and the information exchange between BS and CPEs, every BS conveys to the coordinator I) its geographic position II) its WRAN resource request, III) its WRAN sensing results. For this protocol we assume that if a BS hears a channel it is available for all the CPEs of the WRAN, which is a plausible hypothesis considering that the channels are those made free by TV broadcasters.

IEEE 802.22 is a time slotted protocol [19]; the operations are spread in a slotted structure composed by frames and superframes. The 802.22 superframe is composed by 16 MAC frames, composed by time slots which are preceded by a preamble and a frame control header (FCH). The coordinator exploits data for resource sharing optimization, in this context the resources are the frames. According to the data and implementing the first step of 2I-RSA, the coordinator determines the WRANs which are allowed to transmit and the respective channels to be occupied. The information elaborated by the coordinator is included in the *inter network channel access map*, named also WRAN CAM, broadcast by the coordinator to the BSs. With the *inter network channel access map* the coordinator communicates to each WRAN during which time slots and which channels its outer CPEs can exploit for the transmission.

To maximize the spectral efficiency, taking into account the spatial diversity, the coordinator has to verify, for each channel, the group of WRANs which are able to transmit simultaneously without causing interference, it has to create the interference map.

Data are collected with the help of *overlay* and *channel table*, the first one exploited to memorize topology information, the second one for the available resources and the spacial diversity. The *inter-network overlay table* and *channel table* have to be updated as soon as new information is received.

The *inter-network overlay table* is a square matrix, with size N, where N is the total number of WRANs in the community. The (i, j) element of the matrix is equal to 1 if the $WRANi$ and the $WRANj$ could interfere each other. While the *channel table* has the number of rows equal to the number of WRANs in the community, and a number of columns equal to the number of channels. The (i, j) element of the *channel table* is equal to 1 if the *channel* j is available for the $WRANi$.

As an example in tab.1 shows an *inter-network overlay table*. For reasons of compactness in tab.1 and in the other tables of this paper the notation $W1$ stands for $WRAN1$, and so on.

Table 1. Inter-network Overlay Table

	W1	W2	W3	W4	W5	W6
W1	1	0	0	1	0	1
W2	0	1	1	1	1	1
W3	0	1	1	0	1	0
W4	1	1	0	1	0	1
W5	0	1	1	0	1	0
W6	1	1	0	1	0	1

The *channel table* is created according to the received sensing information. Moreover, we suppose to have three available channels with different coverage area, which allow to build up the *channel table* shown in tab.2. As an example, we supposed that *channelA* is available only for $WRAN1$, $WRAN4$ and $WRAN6$, *channelB* is available only for $WRAN2$, $WRAN3$ and $WRAN5$, while the third channel covers all the network.

Table 2. Channel Table

	ChA	*ChB*	*ChC*
W1	1	0	1
W2	0	1	1
W3	0	1	1
W4	1	0	1
W5	0	1	1
W6	1	0	1

According to *channel* and *inter-network overlay table*, the WRAN header is able to elaborate the sets of WRANs which could transmit simultaneously without interfering each other. In the following these WRAN groups are called clusters. The procedure to compute the clusters is:

- for each channel the coordinator has to examine which are the WRANs allowed to use it; i.e. for each *channel table* column the coordinator checks the elements marked with 1.
- Considering only this WRAN group the coordinator computes the non-overlapped WRAN sets.

In the exposed example $WRAN1$, $WRAN4$, and $WRAN6$ are overlapped among them, then only one of these can occupy the *channelA*, simultaneously. While more users can transmit simultaneously using the *channelC* without interfering each other. In tab.3 are shown all the possible channel assignments. Precisely, each column of the table is referred to a specific channel. Each row shows the possible groups of $WRAN$s which can transmit simultaneously, without interfering each other, occupying the same channel, which is indicated on the top of the column.

Table 3. WRAN Cluster

$Ch-A$	$Ch-B$	$Ch-C$
W1	W2	W1-W2
W4	W3	W1-W3
W6	W5	W1-W5
-	-	W3-W4
-	-	W3-W6
-	-	W4-W5
-	-	W5-W6

The coordinator exploits these tables for resource sharing optimization. It creates the CAM, which indicates WRANs allowed to transmit at a given moment. The map is created selecting for each channel a single cluster which is allowed to occupy it for a superframe. The choice is taken with the aim to assign resources proportionally to the requests, taking into account the spatial diversity. Note that the protocol is dynamic because CPEs update information every two seconds, whenever an environment change is detected, as PU transmission or new requests. The exploited resource sharing algorithm takes into account two different aspects: the resource already assigned to the user and the fairness of the allocation. The goal is to take the fairest choice among the ones which maximize the total throughput of the network community. In the Appendix section a mathematical description of the algorithm is introduced, to provide a formal justification for the choice of the algorithm; while in this section an intuitive exposition is suggested.

Each BS communicates to the coordinator the resource request, to each BS_i a req_i is associated, where req_i is the amount of data that BSi needs to transmit. In the exposed case the WRAN set which can transmit simultaneously without producing interference are illustrated in Tab.3. As an example in Tab.3, 7 different combinations are individuated for *channelC*, namely [W1-W2], [W1-W3], [W1-W5], [W3-W4], [W3-W6], [W4-W5], [W5-W6]. To define the *channelC* access map the coordinator must select one of the above sets to assign the resource, for each slot. The criteria introduced in order to chose the best cluster among the available group for each channel, is the one which maximizes the y function:

$$y = \sum_{i=1}^{N} p_i \ln(n_i + 1). \tag{1}$$

Where n_i is the amount of data already transmitted by $WRANi$, with $n_i \geq 0$. While, p_i, the transmission probability, is a factor that takes into account the amount of data that the single user desires to send; p_i is computed in the Appendix.

The function y was build as in eq.3 because it increases when the resources are assigned to the WRANs with higher requests: they have higher p_i, so increasing the corresponding n_i the global value of y will increase too. Moreover y has a logarithmic growth, for this reason y raises very fast with the first assigned superframes to the users with higher p_is. Subsequently, in the sum, the contributes of these will raise more slowly, because it is a logarithmic function, then, for increasing y, it is necessary to assign resources to the WRANs with lower p_is. Moreover, as shown in the appendix, in the optimal solution, n_i is proportional to the related p_i, this confirms the fairness of the proposed method.

The coordinator creates the WRAN CAM choosing for each time slot, for each channel the WRAN cluster which is enabled for transmission in that period. The coordinator dispenses the channel access map to each BS of the community. With this transmission the inter-RSA concludes; now the intra WRAN transmission has to be coordinated.

3.2. Intra Network-RSA

The BS is the coordinator of its WRAN. It has to disposes the channel access map of its WRAN, the *intra network channel access map*, taking into account the CPE requests and the *inter network channel access map*. As previously explained, the inter network interference problem is connect to the CPEs in the outer zone of the WRAN, while the inner zone CPEs do not interfere with users of neighbouring WRANs. Then, the BS, creating the CAM for its cell, has to schedule the resource avoiding interference intra WRAN and inter WRANs, at this aim the BS conducts the second step of the 2I-RSA. The BS has all to implement the Intra-RSA: the CPEs' request and position, channel availability and WRAN CAM. The BS builds the *overlay* and *channel table*, for its WRAN, and consequently the table for the CPE clusters which are allowed to transmit simultaneously without interfering, as in table 3, but referred to the CPEs. In the subsequent phase the BS, exploiting the y function, chooses the CPE cluster allowed to transmit for each time slot and for each channel. Differently by the inter-RSA, in computing final y value, there is another constraint, which is represented by the WRAN CAM. Calculating the y value the BS has to exclude the outer CPEs if they are not allowed to transmit according to the *inter network channel access map*. That is why in figure.2 the input data for computing CPE MAP are CPE clusters, CPE requests and WRAN CAM.

4. Simulation Results

In this section the performance of the 2I-RSA are evaluated.

In particular 2I-RSA is compared with the previous version, P2P-RSA [3], at the aim to demonstrate that the new version is equally fair. Moreover the proposed method is compared with another algorithm to prove that it is more fair without losing performance in terms of network throughput. The comparing protocol, named self-adapting interference map building protocol (SIMBP), exploits the interference map to maximize the total throughput of the network [5]. SIMBP is introduced as comparison protocol because it is used in a 802.22 network for multiple channel allocation, as 2I-RSA.

In the simulation scenario we have supposed a wide area covered by five overlapped WRANs. In the multichannels environment, the number of available channels is varied in the range [1,3]. The number of nodes in the single network is increased, with step of 10 units, in the range [10:30]. Results have been computed supposing CPE request and position randomly chosen.

Since SIMBP and P2P-RSA are exploited for resource sharing in a single WRAN, the showed results are refereed to a WRAN of the scenario, but while 2I-RSA is an intra network collision free protocol, the other cited protocols suffer of interference for the overlapped CPEs.

The results obtained simulating 2I-RSA are illustrated by red lines, while the P2P-RSA results with green lines, and SIMBP with yellow.

In figure3 the Fairness Index (FI) is computed whit the use of the Jain's index, for more details see [8]. The index was computed varying the number of available channels and users in the WRAN.

The results illustrated in figure3 show that the proposed method is more fair then SIMBP, and also compared with P2P-RSA the performance is not decreased. Note that the maximum value of Jainś index is 1, so figure3 demonstrates that the proposed protocol is a really fair resource sharing method.

In the same scenarios we computed the network capacity, as the total throughput of a WRAN. The results are shown in figure4

Figure4 shows that the network capacity grows increasing the number CPEs and channels. It is important to notice that, exploiting 2I-RSA, the total

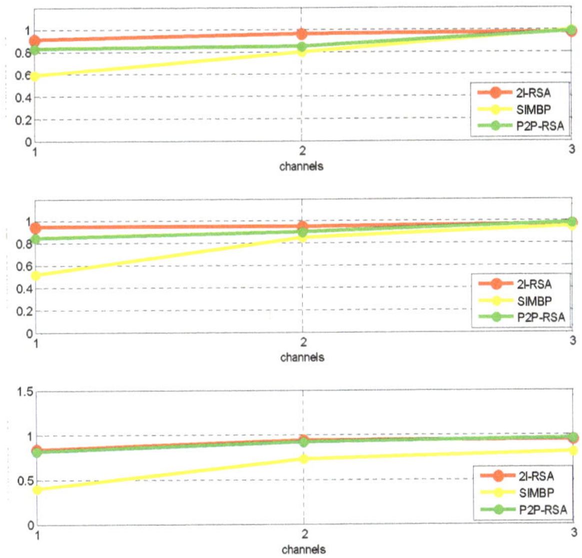

Figure 3. FI evaluated with different algorithms

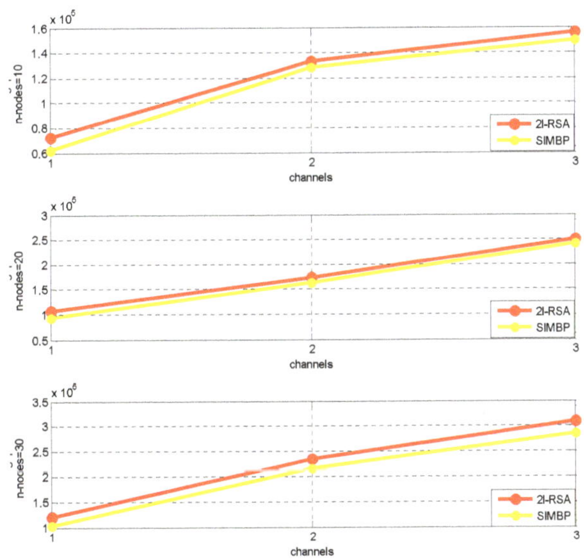

Figure 4. WRAN Capacity

capacity of the network is not less than the one obtained with SIMBP, this is because 2I-RSA is not affected by inter network interference.

The purpose of our algorithm was to create a fair resource sharing protocol, the showed results demonstrate that the goal is reached without losing performance in term of network capacity.

5. Conclusions

We proposed a *inter/intra resource sharing algorithm*, named 2I-RSA, for resource allocation inter and intra 802.22 P2P WRANs. The procedure is divided into two steps, in the first there is a propagation of informations intra and inter networks, to collect all data to build the inter network interference map. This first interference map is used to avoid collision among CPE overlapped in two different WRANs. The allocation resource intra network is managed by the BS, which according with inter network-IM, CPE requests and geographic positions, builds the intra network channel access map. The proposed protocol is an evolution of the previous P2P-RSA,[3]. 2I-RSA compared with P2P-RSA presents similar values of FI, in addition it solves interference problems with neighbouring cells.

Appendix

In this Section the resource sharing algorithm is explained in detail, demonstrating that it allows the best resource distribution.

Note that the algorithm is used to schedule resources or among CPEs, in this case the BS implements the algorithm, or among WRAN, here the coordinator implements the RSA. In this section we refer to the inter RSA, so we named WRAN and coordinator, but the same line of reasoning is applicable to the intra-RSA with CPEs and BS.

The algorithm aims to schedule in a fairness way the transmission. To achieve this, for each frame and for each available channel, the coordinator considers the WRAN clusters which may transmit simultaneously, see table3, and it chooses the fairest option among the ones which maximize the assigned resources.

Each BS, BS_i, estimates the total amount of data which its CPEs need to transmit, named $request_i$, which is the number of frames requested by $WRANi$; $request_i$ is computed according to the channel bandwidth and the amount of data to be transmitted by $WRANi$. According to the requests the coordinator computes the transmission probability p_i, for each $WRANi$, by using the following formula:

$$p_i = \frac{request_i}{\sum_{j=1}^{N} request_j}, \tag{2}$$

where N is the number of WRANs in the scenario. All the p_i of the community are included in the probability vector, where the *ith* element is the transmission probability of the WRAN*i*.

To choose the WRAN cluster, the state_vector, namely s_v, is used. It is a vector of N elements, where the *ith-element* indicates how many frames have been totally assigned to the $WRANi$. As an example, vector [3,3,0,0,0,0] represents the state where 3 frames have been assigned to $WRAN1$ and $WRAN2$. Specifically, the best s_v is the one which maximizes y, where:

$$y = \sum_{i=1}^{N} p_i \ln(n_i + 1), \tag{3}$$

with $n_i \geq 0$, where n_i is the number of frames assigned to $WRANi$. In practice, the n_i value introduced in the eq.3 is the *ith-element* of the s_v.

The RSA goal is to maximize y in eq.3; in the following is demonstrated that the optimal solution is obtained when $n_i = n \cdot p_i$, where n is the total number of assigned frames. In other words, in the optimal solution n_i is proportional to the related p_i, according to the WRAN requests.

At the beginning of the assignment process the s_v is a null vector with length equal to the number of the community members. For the example of tab.1 it is s_v=[0,0,0,0,0,0], because there are 6 WRANs. Every time that the coordinator increases the number of frame where a WRAN can exploit a channel, the *ith* element of the s_v is incremented of a unit. For each available channel and for each superframe, the coordinator has to choose among the combinations of tab.3 the one that gives the s_v which maximizes the function of eq.3.

To choose the WRAN clusters for the transmission, a greedy algorithm is implemented. Starting from the channel with larger band, the coordinator evaluates all the possible finals_vs which could be chosen according to the cluster of tab.3. For each entry of tab.3, the result of eq.3 must be calculate, evaluating the s_v and computing the corresponding y value, and finally choose the state which returns as result the highest value of y. In this way the s_v is updated according with the selected cluster.

The criterion on witch RSA is based is to maximize the eq.3, in particular we prove that the optimal solution is obtained when $n_i = n \cdot p_i$, where n is the total number of frames assigned. This means that, in case of p_i all equal among them, the algorithm chooses the state_vector where n_i approaches more to the mean value $n_i = n/N$. In general, in the optimal solution n_i is proportional to the related p_i, i.e. respects the resource request of $WRAN_i$.

To show the optimality condition we assume the following approximation:

$$\sum_{i=1}^{N} p_i \ln(n_i + 1) \approx \sum_{i=1}^{N} p_i \ln(n_i). \tag{4}$$

Eq.4 is true for $n_i > 0$; note that $n_i = 0$ implies $ln(n_i + 1) = 0$ and thus it does not give a contribution to the sum. So we can consider $n_i \geq 1$.

To show the optimality condition eq.3 can be rewritten in the following way :

$$\sum_{i=1}^{N} p_i \ln(n_i + 1) = \sum_{i=1}^{N} p_i[\ln(n_i + 1) + \ln n_i - \ln n_i]. \tag{5}$$

Eq.5 can be rewritten as:

$$\sum_{i=1}^{N} p_i \ln(n_i+1) = \sum_{i=1}^{N} p_i \ln n_i + \sum_{i=1}^{N} p_i \ln \frac{n_i+1}{n_i}. \quad (6)$$

It is possible to observe that the following inequality is always true:

$$\sum_{i=1}^{N} p_i \ln \frac{n_i+1}{n_i} \leq 1. \quad (7)$$

In particular, the above sum has limit 0 when n_i increases. The second term of eq.6 is negligible and consequently:

$$\sum_{i=1}^{N} p_i \ln(n_i+1) \approx \sum_{i=1}^{N} p_i \ln(n_i). \quad (8)$$

Let us now prove the following inequality:

$$\sum_{i=1}^{N} p_i \ln(n_i) \leq \sum_{i=1}^{N} p_i \ln(np_i), \quad (9)$$

and in particular $\sum_{i=1}^{N} p_i \ln(n_i)$ is maximized when $n_i = np_i$.

Now we consider:

$$\sum_{i=1}^{N} p_i \ln(n_i) - \sum_{i=1}^{N} p_i \ln(np_i), \quad (10)$$

which can be written as:

$$\sum_{i=1}^{N} p_i \ln \frac{n_i}{np_i}. \quad (11)$$

Given that that $\ln y \leq y - 1$, see [20], we obtain:

$$\sum_{i=1}^{N} p_i \ln \frac{n_i}{np_i} \leq \sum_{i=1}^{N} p_i [\frac{n_i}{np_i} - 1]. \quad (12)$$

The second member of the inequality is equal to 0, in fact:

$$\sum_{i=1}^{N} p_i [\frac{n_i}{np_i} - 1] = \sum_{i=1}^{N} p_i \frac{n_i}{np_i} - \sum_{i=1}^{N} p_i = 0. \quad (13)$$

Then eq.12 can be written as:

$$\sum_{i=1}^{N} p_i \ln \frac{n_i}{np_i} \leq 0; \quad (14)$$

We proved that:

$$\sum_{i=1}^{N} p_i \ln(n_i) \leq \sum_{i=1}^{N} p_i \ln(np_i), \quad (15)$$

which means that first and second member become equal when $n_i = n \cdot p_i$, i.e. $\sum_{i=1}^{N} p_i \ln(n_i)$ is maximized when $n_i = n \cdot p_i$.

6. Acknowledgements

This work was supported by the PON projects (RES NOVAE, ERMES- 01-03113, DSS-01-02499 and EURO6-01-02238) funded by the Italian MIUR and by the European Union (European Social Fund).

References

[1] Khaled and W. Zhang, "Cooperative communications for cognitive radio networks," *Proceedings of the IEEE*, vol. 97, pp. 878–893, May 2009.

[2] M. Hani, H. Salameh, Y. Jararweh, and A. Bousselham, "Traffic-aware self-coexistence management in IEEE 802.22 WRAN systems," in *GCC Conference and Exhibition (GCC), 2013 7th IEEE*, pp. 507–510, Nov 2013.

[3] C. Passiatore and P. Camarda, "A P2P resource sharing algorithm (P2P-RSA)for 802.22b networks," in *Proc. of International Conference on Context-Aware Systems and Applications (ICCASA)*, (Dubai, United Arab Emirates), Oct. 2014.

[4] J. Shimonovich, A. S. Baruch, and S. S. Shitz, "Cognitive cooperative communications on the multiple access channel," in *Information Theory Workshop (ITW), 2013 IEEE*, pp. 1–5, Sept 2013.

[5] H. Shi, R. Prasad, V. Rao, and I. Niemegeers, "Procedure to build interference map in peer to peer ieee 802.22 networks," in *Global Communications Conference (GLOBECOM), 2012 IEEE*, pp. 1103–1108, Dec 2012.

[6] H. Shi, R. Prasad, and I. G. M. M. Niemegeers, "An intra-cell peer to peer protocol in IEEE 802.22 networks," in *GLOBECOM Workshops (GC Wkshps), 2011 IEEE*, pp. 533–537, Dec 2011.

[7] H. Shi, R. Prasad, V. Rao, and I. Niemegeers, "Fairness and network capacity trade-off in P2P IEEE 802.22 networks," in *Global Communications Conference (GLOBECOM), 2012 IEEE*, pp. 1435–1440, Dec 2012.

[8] C. D. . H. W. Jain, R., "A quantitative measure of fairness and discrimination for resource allocation in shared computer systems," tech. rep., September 1984.

[9] [Online]. Available at: http://www.ieee802.org/22/.

[10] C.-W. Pyo, X. Zhang, C. Song, M.-T. Zhou, and H. Harada, "A new standard activity in IEEE 802.22 wireless regional area networks: Enhancement for broadband services and monitoring applications in tv whitespace," in *Wireless Personal Multimedia Communications (WPMC), 2012 15th International Symposium on*, pp. 108–112, Sept 2012.

[11] M.-T. Zhou, C. Song, C. Sum, and H. Harada, "Energy/bandwidth-saving cooperative spectrum sensing for two-hop wran," in *Cognitive Radio Oriented Wireless Networks and Communications (CROWNCOM), 2014 9th International Conference on*, pp. 51–56, June 2014.

[12] T. sub Kim, S. joon Lee, S. Chhorn, C. ho Cho, and S. Ryu, "Resource allocation and power control scheme for interference avoidance in lte-advanced device-to-device communication," in *Computing and Convergence Technology (ICCCT), 2012 7th International Conference on*, pp. 1201–1204, Dec 2012.

[13] N. Tadayon and S. Aissa, "Modeling and analysis of cognitive radio based IEEE 802.22 wireless regional area

networks," *Wireless Communications, IEEE Transactions on*, vol. 12, pp. 4363–4375, September 2013.

[14] C. Passiatore and P. Camarda, "A centralized inter-network resource sharing (CIRS) scheme in IEEE 802.22 cognitive networks," in *Ad Hoc Networking Workshop (Med-Hoc-Net), 2011 The 10th IFIP Annual Mediterranean*, pp. 17–24, June 2011.

[15] S.-H. Park, Y. Kim, and J. S. Hwang, "An efficient algorithm for leader-election in synchronous distributed systems," in *TENCON 99. Proceedings of the IEEE Region 10 Conference*, vol. 2, pp. 1091–1094 vol.2, Dec 1999.

[16] W. Hu, D. Willkomm, M. Abusubaih, J. Gross, G. Vlantis, M. Gerla, and A. Wolisz, "Cognitive radios for dynamic spectrum access - dynamic frequency hopping communities for efficient ieee 802.22 operation," *Communications Magazine, IEEE*, vol. 45, pp. 80–87, May 2007.

[17] "IEEE standard for information technology– local and metropolitan area networks– specific requirements– part 22: Cognitive wireless ran medium access control (mac) and physical layer (phy) specifications: Policies and procedures for operation in the tv bands," *IEEE Std 802.22-2011*, July 2011.

[18] S. Sengupta, S. Brahma, M. Chatterjee, and S. Shankar, "Enhancements to cognitive radio based IEEE 802.22 air-interface," in *2007 IEEE International Conference on Communications*, pp. 5155 – 5160, June 2007.

[19] H. Shi, R. Venkatesha Prasad, V. Rao, and I. Niemegeers, "Adapting IEEE 802.22 OFDMA system for P2PWRANs," in *Global Communications Conference (GLOBECOM), 2013 IEEE*, pp. 974–979, Dec 2013.

[20] S. Benedetto, E. Biglieri, and V. Castellani, *Digital Transmission Theory*. Prentice Hall, 1987.

15

Design and optimization of the Automatic Brown PlantHopper (BPH) light trap surveillance network

Huong Hoang Luong[1,*], Thanh Hai Nguyen[2], Hung Huu Huynh[3] and Hiep Xuan Huynh[4]

[1]CUSC-CTU, Cantho, Vietnam
[2]Ministry of Education and Training, Vietnam
[3]The University of Danang, Vietnam
[4]DREAM-CTU/IRD, CICT-CTU, Cantho, Vietnam

Abstract

Communicating coverage of automatic BPH light trap surveillance network characterizes how well an area is monitored or tracked by automatic light traps. Connectivity is an important required that shows how nodes in a automatic BPH light trap surveillance network (BSNET) can effectively communicate. Some areas in the deployment region are more important than other areas and need to be covered. In this paper, we propose a new approach based hexagonal cellular automata to find the automatic light trap node distribution. This approach are ensure the deployment region that maximizes the coverage area of BSNET, and preserve connectivity between nodes provided.

Keywords: light trap, BPH, surveillance network, optimization, optimal- design, hexagonal cellular automata.

1. Introduction

The light trap surveillance network [1] in Mekong Delta region is one kind of representative sampling applying for the geographical region. The light trap surveillance network that can capture multiple kinds of insects, especially BPH, and which data (the density of insects per trap) is collected and analyzed daily. The light trap surveillance network is deployed in the experiments where the BPH trapped density is considered as monitoring called BPH light trap surveillance network (BSNET). BSNET is a spatial sampling network applying for the geographical region.

The light trap [19] is one kind of passive trap helping to catch only the mature insects, and it operates only at night. A light trap uses light as an attraction source [20]. Light traps depend on the positive phototactic response of the insects, physiological as well as abiotic environmental factors which can influence the behavior [21]. Many kinds of insect will be caught and counted every day to observe the current density of them. BPH monitoring process is done manually. To automate the process of monitoring BPHs, a network of automatic BPH light traps need building. An automatic BPH light trap includes some functions such as detecting the BPHs and counting the number of BPHs in the trap. Also, the automatic BPH light trap can trasmit data to other(s). An automatic BPHs light trap [2] was equipped with light source, tray, a camera, communication devices, some sensors and a power. The camera is programmed to capture the images from tray. Also, it can recognize the BPHs and count the number of BPHs in the image. The sensors includes temperature, light, humidity, wind speed,and wind direction. The communication devices which use radio are used to transmit or receive data.

In this paper, we propose a new approach based hexagonal cellular automata[3] to find the automatic light trap node distribution. This approach are ensure the deployment region that maximizes the coverage area of BSNET, and preserve connectivity between nodes provided

This paper contains 6 sections. Some related works are introduced in the next section. Automatic Brown PlantHopper surveillance network is presented in the section 3. Section 4 will decscribe how to optimize the

*Corresponding author. Email: lhhuong@ctu.edu.vn

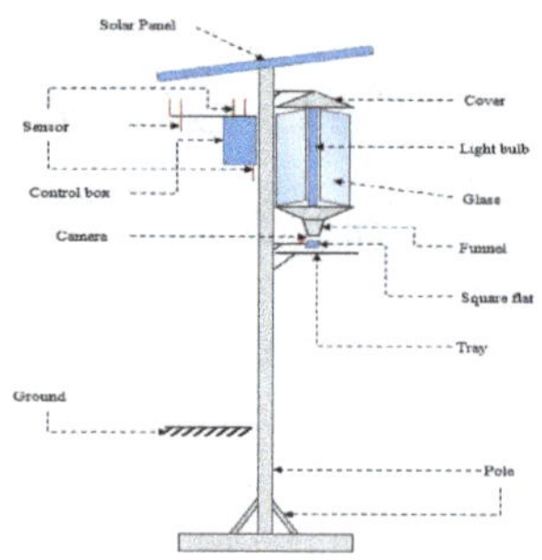

Figure 1. An automatic light trap

light trap position for Brown PlantHopper surveillance network (This method is called OBSNET). Section 5 will introduce some experimental results by applying the new approach. The last section summarizes the contribution and suggests some researches in the future.

2. Related works

The surveillance network is applied in many domain of environment and ecological research such as in the agricutural management[4], in the fishery surveillance, and in the forest management[1] [5]. Light traps are used to monitor the kinds of insect in the agricutural such as BPHs.

Optimization for wireless sensor network or particular light trap network is an important research. In fact, there are many related researches such as layout optimization [8], optimization for energy [9], optimization for coverage - connectivity - topology... [10], schemes optimization [11], optimizing for environment surveillance network [12], and etc [13] [14] [15].

In optimal design, optimization for location wireless sensor network or light trap network that ensures the network is coverage or connectivity and so on, which is a popular research. Many researches for that are presented in [12] [16] [17] [18] [10].

3. Automatic Brown PlantHopper surveillance network

An automatic BPH light trap surveillance network is a graph *G=(V, E)*. This graph built from a set of vertices $V = \{v_1, v_2, ..., v_n\}$ and the set of edges $E = \{e_1, e_2, ..., e_m\}$. The vertice v_i with $i \in \{1..n\}$ is an automatic light trap. The edge e_k with $k \in \{1..m\}$, $i \in \{1..n\}$, $j \in \{1..m\}$ is an edge between two vertices v_i and v_j. The weights of the edges are defined by W=$\{w_1, w_2, ..., w_m\}$ where the value of w_k is given by distance function $f_d(v_i, v_j)$.

Fig. 2 illustrates the logical graph of a light trap network where the black dots mean the vertices in V and the red lines mean the edges in E. The graph contains 9 vertices $V = \{v_1, v_2, v_3, v_4, v_5, v_6, v_7, v_8, v_9\}$ and 10 edges $E = \{e_1 = e(v_1, v_2), e_2 = e(v_1, v_4), ..., e_9 = e(v_6, v_7), e_{10} = e(v_7, v_8)\}$.

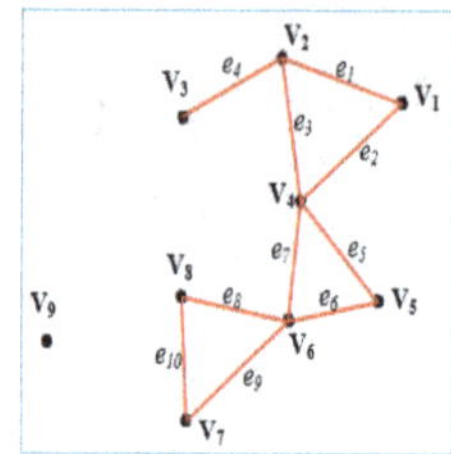

Figure 2. A light trap network is presented as a graph

Each node of light trap network has a communication range that is indicated by a circle with radius *r*. Conditions to define existence of an edge are introduced as following:

Definition 1 (Established edge): An edge is established if and only if the distance between a pair of vertices is less or equal to the minimum value of their radius - $f_d(v_i, v_j) \leq min(r_i, r_j)$.

Definition 2 (Unestablished edge): An edge is not established if distance between a pair of vertices is greater than the minimum value of their radius - $f_d(v_i, v_j) > min(r_i, r_j)$.

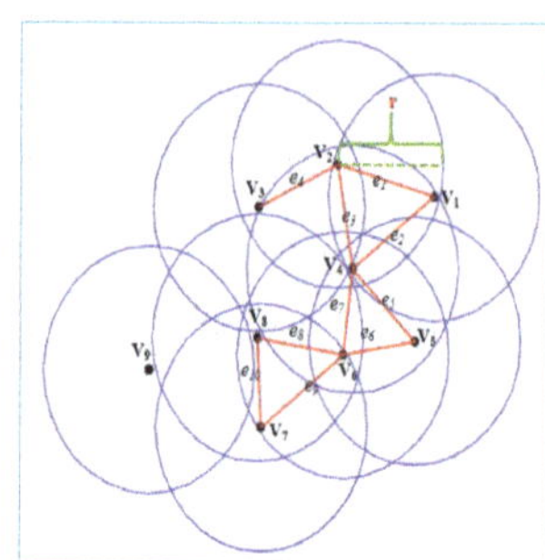

Figure 3. The communication range of light traps that is used to establish the egdes between the light traps

In the Fig. 3, the graph contains 1 subgraph and an isolated node. The sub graph consists of 8 nodes since distaces among these nodes are less than the radius r while the vertex v_9 is an isolated node because all distance values between it to others are insufficient to the definition 1.

There are some main factors in deployment of BSNET includings the positions where to place the automatic light traps so that the number of light traps is minimum and the network must be connected and coverage the deployment region. In the next section, we will present a new approach to optimize connectivity for BPH light trap surveillance network (The light trap network which is created by using this approach is called *Optimized BPH Surveillance Network*, contracted *OBSNET*).

4. Optimization of the Automatic Brown PlantHopper (BPH) light trap surveillance network

The BSNET will be deployed based on hexagonal cellular automata [22]. In [23], this paper specifies the condition that ensures the coverage of the region and guarantees network connectivity [24] [22] [25] [26] [27] [28] [29]. If $R \geq r$ and $0 \leq \frac{R}{r} \leq \frac{1}{2}3^{\frac{3}{4}}$, the hexagonal cellular automata is the best deployment, it ensured the region is full coverage, the network is connected and it requires the minimum number of light trap nodes. Otherwise, if $R \geq \sqrt{3}r$, the triangle lattice is the optimal deployment pattern to ensure full region coverage and network connectivity.

To determine the automatic light trap node distribution is initiated by placing a light trap in the center hexagon cell. The others will be set based on the first light trap. For example, the first light trap is located at (x,y) in Euclidean space, the neighbor light traps are located at $(x, y \pm \sqrt{3}r)$, and $(x \pm 1.5r, y \pm \frac{\sqrt{3}r}{2})$. Through the recursive this construction, we not only determine the position for all the light traps in the surveillance region with the minimum number of the light traps but also ensure the surveillance region that is full coverage about the communication.

There are two cases for optimizing the BSNET. In the first case, the deployment region will be devided into smaller units based on some conditions such as river, road, province, district and so on. After that, the biggest unit will be considered and hexagon cell at this unit will be created. Also, a hexagon grid will be created based on the first hexagon cell. The light traps will be located at the center of the hexagon cell. The pseudo-code for this case is presented in Alg. 1

Algorithm 1: Optimizing the BSNET for the first case

Data: Deployment Region
Result: Light trap network
begin
Divide deploytment region into smaller unit;
Get the biggest unit;
Let w is the width of the biggest unit;
Let c is center coordinates of the biggest unit;
list<hexagon>⟵ gridBuilder(c,w);
list<lighttrap> ⟵ trapBuilder(list<hexagon>);
network ⟵ networkBuilder(list<lighttrap>);
return network;
end

In the second case, a hexagon grid will be created by using the same method of the first case. If there are more than an unautomated light traps in a hexagon cell, build the unautomated light trap which is nearest from center of the hexagon cell to become the automatic light trap. After that, if the BSNET is not connected or not covered the deployment region about the communication, a light trap will be added at the center of the blank hexagon cell. Then, the connectivity will be checked again. If the BSNET is still not connected, move the light trap in the hexagon cell which is not connected with the network to the center of that hexagon cell or intersection of communication range between two automatic light traps (choose the nearest point). The pseudo-code for the second case is presented in Alg. 2

5. Experiment

5.1. Data used

The data of experiment is a GIS map data of the *Hau Giang* province at administrative levels including province, district, and commune. The data is stored as a table includings id, name (province, district, commune), shape length, shape area and so on (Fig. 4).

id_province	name_province	id_district	name_district	id_commune	name_commune	type	shape_leng	shape_area
38254	Hau Giang	103164	Chau Thanh	34429	Nga Sau	Townlet	0.142665	0.00088672
38254	Hau Giang	103164	Chau Thanh	34428	Dong Thanh	Commune	0.172407	0.00093227
38254	Hau Giang	103164	Chau Thanh	34427	Dong Phuoc A	Commune	0.200755	0.00139623
38254	Hau Giang	103164	Chau Thanh	34426	Dong Phuoc	Commune	0.209359	0.00180681
38254	Hau Giang	103164	Chau Thanh	34432	Phu Huu	Commune	0.165654	0.00142994
38254	Hau Giang	103164	Chau Thanh	34433	Mot Ngan	Townlet	0.130744	0.00067739

Figure 4. Data of Hau Giang province

The position of the light traps are stored in the plain text with xml format (*.gpx) that are used as input data. Fig. 5 presents the structure of the data with three types of information including date, coordinate of the light trap (longitude, latitude), and name. This file is created by using NetGen platform (a platform is developed by Brest university - France) [30]. Also, an abstract network of the light traps for BPH surveillance region at *Hau Giang* province was generated from NetGen [30].

```
<?xml version="1.0"?>
<gpx version="1.0" creator="NetGen for Hau Giang">
  <metadata>
    <name>Hau Giang's sensors</name>
    <desc>Sensors in city: Hau Giang, Vietnam</desc>
  </metadata>
    <wpt lat="9.957438" lon="105.743126">
        <sym>sound</sym>
        <cmt>title1</cmt><time>2013-031916:02:42</time>
        <name>Node:60</name>
        <desc>Noise:</desc>
    </wpt>
```

Figure 5. The position of the light traps in the xml format

5.2. OBSNET tool

We have developed the OBSNET tool in GAML [31] that enables to optimize the number of the light traps needed and their positions. OBSNET tool enables to

Algorithm 2: Optimizing the BSNET with existing unatomatic light traps

```
Data: Deployment region, unautomatic light trap
      position
Result: light trap network
begin
    Divide deploytment region into smaller unit;
    Get the biggest unit;
    Let w is the width of the biggest unit;
    Let c is center coordinates of the biggest unit;
    list<hexagon> ⟵ gridBuilder(c,w);
    foreach cell in the list < hexagon > do
        if there are more than unautomated light trap
        in a hexagon cell then
            Build the nearest unautomated light trap
            from center to automatic light trap;
        end
    end
    Build the automatic light trap network;
    repeat
        if automatic light trap network is not
        connectivity then
            Find all isolated light trap;
            repeat
                foreach lt in
                list < isolated_lighttrap > do
                    move it to center of hexagon cell
                    or intersection of communication
                    range;
                end
            until automatic network is connectivity;
        end
        if automatic light trap network is not coverage
        then
            foreach cell in list < hexagon > do
                if no light trap in a cell then
                    Create a light trap at the center of
                    the cell;
                end
            end
        end
        Build the automatic light trap network;
    until automatic light trap network is connectivity
    and coverage;
    return automatic network;
end
```

show the GIS map data, determine the position of the light trap on a map, create and display a hexagon grid on map,and build the network. Besides, OBSNET tool is also used to determine the communication range for automatic light trap based on honeyComb network.

5.3. Experiment 1: optimizing the light trap position for BPH surveillance network on the surveilance region without existing unautomated light trap

The requirement for this experiment must create a automatic BPH light trap network for Hau Giang province. First, the experiment will display the gis map data of Hau Giang province as communes. Then, it will determine the biggest commune on the map and construct the hexagon grid based on that commune. The result shown as Fig. 6.

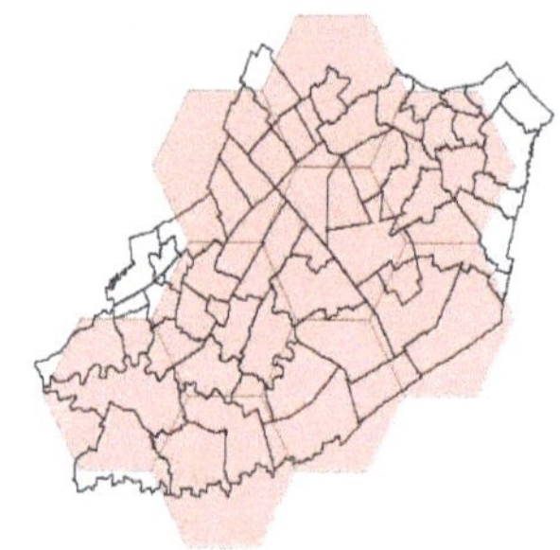

Figure 6. Hexagon grid for Hau Giang province

In Fig. 6, we obtain a hexagon grid with 9 hexagons. Each hexagon has a radius with 8,842 (m). Therefore, the minimum communication range is proposed 8,842*$\sqrt{3}$= 15,315 (m).

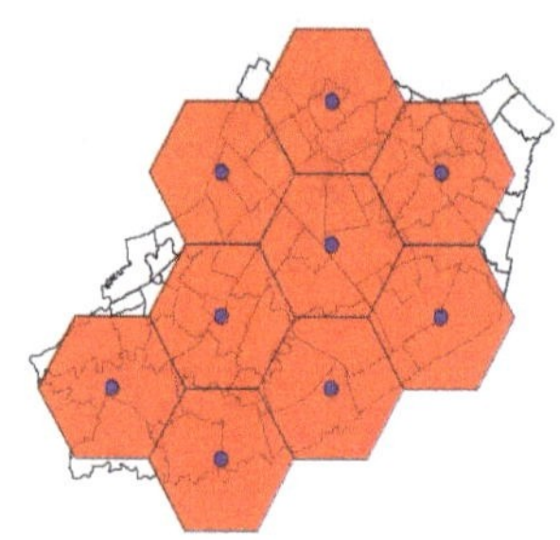

Figure 7. Light trap position in hexagon grid for Hau Giang province

After building the hexagon grid, place a automatic light trap at the center of hexagon (blue circle). The result shown as Fig. 7. The communication range of the automatic light traps are shown as yellow circle (Fig. 8)

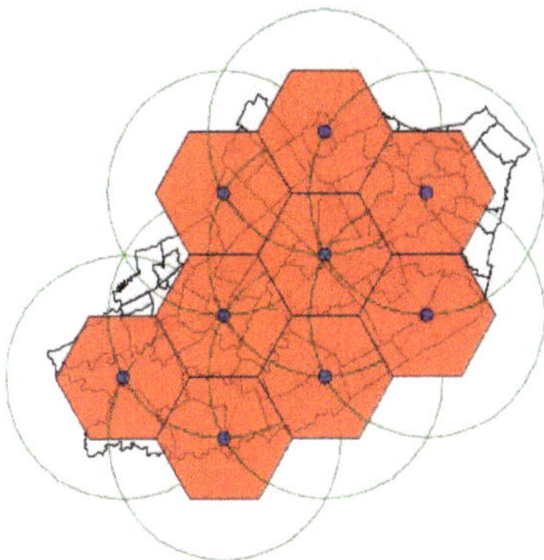

Figure 8. Light trap communication range in Hau Giang

Fig. 9 is present the OBSNet for automatic light traps in Hau Giang province.

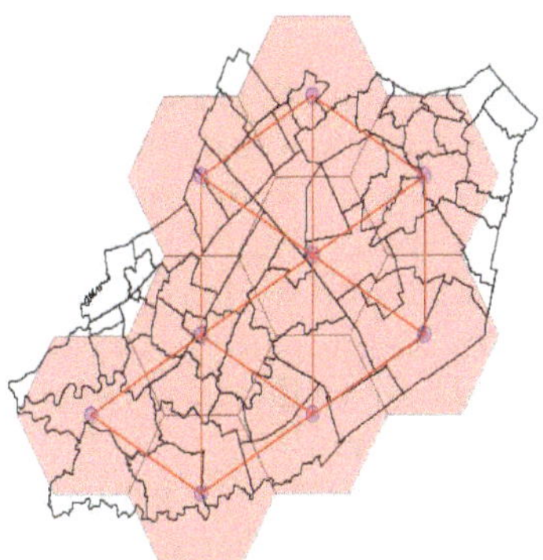

Figure 9. The OBSNET for the automatic light traps in Hau Giang

The logical network of the OBSNet in Hau Giang province is present as below:

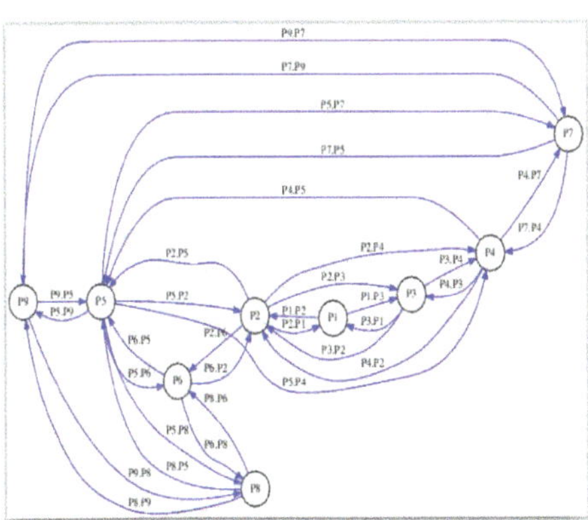

Figure 10. The logical network for the automatic light traps in Hau Giang

5.4. Experiment 2: optimizing the light trap position for BPH surveillance network with existing unautomated light trap

In this experiment, we will build the hexagon grid on the surveillance region that have some existing unautomated light traps. First, we need to consider to build some unautomated light traps to become automatic light traps. Second, we will build the network. If the network is not connected (there are some isolated automatic light traps), these automatic light traps will be considered moving to a new location. The hexagon grid on the surveillance region with existing unautomated light trap is shown as in Fig. 11. There are two cases about the unautomated light trap position. They are the unautomated light trap is located inside or outside the hexagon grid.

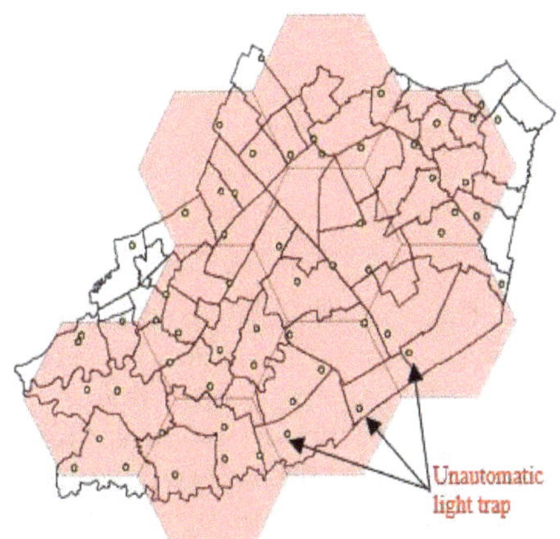

Figure 11. The hexagon grid on Hau Giang with existing non-auto light traps

Then, skip all the unautomated light traps outside the hexagon grid. After that, we will traverse every hexagon cell in hexagon grid, and get the unautomated light trap nearest from the center of the cell and skip all the others. The result is shown as in Fig. 12

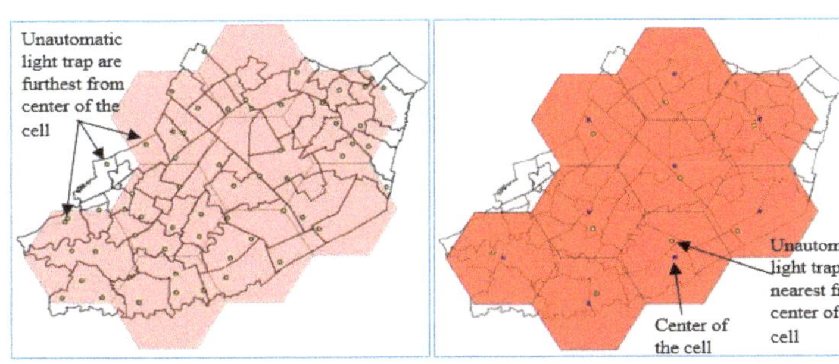

Figure 12. Get unautomated light trap is nearest from center of the hexagon cell

The nearest unautomated light traps from center of the hexagon cell will become an automatic light trap with communication range that is calculated in the experiment 1. Now, we will build the network for the automatic light traps (Fig. 13).

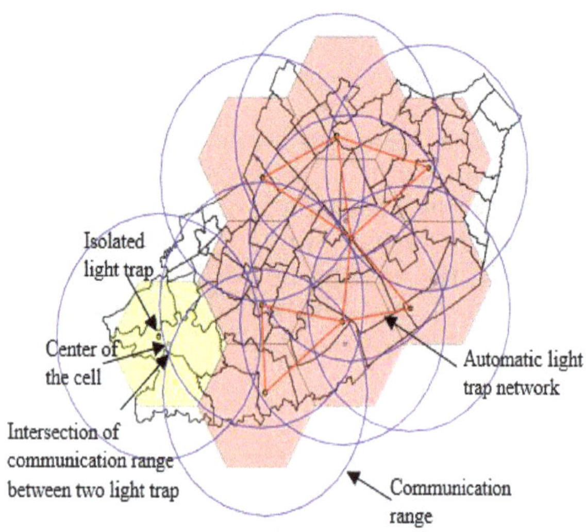

Figure 13. Skip the unautomated light trap

In the Fig. 13, there is an isolated automatic light trap, so the network is not connected. Therefore, we must move the isolated to new position that helps the network connect. There are two new positions including the center of the cell and the intersection between two communication of two automatic light traps in neighbor cells. In this experiment, we will choose the nearest position that helps network connect. It is the intersection between two communication ranges.

The Fig. 14 shows the network after moving the isolated light trap to new position (intersection between two communication ranges of two automatic light traps).

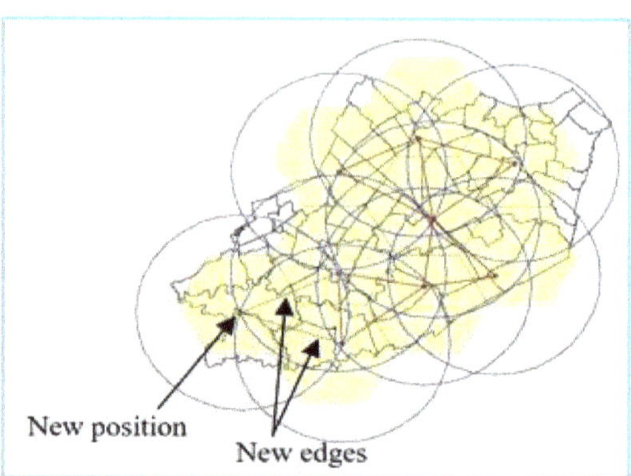

Figure 14. OBSNet is built after moving the isolated automatic light trap to new position

The logical network for OBSNet after moving the isolated automatic light trap to new position (Fig. 15).

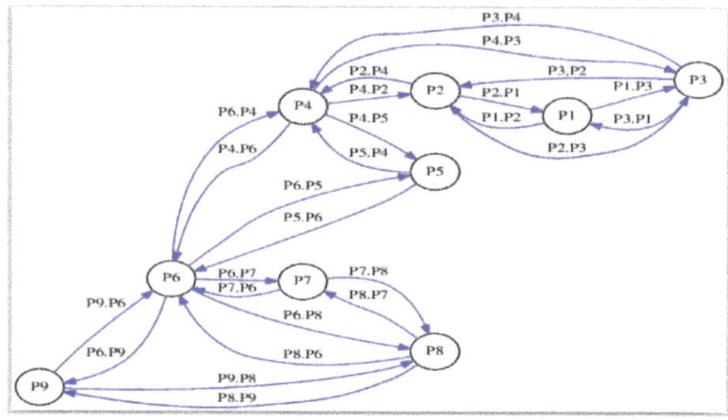

Figure 15. Logical network for OBSNet

6. Conclusion

The research on the optimization for automatic Brown PlantHopper surveillance network is one of the important trends in the environment and ecological research. This trend solves some questions such as where light traps are placed, how to fully cover the surveillance region and so on. Therefore, we propose a new approach to design and optimize the light trap position for BPH surveillance network based on hexagonal cellular automata.

Building the hexagon grid and network helps to determine the number of light traps needed and their positions. The result of the network model is deployed in Hau Giang province, a province in Mekong Delta. Based on the experiment results, we can deploy the OBSNET in the Mekong Delta region.

The experiment results show the effects of OBSNET based on hexagonal cellular automata. Using this method not only helps to optimize the light trap position for BPH surveillance network but also saves the cost in actual deployment. Actual data is used to validate the correctness of the OBSNET.

6.1. Acknowledgements

The work of this paper was in the scope of the "OBSNET - Optimizing the Brown Planthopper surveillance network using automatic light traps in the Mekong Delta region" project of the Ministry of Education and Training, Vietnam.

References

[1] Michael, K.M.M., Steen, S.M., (2006) Sampling Methods, Remote Sensing and GIS Multiresource Forest Inventory. In *Springer-Verlag Berlin Heidelberg*.

[2] Ky, N.M., Bao, L.H., Tuyen, T.P., Hieu, P.T.M., Chien, H.V., Pottier, B. and Hiep, H.X., (2015) Automatic light-trap using sensors. In *Proceeding in National Conference on*

Information and Communication Technology 2015, Can Tho University, Vietnam.

[3] Wang, B. (2011) Coverage problems in sensor networks: A survey. In *ACM Comput. Surv.*, vol. 43, no. 4, 1-53. doi: 10.1145/1978802.1978811

[4] Marcin, K.R. W. (2005) Ps sampling versus stratified sampling comparison of efficiency in agricultural surveys. In *Statistics in transition*, vol. 7, no. 1, 5-12.

[5] Talvitie, M., Leino, O., and Holopainen, M. (2006) Inventory of sparse forest populations using adaptive cluster sampling. In *The Finnish Society of Forest Science, Silva Fennica*, 0037-5330.

[6] Warrick, A.W. and Myers, D.E. (1987) Optimization of sampling locations for variogram calculations. In emphWater Resources Research, vol. 23, no. 3,496-500, 1987. doi: 10.1029/WR023i003p00496

[7] Brus, D.J. and Heuvelink, G.B. (2007) Optimization of sample patterns for universal kriging of environmental variables. In *Geoderma*, vol. 138, no. 12, pp. 86-95.

[8] Jourdan, D. and Weck, O. (2004) Layout optimization for a wireless sensor network using a multi-objective genetic algorithm. In *Vehicular Technology Conference*, vol. 5, 2004, 2466-2470.

[9] Lee, J.H. and Moon, I. (2014) Modeling and optimization of energy efficient routing in wireless sensor networks. In *Applied Mathematical Modelling*, vol. 38, no. 78, 2280-2289.

[10] Gogu, A., Nace, D., Dilo, A. and Mertnia, N. (2011) Optimization problems in wireless sensor networks. In *International Conference on Complex, Intelligent and Software Intensive Systems (CISIS)*, 302-309.

[11] Niewiadomska, S.E. and Marks, M. (2009) Optimization schemes for wireless sensor network localization. In *Int. J. Appl. Math. Comput. Sci.*, vol. 19, no. 2, 291-302.

[12] Viet, T.X., Hiep, H.X., Minh, L.N. and Drogoul, A. (2013) Optimizing an environmental surveillance network with gaussian process entropy - an optimization approach by agent-based simulation. In *7th International KES Conference on Agents and Multi-agent Systems - Technologies and Applications*, Vietnam.

[13] Aio, W., Mateus, G., and Quintao, F. (2007) Optimization issues and algorithms for wireless sensor networks with mobile sink. In *Proc. INOC*.

[14] Iqbal, M., Naeem, M., Anpalagan, A., Ahmed, A. and Azam, M. (2015) Wireless sensor network optimization: Multi-objective paradigm. In *Sensors*, vol. 15, no. 7, 17572.

[15] Madan, R., Cui, S., Lall, S. and Goldsmith, A. (2007) Modeling and optimization of transmission schemes in energy-constrained wireless sensor networks. In *IEEE/ACM Transactions on Networking*, vol. 15, no. 6, 1359-1372.

[16] Gopakumar, A. and Jacob, L. (2008) Localization in wireless sensor networks using particle swarm optimization. In *International Conference on Wireless, Mobile and Multimedia Networks*, 227-230.

[17] Mollaebrahim, G.P., Shahbazian, R. and Ghorashi, S.A. (2014) Localization in Wireless Sensor Networks Using Quadratic Optimization. In *ArXiv e-prints*.

[18] Monica, S. and Ferrari, G. (2013) Particle swarm optimization for auto-localization of nodes in wireless sensor networks. Adaptive and Natural Computing Algorithms. In Tomassini, M., Antonioni, A., Daolio, F. and Buesser, P. *Lecture Notes in Computer Science*, Eds. Springer Berlin Heidelberg, vol. 7824, 456-465.

[19] Vaithilingam, D.C. (2013) Role of insect light trap in organic agriculture. In *Romvijay Biootech Private Limited.*

[20] Linker, D.B.O.H.M. and Barbercheck, M.E. (2009) Insect management on organic farms.

[21] Brown planthopper: Threat to rice (1979). In *International Rice Research Institude.*

[22] KhouïňĄ, I., Minet, P., Laouiti, A. and Mahfoudh, S. (2014) Survey of Deployment Algorithms in Wireless Sensor Networks: Coverage and Connectivity Issues and Challenges. In *International Journal of Autonomous and Adaptive Communications Systems (IJAACS)*, 24.

[23] Baltzis, K.B. (2011) Hexagonal vs circular cell shape: a comparative analysis and evaluation of the two popular modeling approximations. In *INTECH Open Access Publisher.*

[24] Sun, Y., Yu, Z., Ge, J., Lin, B. and Yun, Z. (2009) On deploying wireless sensors to achieve both coverage and connectivity. In *Proceedings of the 5th International Conference on Wireless Communications, Networking and Mobile Computing*(IEEE Press),3369-3372.

[25] Ghosh, A. and Das, S.K. (2008) Coverage and connectivity issues in wireless sensor networks: A survey. In *Pervasive and Mobile Computing*, vol. 4, no. 3, 303-334.

[26] Zhu, C., Zheng, C., Shu, L. and Han, G. (2012) Review: A survey on coverage and connectivity issues in wireless sensor networks. In *J. Netw. Comput. Appl.*, vol. 35, no. 2, 619âĂŞ632. doi: 10.1016/j.jnca.2011.11.016

[27] Zhang, H. and Hou, J.C. (2005) Maintaining sensing coverage and connectivity in large sensor networks. In *Ad Hoc and Sensor Wireless Networks*, vol. 1, no. 1-2.

[28] Bai, X., Kumar, S., Xuan, D., Yun,Z. and Lai, T.H. (2006) Deploying wireless sensors to achieve both coverage and connectivity. In *Proceedings of the 7th ACM International Symposium on Mobile Ad Hoc Networking and Computing*, 131-142. doi: 10.1145/1132905.1132921

[29] Wang, Y.c., Hu, C.C. and Tseng, Y.C. (2005) Efficient deployment algorithms for ensuring coverage and connectivity of wireless sensor networks. In *WICON, S. Imreand J. Crowcroft, Eds. IEEE Computer Society*,114-121.

[30] netgen website, A generator of concurrent systems, `http://wsn.univ-brest.fr/NetGenNews` (accessed on 21 December 2015)

[31] gama website, Modeling and simulation platform, `https://github.com/gama-platforms` (accessed on 21 December 2015)

16

Book Titled *Autonomic Networking-on-Chip: Bio-Inspired Specification, Development, and Verification*: An Introduction

Phan Cong Vinh

Nguyen Tat Thanh University, 300A Nguyen Tat Thanh street, Ward 13, District 4, Ho Chi Minh City, Vietnam

Abstract

Despite the growing mainstream importance and unique advantages of autonomic networking-on-chip (ANoC) technology, Autonomic Networking-On-Chip: Bio-Inspired Specification, Development, and Verification is among the first books to evaluate research results on formalizing this emerging NoC paradigm, which was inspired by the human nervous system. The third book in the Embedded Multi-Core Systems series from CRC Press, this is an advanced technical guide and reference composed of contributions from prominent researchers in industry and academia around the world. A response to the critical need for a global information exchange and dialogue, it is written for engineers, scientists, practitioners, and other researchers who have a basic understanding of NoC and are now ready to learn how to specify, develop, and verify ANoC using rigorous approaches.

Keywords: ANoC, BioChipNets, Bio-inspiration

1. Introduction

A new networking-on-chip paradigm, which is seen as a cutting-edge approach to network-on-chip (NoC), is currently on the spot as one of the priority research areas: autonomic networking-on-chip (ANoC), which is inspired by the human autonomic nervous network.

ANoC is a networking-on-chip paradigm able to realize its self-* functionality such as self-organization, self-healing, self-configuration, self-optimization, self-protection, and so on whose context awareness is used to control networking functions dynamically. The overarching goal of ANoC is to realize intelligent networks-on-chips that can manage themselves without direct human interventions. Meeting this grand challenge of ANoC requires a rigorous approach to ANoC and the notion of self-*. To this end, taking advantage of formal engineering methods we will establish, in this book, formal and practical aspects of ANoC through specifying, refining, programming and verifying ANoC and its self-*. All of these are to achieve foundations and practice of ANoC.

From the above characteristics, novel approaches of specification, refinement, programming and verification are arising in formal engineering methods for ANoC. Therefore, new methodologies, programming models, tools and techniques are imperative to deal with the impact of ANoC and its self-* mentioned above on emerging intelligent networks-on-chips.

The book [1] is a reference for readers who already have a basic understanding of NoC and are now ready to know how to bio-inspiredly specify, develop and verify ANoC using rigorous approaches. Hence, the book includes both theoretical contributions and reports on applications. To keep a reasonable tradeoff between theoretical and practical issues, a careful selection of the chapters was completed, on one hand, to cover a broad spectrum of formal and practical aspects and, on the other hand, to achieve as much as possible a self-contained book.

Formal and practical aspects will be presented in a straightforward fashion by discussing in detail the

★Please ensure that you use the most up to date class file, available from EAI at http://doc.eai.eu/publications/transactions/latex/

*Corresponding author. Email: pcvinh@ntt.edu.vn

Figure 1. Cover of the book

necessary components and briefly touching on the more advanced components. Therefore, bio-inspired specification, development and verification demonstrating how to use the formal engineering methods for ANoC will be described using sound judgment and reasonable justifications.

This book, with chapters contributed by prominent researchers from academia and industry, will serve as a technical guide and reference material for engineers, scientists, practitioners and researchers by providing them with state-of-the-art research findings and future opportunities and trends. These contributions include state-of-the-art architectures, protocols, technologies, and applications in ANoC. In particular, the book covers existing and emerging research issues in bio-inspired on-chip networks (BioChipNets). The book has eight chapters addressing various topics from specification to implementation of ANoC based on rigorous interdisciplinary approaches.

2. Contents of the Book

Chapter 1 by M. Bakhouya presents an overview of state-of-the-art approaches for ANoC. Unlike design-time approaches in which all parameters/protocols are optimized/selected at design time targeting a specific application, in run-time approaches the adaptation process is running continuously to evolve the system. Based on this state-of-the-art review, the self-* capabilities are considered in the context of ANoC. These capabilities can provide the core scheme to develop ANoC at application, communication and architecture levels. An approach inspired by a biological immune system toward developing ANoC with these self-* capabilities is introduced. One aspect of the self-healing capability is partially developed, and some preliminary results are reported and showed that the adaptation is energy and latency efficient.

Chapter 2 by A. A. Morgan, H. Elmiligi, M. W. El-Kharashi, and F. Gebali discusses a GA-based multi-objective technique for autonomous on-chip network architecture optimization. This technique considers four NoC metrics: power, area, delay, and reliability. The models of the four metrics are discussed. The formulation of the fitness function is then presented. Finally, the GA representation of the problem is explained. The optimization can be carried out for power, area, delay, reliability or the four of them according to weight factors supplied by the designer. Results show that the technique is an efficient way to compromise between different NoC metrics. Moreover, the running time of the technique makes it more suitable for ANoC than previous architecture optimization techniques.

Chapter 3 by K. Latif, A. M. Rahmani, T. Seceleanu, and H. Tenhunen discusses an autonomous PVS-NoC architecture where an ideal tradeoff among virtual channel utilization, system performance and power consumption is presented. The virtual channel buffers are shared between two input ports in PVS-NoC. Apart from system performance, the architecture is also fault tolerant. In case of any link failure, the VC buffers are used by the other physical channel sharing the VC buffers, and system performance is not affected severely. The architecture is simulated with synthetic and real application traffic patterns. The performance is compared with typical VC-based NoC architecture and FVS-NoC. The PVS-NoC architecture shows significant improvement in system throughput without significant power consumption overhead.

Chapter 4 by J. Jaros and V. Dvorak discusses an evolutionary design that is able to produce optimal or near optimal communication schedules comparable to or even better than those obtained by a conventional design for the networks sizes of interest. Moreover, evolutionary design reduces many drawbacks of present techniques and invents still unknown schedules for an arbitrary topology and scatter/broadcast communication patterns.

Chapter 5 by P. C. Vinh breaks new ground in dealing with the *core-to-core* and *agent-based networking techniques* for ANoC using a categorical approach of tasks and data parallel processing – the firm formal method applicable to a wide variety of BioChipNets. The major contribution of the chapter is to propose some applied categorical structures of tasks parallel and data parallel for parallel processing on BioChipNets that has never been tackled thoroughly in this emerging field. By this approach, category theory is applied to deal in an abstract way with algebraic objects and relationships between them for specifying tasks parallel and data parallel on BioChipNets. The chapter shows that for specifying, analyzing and verifying tasks and data parallelism, the categorical approach becomes much better approaching than other ones in theory of algebras. From an applicative aspect of the approach, moreover, formalizing tasks parallel and data parallel on BioChipNets is a categorical specification of middleware that can be used to develop implementations for core-to-core and agent-based networking.

Chapter 6 by L. Guang, J. Plosila, J. Isoaho, and H. Tenhunen presents a novel design approach, hierarchical agent-monitored SoC (HAMSoC), by elaborating its formal specification framework and demonstrating a design example of hierarchical power monitoring on NoCs. This approach provides scalability in terms of both design effort and physical overhead. The formal specification framework of HAMSoC enables efficient system-level specification and modeling for early-stage development. It adopts a highly abstracted formal language to specify exclusively the monitoring operations and parameters, with proper exposure of SW/HW interfaces. The FSM model is used for state transition analysis in monitoring operations. The specification is transformed into generic formal languages so that existing verification tools can be used for further system development.

Chapter 7 by L. Petre et al. discusses a middleware language based on formal methods that has been developed for real networks to be adapted to applications running on NoC. The chapter presents how to place modules that communicate often in each other's vicinity for efficiency. This becomes even more interesting if the NoC consists of vertically stacked layers, commonly referred to as 3-D NoCs. Furthermore, three dynamic alternatives for replacing an application are analyzed, some of which are centralized and others are distributed.

Finally, Chapter 8 by H. Zakaria, E. Yahya, and L. Fesquet discusses a survey on different problems facing designers over the nanometric era. Analyses and solutions to each of these challenges are presented that enable the design of a self-adaptable SoC. GALS systems are considered and some proposed solutions are also applicable to simpler designs.

3. Remarkable Features of the Book

This title is the third book in a series on Embedded Multi-Core Systems (EMS) published by CRC Press. The EMS series is under the editorial supervision of the Integrated Microsystems Research Group, which is one of the research groups of the Electrical and Computer Engineering Department (ECE), Faculty of Engineering at the University of Victoria (UVic) in Canada.

This book has the following remarkable features:

- Provides a comprehensive reference on ANoC
- Presents state-of-the-art techniques in ANoC
- Formally specifies, develops and verifies bio-inspired on-chip networks
- Includes illustrative figures facilitating easy reading
- Discusses emerging trends and open research problems in ANoC

The book serves as a comprehensive and essential reference on autonomic networking-on-chip and is intended as a textbook for senior undergraduate- and graduate-level courses. It can also be used as a supplementary textbook for undergraduate courses. The book is a useful resource for students and researchers to learn autonomic networking-on-chip. In addition, it will be valuable to professionals from both academia and industry and generally has instant appeal to people who would like to contribute to autonomic networking-on-chip technologies.

We highly hope you enjoy reading the book.

4. Reviews and Testimonials

"The FIRST Book to Assess Research Results, Opportunities, & Trends in BioChipNets".

– CRC Press

"The book has come into being from a fine collection of chapters emphasizing the multidisciplinary character of investigations from the point of view of not only the ANoC field involved but also the formal methods. The book, which is specially dedicated to reporting on recent progress made in the field of bio-inspired on-chip networks (BioChipNets), will usefully serve as a technical guide and reference material for engineers, scientists, practitioners, and researchers by providing them with state-of-the-art research results and future opportunities and trends. To the best of my knowledge, this is the first book that presents achievements and findings of ANoC research covering the full spectrum of formalizing BioChipNets. These make the book unique and, in more than one respect, a truly valuable source of information that may be considered a landmark in the progress of ANoC".

– Petre Dini, Advisory Committees Board Chair of IARIA

References

[1] P.C. Vinh(2011) *Autonomic Networking-on-Chip: Bio-Inspired Specification, Development, and Verification* (CRC Press), 1st ed.

17

Representing and Reasoning with the Internet of Things: a Modular Rule-Based Model for Ensembles of Context-Aware Smart Things

S. W. Loke[1,*]

[1]Department of Computer Science and Information Technology, La Trobe University, Kingsbury Drive, VIC 3086, Australia

Abstract

Context-aware smart things are capable of computational behaviour based on sensing the physical world, inferring context from the sensed data, and acting on the sensed context. A collection of such things can form what we call a *thing-ensemble*, when they have the ability to communicate with one another (over a short range network such as Bluetooth, or the Internet, i.e. the Internet of Things (IoT) concept), sense each other, and when each of them might play certain roles with respect to each other. Each smart thing in a thing-ensemble might have its own context-aware behaviours which when integrated with other smart things yield behaviours that are not straightforward to reason with. We present Sigma, a language of operators, inspired from modular logic programming, for specifying and reasoning with combined behaviours among smart things in a thing-ensemble. We show numerous examples of the use of Sigma for describing a range of behaviours over a diverse range of thing-ensembles, from sensor networks to smart digital frames, demonstrating the versatility of our approach. We contend that our operator approach abstracts away low-level communication and protocol details, and allows systems of context-aware things to be designed and built in a compositional and incremental manner.

Keywords: smart artifacts; context-aware artifacts; operator language; compositionality; Internet-of-Things (IoT)

Introduction

Due to developments in pervasive computing and embedded systems, there has been a growing number of types and varieties of *smart things*, which are capable of computationally based behaviours, communicating with each other over the Internet[1] or over short range networks, and sensing the physical world [4, 30]; they can have context-aware behaviours [41], and are becoming even cloud-enabled [22]. These things can then intelligently respond to the context or situation they sense, via some computation. Smart things might also be everyday objects (e.g., vases, books, picture frames, furniture, clothings, etc)[2] endowed with embedded processors, intelligent sensors, mobile devices with sensors, appliances [26], or simply new types of devices [48].

A number of paradigms have been employed to program such things, including Web service based methods,[3] that assume each device has an embedded Web server that receives and responses to invocations, UPnP service styles,[4] JINI,[5] and peer-to-peer based styles such as JXTA,[6] and various visual editors such as [21, 25, 38]. For some time now, we have seen the application of rule-based programming for behaviours of collections of smart things, such as [24, 25, 49, 50],[7]

*Corresponding author. Email: s.loke@latrobe.edu.au

[1]http://www.theinternetofthings.eu

[2]The e-gadgets model view everyday objects as components, see http://www.extrovert-gadgets.net/

[3]See http://docs.oasis-open.org/ws-dd/ns/dpws/2009/01 and [31]

[4]See http://www.upnp.org/

[5]See http://www.jini.org/wiki/Main_Page

[6]See https://jxta.dev.java.net/

[7]Others are https://www.onx.ms/ and http://appinventor.mit.edu/

though such work generally do not consider reasoning with groups of rules. Basically, context information gathered via sensors are reasoned with and appropriate action taken based on these rules.

The general structure of smart things, sensing, reasoning and acting [33] corresponds to context-aware mobile computing systems as early as 1994 [43], and in fact, many context-aware systems do have that general structure.

However, when a collection of such smart things are put together, and if they can sense and react to each other, it is not easy to reason about the collective behaviour of the ensemble, and it is not easy to determine what will happen as a result of combining such smart things.

Following the design philosophy first introduced in [36] and also used in [35], in this paper, we elaborate on an approach (and formalism), which we call SIGMA, to represent compositions (or collections) of smart things that can sense their environment and context and act in response to what they sense. The formalism comprises a language of operators, inspired by modular logic programming [12], for composing systems, the advantage of using operators being to encapsulate the cooperation between systems, leaving it to the implementation to realize that cooperation. The compositions are not necessarily physical compositions of the things (e.g., if the things are vases, a composition here does not necessarily mean gluing the things together), but rather, are compositions of their representations. Such compositions specify how things may cooperate in their sensing and acting; we call a collections of things that can communicate with each other, and whose behaviours are composed, a *thing-ensemble*. Note that our view of cooperation here is the idea of a set of devices working together complementarily for a purpose, and coordinated to achieve an effect via a set of operators with predefined semantics, not autonomous cooperation as in multiagent systems research. It should be noted that a thing-ensemble is, hence, different from an arbitrary set of appliances or sensors, in that we assume that the things in the ensemble can interact with each other, in the way we describe later.

We show how a variety of thing-ensembles can be represented, ranging from the smart living room applications, sensor networks to smart digital frames, and their behaviours specified. We use the term *ensembles* to emphasize that the things are in some kind of coordinated relationship with each other in order to fulfill their own tasks or end-user specified tasks, and things typically play different roles in such a cooperation. Moreover, this paper deals with the formalism to specify such compositions but not the implementation details.

The twofold purpose of our formalism for representing and reasoning with thing-ensembles is as follows:

- *Abstraction*. We present a set of operators that provides well-defined compositions of smart context-aware things. This enables reasoning about what collective behaviours could emerge when an ensemble of such smart things interact or are composed. While this may seem constraining (restricted to the collection of operators we have), we present a small set of operators which can be used to model interactions in a range of scenarios, showing that even a small set of operators can be quite expressive and useful.

- *Incremental and compositional design for the Internet of Things*. We also argue for the need of a design approach that provides a means to build or extend thing-ensembles in an incremental manner, i.e., new things can be added to a collection, and new composed behaviours attainable without considerable developer effort, or considerable change to the existing thing-ensemble. The idea is that smart context-aware behaviours of a collections of smart things (and/or computer devices) can be effectively "grown" over time. While in a different context, a parallel of this idea is McCarthy's notion of *elaboration tolerance* (especially, additive types of elaborations), which is "the ability to accept changes to a person's or a computer program's representation of facts about a subject without having to start all over."[8] In our case, it is the addition of new smart things (e.g., new computational nodes, storage, sensors, appliances, etc) to an existing system (a collection of smart things), without needing to start over or redesigning the whole system. Our examples in a range of scenarios demonstrate this idea where the behaviour of each thing is simply modelled as a module of rules. There are pragmatic advantages of this since if old devices can be combined with newly added devices, then reuse of the old devices in the presence of the new is facilitated, or if newly added devices can override (or suppress) the functionality of older devices in a well-defined manner, then the new behaviour can be better understood.

The contributions of this paper are twofold:

- Theoretically, the SIGMA formalism serves as a basis for a unified model of cooperating thing-ensembles that act on an understanding of their situations. We seek to expose and formalize the

[8] http://www-formal.stanford.edu/jmc/elaboration/elaboration.html

underlying structure of a diverse range of thing-ensembles, which might seem rather different on the surface, but in fact, have common structure and behaviour. The fact that these operators can be defined which capture how different systems of context-aware things work together in different contexts suggests the generality of the operators and emphasizes the common features of these systems.

- Practically, the SIGMA formalism (perhaps with syntactic sugar if needed) can be used in two ways:
  - as a *specification language* for designing context-aware IoT applications [41, 47, 58]; the formalism encourages a modular approach to the problem, and also allows incremental development where an existing system can be modelled and then, add-ons can be represented and combined with the existing system via suitable operators; a smart thing can be modelled as a rule-base but then implemented in hardware, or it can be a software-programmable device - our approach is agnostic to this, and
  - as a *scripting language* for programming behaviours of cooperating context-aware things, where the combined behaviour is embedded within the operators, thereby providing a high-level of abstraction, without the programmer needn't to deal with underlying protocols; but an interpreter for the operators would need to be implemented first.

In the next section, we present a brief overview of operators we will use in composing behaviours of things. Then, we will describe scenarios of thing-ensembles. Related work is then reviewed, and finally, we conclude with future work.

SIGMA Operators for Composing Context-Aware Systems

Context-Aware Systems

We use the operator language and abstract model of context-aware systems from [36], but extend the repertoire of operators here. We use the term *context-aware system* to refer to three categories of systems:

- systems comprising only of sensors,
- systems with sensors and context interpretation components, and
- systems with sensors, context interpretation components, and situation reasoners.

The above components are explained below.

The general model of a context-aware system R is a triple of the form (Σ, Π, Θ). Σ is a finite set of sensors where each sensor σ_i is a function which senses a part of the world (in which the sensor is situated at the time) $W \in \mathbf{W}$ and produces a set of sensor readings $G \in \mathbf{G}$, i.e. $\sigma_i : \mathbf{W} \rightarrow \mathbf{G}$. We let $\mathbf{W}$ and $\mathbf{G}$ remain opaque as our discussions do not require their details, but explicit and precise definitions could be given for them; also, in our case, we assume that the sensors Σ are attached to (or part of) the particular smart thing (e.g., a vase embedded with touch sensors and a microphone). Often, we will write $reading(\sigma_i) = G$ to denote when reading from the sensor σ_i has value G.

In this paper, a context-aware system can map to a thing with sensing, networking and computational capabilities. Or a context-aware system might comprise a thing and associated software and hardware that tracks the thing and provides context-aware behaviour associated with the thing.

The interpreter Π can then be defined as a mapping from sensor readings $\mathbf{G}$ to some context (e.g., a symbolic location such as a room number) $C \in \mathbf{C}$ (which we assume are concepts grounded in some ontology such as SOUPA[9] and CONON [52], or other ontologies as surveyed in [7, 57]), i.e., $\Pi \subseteq (\mathbf{G} \times \mathbf{C})$. So, given $W \in \mathbf{W}$, and suppose $\Sigma = \{\sigma_1, \dots, \sigma_n\}$, and $\sigma_1(W) = G_1$, ..., $\sigma_n(W) = G_n$ (for $G_i \in \mathbf{G}$), then Π can be applied to interpret each G_i to obtain a set of contexts $\{C_1, \dots, C_n\}$, denoted by $\Pi(G_i) = C_i$ (taken here to mean $(G_i, C_i) \in \Pi$). We will also often write the mapping $(G_i, C_i) \in \Pi$ as a rule of the form $(G_i \Rightarrow C_i) \in \Pi$, for readability.

The situation reasoner Θ is a relation mapping sets of contexts or situations to situations, i.e. $\Theta \subseteq (\wp(\mathbf{C} \cup \mathbf{S}) \times \mathbf{S})$, where $\mathbf{S}$ is a set of situations (again, which we assume grounded in some ontology). We will also often write the mapping $(\{C_1, \dots, C_k, S_1, \dots, S_l\}, S) \in \Theta$ as $(\{C_1, \dots, C_k, S_1, \dots, S_l\} \Rightarrow S) \in \Theta$, for readability.

Examples of aggregating context to infer situations can be found in the literature [2, 16, 56, 57]. Our model makes the distinction between the notions of context and situation following the work such as [17, 32], where context information is used to "characterize the situation of an entity," modelled via Θ. While we do not commit to any ontology here, in practice, we assume that systems do return inferred contexts or situations using concepts from a "well-known" ontology in order to facilitate interoperability.

Moreover, we define the relation denoted by $\vdash$ between systems and pairs of the form (W, S) where $W \in \mathbf{W}$ and S is some situation, such that $R \vdash (W, S)$ if and only if R recognizes S when sensing part of the world W, or alternatively, if one places R in the part of

[9] http://pervasive.semanticweb.org/soupa-2004-06.html

the world W where S is happening, S will be detected by R. R may be in a world W where S is not happening, but as soon as S happens, it detects it.

A situation that is recognized by a system is then either computed from contexts (recognized by some sensors and the interpreter) and other situations (if any) via Θ. This meaning of the relation $\vdash$ can be expressed recursively as follows in a rule written in the form $\frac{premises}{conclusion}$:

$$\frac{\begin{array}{l}(\{C_1,\ldots,C_m,S_1,\ldots,S_k\} \Rightarrow S) \in \Theta, \\ \text{for some } m, \text{ where for each } C_i,\ i \in \{1,\ldots,m\}, \\ \Pi(\sigma(W)) = C_i, \text{ for some } \sigma \in \Sigma, \\ \text{and for some } k, \text{ where for each } S_i,\ i \in \{1,\ldots,k\}, \\ (\Sigma,\Pi,\Theta) \vdash (W,S_i)\end{array}}{(\Sigma,\Pi,\Theta) \vdash (W,S)} \ (one-system)$$

We represent context recognition by a system R, i.e., $R \vdash (W,C)$ for some context C, where $\Pi(\sigma(W)) = C$, for some $\sigma \in \Sigma$.

Sigma Operators on Context-Aware Systems

The Sigma operators on context-aware systems form a small language of Sigma expressions, given by the following EBNF syntax:

$$\begin{array}{rcl} Q & ::= & R \mid Q+Q \mid Q \ominus Q \mid Q \oslash Q \\ E & ::= & Q \mid E \oplus E \mid E \otimes E \end{array}$$

The Sigma expressions defined by Q only are called *Q-expressions* and those defined by E are called *E-expressions*. The idea is that, given a thing-ensemble, i.e., a collection of context-aware things, the Sigma operators can be used to define different compositions of things from this collection. For example, given a thing-ensemble of three system $\{R_1,R_2,R_3\}$, different Sigma expressions can be defined over them, such as $R_1+R_2+R_3$ and $R_1 \oplus (R_2+R3)$. Moreover, for these expressions to be meaningful, there must be some way for the R_is to interact with each other, in the way that we define below. We distinguish between the thing-ensemble and the Sigma expressions defined over (perhaps some subset of the) things in the thing-ensemble, since each Sigma expression describes only *one way (among many) in which the things can work together.*

Informally, the meaning of the operators are as follows, extending our relation $\vdash$ given above. The union of two context-aware systems used in attempting to recognize C (or a situation S), denoted by $R_1 \oplus R_2 \vdash C$, means that context C (or situation S with $R_1 \oplus R_2 \vdash S$) is recognized either using R_1 or R_2, succeeding if either succeeds. Note that this can be nondeterministic but one could employ short-circuit evaluation in practice, that is, try R_1 first and only on failure try with R_2.

Below, C can be replaced by S. The *intersection* of two context-aware systems used in attempting to recognize C, denoted by $R_1 \otimes R_2 \vdash C$, means that context C is recognized using both R_1 and R_2, succeeding if both succeeds.

The *tight-union* of two context-aware systems used in attempting to recognize C, denoted by $R_1 \otimes R_2 \vdash C$, means that context C is recognized using both R_1 cand R_2 working in a tightly coupled combined manner (as we define below).

The *restriction* of two context-aware systems used in attempting to recognize C, denoted by $R_1 \ominus R_2 \vdash C$, means that context C is recognized using rules from $R_!$ and R_2 with joint conditions (as we define below). The idea is that the rules in the situation reasoner would have joint conditions. For example, if $R_1 = (\Sigma_1,\Pi_1,\Theta_1)$ and $R_2 = (\Sigma_2,\Pi_2,\Theta_2)$, whenever $(\mathbf{S_1} \Rightarrow S) \in \Theta_1$, where $\mathbf{S_1} \subseteq (\mathbf{S} \cup \mathbf{C})$ and $S \in \mathbf{S}$, and also $(\mathbf{S_2} \Rightarrow S) \in \Theta_2$ (in other words, both Θ_1 and Θ_2 have a rule for S), then we join their conditions when applying the rules to infer S, i.e. we must use joint condition rules of the form $(\mathbf{S_1} \cup \mathbf{S_2} \Rightarrow S)$ to infer S. This generalizes to three or more systems, say with R_3 having $(\mathbf{S_3}, S) \in \Theta_3$, we have $(\mathbf{S_1} \cup \mathbf{S_2} \cup \mathbf{S_3} \Rightarrow S)$. Note that this is different from intersection in that with intersection of R_1 and R_2, each independently try to infer S, whereas with restriction, R_1 and R_2 can share sensors and context interpretation, and even situation reasoning rules, but at any step, the rules for a corresponding situation must be joined in their conditions.

The *overriding union* of one context-aware system by another in attempting to recognize C, denoted by $R_1 \oslash R_2 \vdash C$, means that the rules of R_1 takes precedence over those of R_2 whenever the same situation is being considered. For example, if $R_1 = (\Sigma_1,\Pi_1,\Theta_1)$ and $R_2 = (\Sigma_2,\Pi_2,\Theta_2)$, whenever $(\mathbf{S_1} \Rightarrow S) \in \Theta_1$, where $\mathbf{S_1} \subseteq (\mathbf{S} \cup \mathbf{C})$ and $S \in \mathbf{S}$, and also $(\mathbf{S_2} \Rightarrow S) \in \Theta_2$ (in other words, both Θ_1 and Θ_2 have a rule for S), then we use the rule from R_1 only, i.e. $(\mathbf{S_1} \Rightarrow S)$, and ignore any others for the same situation (of the form $(\mathbf{S_2} \Rightarrow S)$) from Θ_2. This operator is inspired by the notion of input suppression from Brook's subsumption architecture [13], as we illustrate below.

The operators $\oplus$ and $\otimes$ corresponds to logical "or" and"and" respectively, and operators in Q-expressions can be rewritten as single systems with contents from the component systems in the way given by the operational semantics of the operators below. In an implementation of these operators, it may not be possible to physically take the actual constituent components and combine them, but in fact, appropriate (e.g., Web service) interfaces can be used by each component to expose their sensors, interpreter and situation reasoner components, and then calls are made to these interfaces to emulate effectively a single virtual system (the virtual system represented by a Q-expression).

The rules in Figures 1 and 2 provide the operational semantics, defining the modified relation $\vdash$ more precisely. Note that in rules $(one-system)$, (tu), (res) and (ovr), m and k could be zero for a given member of Θ, in which case, there is no context or situation in the antecedent.

$$\frac{\begin{array}{l}(\{C_1,\ldots,C_m,S_1,\ldots,S_k\} \Rightarrow S) \in \Theta,\\ \text{for some } m \text{, where for each } C_i, i \in \{1,\ldots,m\},\\ \Pi(\sigma(W)) = C_i \text{, for some } j \text{ and } \sigma \in \Sigma,\\ \text{and for some } k \text{, where for each } S_i, i \in \{1,\ldots,k\},\\ (\Sigma,\Pi,\Theta) \vdash (W,S_i)\end{array}}{(\Sigma,\Pi,\Theta) \vdash (W,S)} \; (one-system)$$

$$\frac{\begin{array}{l}(\{C_1,\ldots,C_m,S_1,\ldots,S_k\} \Rightarrow S) \in \Theta \cup \Theta',\\ \text{for some } m \text{, where for each } C_i, i \in \{1,\ldots,m\},\\ \Pi(\sigma(W)) = C_i \text{ or } \Pi'(\sigma(W)) = C_i \text{, for some } \sigma \in (\Sigma \cup \Sigma'),\\ \text{and for some } k \text{, where for each } S_i, i \in \{1,\ldots,k\},\\ ((\Sigma,\Pi,\Theta) + (\Sigma',\Pi',\Theta')) \vdash (W,S_i)\end{array}}{((\Sigma,\Pi,\Theta) + (\Sigma',\Pi',\Theta')) \vdash (W,S)} \; (tu)$$

$$\frac{(\bigcup_{i=1}^{n} \Sigma_i, \bigcup_{i=1}^{n} \Pi_i, \bigcup_{i=1}^{n} \Theta_i) \vdash (W,S)}{Q \vdash (W,S)} \; (gtu)$$

where $n > 1$ and $Q = (\Sigma_1,\Pi_1,\Theta_1) + \ldots + (\Sigma_n,\Pi_n,\Theta_n)$

$$\frac{\begin{array}{l}(\{C_1,\ldots,C_m,S_1,\ldots,S_k\} \Rightarrow S) \in \Theta \mid \Theta',\\ \text{for some } m \text{, where for each } C_i, i \in \{1,\ldots,m\},\\ \Pi(\sigma(W)) = C_i \text{ or } \Pi'(\sigma(W)) = C_i \text{, for some } \sigma \in (\Sigma \cup \Sigma'),\\ \text{and for some } k \text{, where for each } S_i, i \in \{1,\ldots,k\},\\ ((\Sigma,\Pi,\Theta) \ominus (\Sigma',\Pi',\Theta')) \vdash (W,S_i)\end{array}}{((\Sigma,\Pi,\Theta) \ominus (\Sigma',\Pi',\Theta')) \vdash (W,S)} \; (res)$$

where $\Theta \mid \Theta' = \{(\mathbf{S_1} \cup \mathbf{S_2} \Rightarrow S) \mid (\mathbf{S_1} \Rightarrow S) \in \Theta \wedge (\mathbf{S_2} \Rightarrow S) \in \Theta'\}$
$\cup \; \{(\mathbf{S_1} \Rightarrow S) \mid (\mathbf{S_1} \Rightarrow S) \in \Theta \wedge ((\mathbf{S} \Rightarrow S) \notin \Theta', for\ any\ \mathbf{S})\}$
$\cup \; \{(\mathbf{S_2} \Rightarrow S) \mid (\mathbf{S_2} \Rightarrow S) \in \Theta' \wedge ((\mathbf{S} \Rightarrow S) \notin \Theta, for\ any\ \mathbf{S})\}$

Figure 1. Overview of $Sigma$ operators and associated rules - part 1.

In the rule for tight-union (tu), we can observe a tighter coupling of the three components of sensors, interpreter and situation reasoner compared to union, in that a tight-union composition behaves as an integrated system at all three levels (white-box integration), whereas the union employs multiple systems but each as a blackbox. While the rule specifies that the union of the corresponding components (union of sensors, union of interpreters and union of situation reasoners) are employed in recognizing a situation

$$\frac{(\bigcup_{i=1}^{n} \Sigma_i, \bigcup_{i=1}^{n} \Pi_i, |_{i=1}^{n} \Theta_i) \vdash (W,S)}{Q \vdash (W,S)} \; (gres)$$

where $n > 1$ and $Q = (\Sigma_1,\Pi_1,\Theta_1) \ominus \ldots \ominus (\Sigma_n,\Pi_n,\Theta_n)$

$$\frac{\begin{array}{l}(\{C_1,\ldots,C_m,S_1,\ldots,S_k\} \Rightarrow S) \in \Theta \triangleleft \Theta',\\ \text{for some } m \text{, where for each } C_i, i \in \{1,\ldots,m\},\\ \Pi(\sigma(W)) = C_i \text{ or } \Pi'(\sigma(W)) = C_i \text{, for some } \sigma \in (\Sigma \cup \Sigma'),\\ \text{and for some } k \text{, where for each } S_i, i \in \{1,\ldots,k\},\\ ((\Sigma,\Pi,\Theta) \oslash (\Sigma',\Pi',\Theta')) \vdash (W,S_i)\end{array}}{((\Sigma,\Pi,\Theta) \oslash (\Sigma',\Pi',\Theta')) \vdash (W,S)} \; (ovr)$$

where $\Theta \triangleleft \Theta' = \{(\mathbf{S_1} \Rightarrow S) \mid (\mathbf{S_1} \Rightarrow S) \in \Theta \wedge (\mathbf{S_2} \Rightarrow S) \in \Theta'\}$
$\cup \; \{(\mathbf{S_1} \Rightarrow S) \mid (\mathbf{S_1} \Rightarrow S) \in \Theta \wedge ((\mathbf{S} \Rightarrow S) \notin \Theta', for\ any\ \mathbf{S})\}$
$\cup \; \{(\mathbf{S_2} \Rightarrow S) \mid (\mathbf{S_2} \Rightarrow S) \in \Theta' \wedge ((\mathbf{S} \Rightarrow S) \notin \Theta, for\ any\ \mathbf{S})\}$

$$\frac{(\bigcup_{i=1}^{n} \Sigma_i, \bigcup_{i=1}^{n} \Pi_i, \triangleleft_{i=1}^{n} \Theta_i) \vdash (W,S)}{Q \vdash (W,S)} \; (govr)$$

where $n > 1$ and $Q = (\Sigma_1,\Pi_1,\Theta_1) \oslash \ldots \oslash (\Sigma_n,\Pi_n,\Theta_n)$

$$\frac{E \vdash (W,S)}{(E \oplus E') \vdash (W,S)} \; (union1)$$

$$\frac{E \nvdash (W,S) \text{ and } E' \vdash (W,S)}{(E \oplus E') \vdash (W,S)} \; (union2)$$

$$\frac{E \vdash (W,S) \text{ and } E' \vdash (W,S)}{(E \otimes E') \vdash (W,S)} \; (intersection)$$

Figure 2. Overview of $Sigma$ operators and associated rules - part 2.

(or context), this happens conceptually, and does not necessarily imply the actual integration of software components. Tight-union has represents a type of interaction among two systems. The generalized form of tight-union to n-ary is given by the rule (gtu). Note that (gtu) with two operands is equivalent to (tu). Restriction represents another type of interaction among two systems, similar to tight-union, but where rules to infer the same situation must be joined in their conditions. The generalized form of restriction to n-ary is given by the rule $(gres)$. Note that $(gres)$ with two operands is equivalent to (res).

Overriding union represents another type of interaction among two systems, similar to tight-union, but where rules to infer the same situation have precedence. The generalized form of overriding union to n-ary is given by the rule $(govr)$. Note that $(govr)$ with two operands is equivalent to (ovr). Note that overriding union is left associative and non-commutative. The two rules of union rely on the success of one or the other - operationally, $(union1)$ would first be used for evaluations. If rule $(union1)$ succeeds, then $(union2)$ is not needed. Note that costs are added up for evaluations. The rules above can be used in a backward-chaining fashion to determine if a particular context or situation currently holds. For example, $(R_1 + R_2) \vdash (W, S)$ determines if situation S holds in world W with respect to the composition $(R_1 + R_2)$, by pattern matching. The rules can also be used in as a query to ask what contexts or situations currently hold (or are happening), if S is a variable instead of a given situation. For example, $(R_1 + R_2) \vdash (W, S?)$ asks what situations may be happening with respect to $(R_1 + R_2)$, resulting in one or more possible instantiations for $S?$ (we denote a variable by an identifier post-fixed by a "?").

A *composite context-aware system* is formed by using the above language of operators to compose context-aware systems.

Action Systems

We define another category of systems called *action systems* which takes situations or context recognized from context-aware system(s) and map those situations or context to actions. Similar to context-aware systems, we can define operators on action systems. An action system M is modelled simply as a relation between sets of context and situations, and actions, i.e. $M \subseteq \wp(\mathbf{S} \cup \mathbf{C} \cup \{\mathtt{false}\}) \times \mathbf{A}$, where $\mathbf{A}$ is a set of actions.[10] The idea is that the context and situations are conditions that must be satisfied before the action is taken. For some action system M, $\mathbf{S} \in \wp(\mathbf{S} \cup \mathbf{C})$, and $a \in \mathbf{A}$, we will write mappings of the form $(\mathbf{S}, a) \in M$ as $(\mathbf{S} \Rightarrow a) \in M$, for readability. We will also often use the constant `false` to represent a condition that can never be satisfied. Actions with the `false` condition will never be performed - this will be useful in compositions to inhibit behaviours as we show later.

SIGMA Operators on Action Systems

We also have SIGMA expressions on collections of action systems. We define four operators on action systems: *a-union* (denoted by "∪"), *a-intersection* (denoted by "∩"), *a-restriction* (denoted by "|"), and *a-overriding* (denoted by "◃"), as follows. Given two action systems M_1 and M_2, $M_1 \cup M_2$ is simplify the set-theoretic union of the corresponding relations, and the intersection $M_1 \cap M_2$ is the set-theoretic intersection of the corresponding relations. The purpose of a-restriction is to cater for cases where different situations or contexts map to the same actions, and the intention is to aggregate (or "and"-ing) the situations or contexts rather than simply "or"-ing them. For example, suppose we have $M_1 = \{(\{s_1, s_2\} \Rightarrow a_1), (\{s_6\} \Rightarrow a_3)\}$ and $M_2 = \{(\{s_3, s_4\} \Rightarrow a_1), (\{s_5\} \Rightarrow a_2)\}$. Then, forming the union gives $\{(\{s_1, s_2\} \Rightarrow a_1), (\{s_3, s_4\} \Rightarrow a_1), (\{s_5\} \Rightarrow a_2)\}$, which means that action a_1 can be triggered either by $\{s_1, s_2\}$ or $\{s_3, s_4\}$. However, suppose we wish to state that a_1 is to be triggered by $\{s_1, s_2, s_3, s_4\}$, or $s_1 \wedge s_2 \wedge s_3 \wedge s_4$. We do this by using *a-restriction*, which is defined as follows:

$$\begin{aligned} M_1 \mid M_2 = \; & \{(\mathbf{S_1} \cup \mathbf{S_2} \Rightarrow a) \mid (\mathbf{S_1} \Rightarrow a) \in M_1 \wedge (\mathbf{S_2} \Rightarrow a) \in M_2\} \\ & \cup \{(\mathbf{S_1} \Rightarrow a) \mid (\mathbf{S_1} \Rightarrow a) \in M_1 \wedge ((\mathbf{S_2} \Rightarrow a) \notin M_2, for\ any\ \mathbf{S_2})\} \\ & \cup \{(\mathbf{S_2} \Rightarrow a) \mid (\mathbf{S_2} \Rightarrow a) \in M_2 \wedge ((\mathbf{S_1} \Rightarrow a) \notin M_1, for\ any\ \mathbf{S_1})\} \end{aligned}$$

Then, when $M_1 = \{(\{s_1, s_2\} \Rightarrow a_1), (\{s_6\} \Rightarrow a_3)\}$ and $M_2 = \{(\{s_3, s_4\} \Rightarrow a_1), (\{s_5\} \Rightarrow a_2)\}$, we have

$$M_1 | M_2 = \{(\{s_1, s_2, s_3, s_4\} \Rightarrow a_1), (\{s_6\} \Rightarrow a_3), (\{s_5\} \Rightarrow a_2)\}$$

The *a-restriction* operator allows a newly added system to possibly inhibit an existing system since it adds new conditions for triggering an action, and is similar to output inhibition from Brook's subsumption architecture [13], as we illustrate below.

The operator *a-overriding* allows the rules of one system to override those of another, but only with regards to actions found in both systems, i.e. suppose M_1 has precedence over M_2, then

$$\begin{aligned} M_1 \triangleleft M_2 = \; & \{(\mathbf{S_1} \Rightarrow a) \mid (\mathbf{S_1} \Rightarrow a) \in M_1 \wedge (\mathbf{S_2} \Rightarrow a) \in M_2\} \\ & \cup \{(\mathbf{S_1} \Rightarrow a) \mid (\mathbf{S_1} \Rightarrow a) \in M_1 \wedge ((\mathbf{S_2} \Rightarrow a) \notin M_2, for\ any\ \mathbf{S_2})\} \\ & \cup \{(\mathbf{S_2} \Rightarrow a) \mid (\mathbf{S_2} \Rightarrow a) \in M_2 \wedge ((\mathbf{S_1} \Rightarrow a) \notin M_1, for\ any\ \mathbf{S_1})\} \end{aligned}$$

Such an operator allows a newly added action system to override the behaviour of a previous system.

A *composite action system* is formed as a composition of action systems using the above operators (e.g., $(M_1 \cup M_2) \cap M_3$).

Context-Aware Action Systems

A *context-aware action system* is then a pair (E, M), comprising a (possibly composite) context-aware system E and an action system M (which also may be composite). But to be meaningful, the situations which trigger actions in M should be infer-able by E, for otherwise, certain actions in M will never be triggered. Hence, for a **valid** context-aware action system, at minimum, we also require that for any $(\mathbf{S} \Rightarrow a) \in M$, for any situation $S \in \mathbf{S}$, we have $E \vdash (W, S)$, for some conceivable W, i.e., as long as there is the right sensor readings (with the judgement of the system designer). We call this the *validation property*.

Expressions formed using the SIGMA operators are called SIGMA expressions, and corresponds to some composite system.

[10] We use relation rather than function for greater generality, allowing non-determinism but do not discuss specifically how to deal with non-determinism here.

Scenarios of Using SIGMA: Thing-Ensembles and SIGMA expressions

In this section, we illustrate the use of our formalism to model a range of systems from different domains. In each example, we first identify component context-ware systems and action systems, and then describe how they can be composed to form the desired system.

An Illustrative Example: Triggering the Television

We consider an example now of using our operators to incrementally build a system. The system we intend to build is an automatic turn TV on system, where a user *Seng* sits down on an armchair (in front of a television) and the television is automatically turned on.

We start with a basic version of the system which has sensors in the armchair to detect Seng's weight and then infer the situation that Seng is on the chair, afterwhich this is mapped to the action of turning on the television.

We can describe this system, calling it $P = (R_p, M)$, where R_p is the triple $(\{weight\_sensor\}, \Pi_p, \Theta_p)$, where

$$('reading(weight\_sensor) = 65' \Rightarrow seng\_on\_chair) \in \Pi_p$$

that is, if the reading on the chair's weight sensor is 65, we interpret this to mean Seng is sitting on the chair, and

$$\Theta_p = \{(\{seng\_on\_chair\} \Rightarrow seng\_ready\_for\_tv)\}$$

which states that Seng sitting on that chair implies he is ready to watch television. And M only has one rule which maps Seng ready for television to the action of turning on the television, i.e.

$$M = \{(\{seng\_ready\_for\_tv\} \Rightarrow turn\_on\_tv)\}$$

Figure 3 depicts this system.

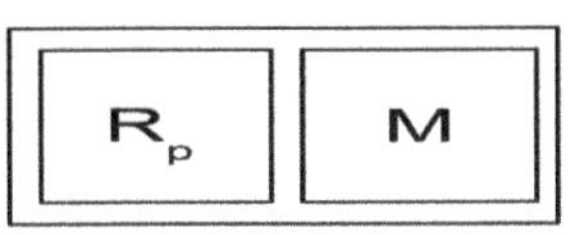

Figure 3. A context-aware action system for the basic version.

Now, there are various shortcomings to such a basic system, namely, when someone else sits on the chair with the same weight as Seng. To further confirm Seng's presence, we can attach an RFID system to the chair that can identify Seng (assume to have an ID of "0101", for example sake) as being nearby. We define this system as $R_q = (\{RFID\_sensor\}, \Pi_q, \Theta_q\})$, where

$$('reading(RFID\_sensor) = 0101' \Rightarrow seng\_detected) \in \Pi_q$$

and

$$\Theta_q = \{(\{seng\_detected\} \Rightarrow seng\_ready\_for\_tv)\}$$

which states that Seng detected by this RFID reader implies he is ready to watch television.

Then, we can form the new composite system $R_p \ominus R_q$, using rule (*gres*), so that:

$$R_p \ominus R_q = (\{weight\_sensor, RFID\_sensor\}, \Pi_p \cup \Pi_q, \Theta_p \mid \Theta_q)$$

where

$$\Theta_p | \Theta_q = \{(\{seng\_on\_chair, seng\_detected\} \Rightarrow seng\_ready\_for\_tv)\}$$

Figure 4 depicts the extended context-aware action system $(R_p \ominus R_q, M)$. If we instead wanted to model suppression, i.e., to have the rules of R_q precede over R_p, we can use $R_q \oslash R_p$ which uses only the rule in Θ_q and ignores that in Θ_p.

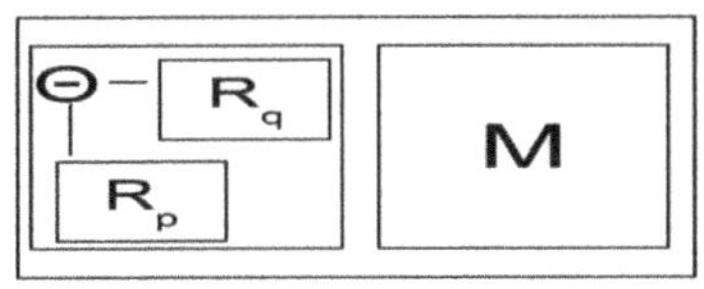

Figure 4. A context-aware action system for the extended version.

Now suppose, a friend of mine invented a head and eye (i.e., gaze) tracking system which can detect if a person is actually looking at the television and determines how long the person does so continuously, and the idea is that a person expresses the intention to watch television by staring at the television continuously for at least 4 seconds.

We provide an abstract description of the system as the following context-aware system $R_r = (\{gaze\_sensor, tv\_sensor\}, \Pi_r, \Theta_r)$, where

$$\{('reading(gaze\_sensor) = 4' \Rightarrow person\_gazing\_tv), ('reading(tv\_sensor) = 0' \Rightarrow tv\_off)\} \in \Pi_r$$

and

$$\Theta_r = \{(\{person\_gazing\_tv, tv\_off\} \Rightarrow person\_want\_tv\_on)\}$$

We also have the action system on the television that uses this trigger M' given by:

$$M' = \{(\{person\_want\_tv\_on\} \Rightarrow turn\_on\_tv)\}$$

Now, considering that it would be useful to combine M' with M in order to better regulate how the television is switched on, we use the operator *a-restriction*, forming

$$M \mid M' = \{(\{seng\_ready\_for\_tv, person\_want\_tv\_on\} \Rightarrow turn\_on\_tv)\}$$

Figure 5 depicts the extended (valid) context-aware action system

$$((R_p \ominus R_q) \oplus R_r, M \mid M')$$

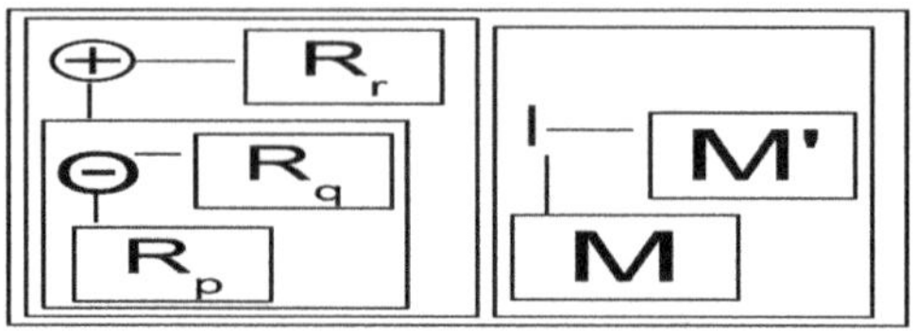

Figure 5. A context-aware action system for the extended (valid) version.

For testing purposes, perhaps we may want to disable M. We can then use *a-overriding* in the composition $M' \triangleleft M$, which only takes the rule in M' into account. Or if we wanted to "or" the triggers for the turning on the television, we would use *a-union*

$$((R_p \ominus R_q) \oplus R_r, M \cup M')$$

A key idea of our operator-based approach is not only that different semantics of composite systems can be easily stated using different expressions, thereby clarifying their differences, but also, that our approach enables incremental extensions to an existing system while preserving the existing systems. We can define a composite system without altering the definitions of its constituent systems. The advantage is, of course, that different compositions can be formed from the same constituent systems.

We an use the *overriding union* on context-aware systems to have a new system suppress the rules of another, or use *a-overriding* or *a-restriction* to effectively inhibit (depending on sensor readings) the action triggers of a system.

Our approach also separates the context-aware system from the action system, in order to provide greater modularity, requiring only that the validation property be satisfied when combining a context-aware system with an action system.

Modelling Sensor Networks

We can use our formalism to model sensor networks, with sensors only having context interpretation, i.e. of the form $(\Sigma, \Pi, \emptyset)$.

Consider a sensor network comprising a collection of nodes (each with a camera and temperature sensor) distributed throughout the fringe of a neighbourhood, the purpose of which is to detect approaching bush fires. We also assume each node has a position.

A node N_i, at position p_i, denoted by $N_i.p_i$ is represented by a context-aware system of the form

$$(\{camera, temp\_sensor\}, \Pi_i, \emptyset)\})$$

where

$$\{('reading(camera) = flame' \Rightarrow fire\_seen), ('reading(temp\_sensor) \geq 70' \Rightarrow hot)\} \subseteq \Pi_i$$

Each node would have its own *camera* and *temp_sensor* though notationally, we don't distinguish among them (if need be, we could use $camera_i$ and $temp\_sensor_i$). Assuming a sensor network $\mathbf{N}$ of n nodes distributed at different locations, we can represent this by the composition $\oplus_{i=1}^{n} N_i(p_i)$ (or we write this as $\oplus\mathbf{N}$, where $\mathbf{N} = \{N_i \mid i \in \{1, \ldots, n\}\}$). And we could pose queries such as $\mathbf{N} \vdash fire\_seen$? which will return true if fire is seen at any one location. A query such as $(N_2 \oplus N_3 \oplus N_6) \vdash fire\_seen$? asks if a fire is seen at nodes 2, 3 or 6. Effectively, we can use our formalism to pose queries to selected sensors or to different parts of the sensor network or modify the semantics slightly to return the node where fire is seen.

However, over the same physical sensor network, one could define different compositions, relating to different possible "behaviours" being sought. For example, what would the composition $\otimes\mathbf{N}$ mean? Physically, the sensor network is not changed, but $\otimes\mathbf{N} \vdash fire\_seen$ is true when fire is seen in all locations.

Extending nodes and the sensor network.

Now, let us consider adding m sensor nodes to the sensor network for nodes N_{n+1} to N_{n+m}. The above compositions then extend to the $n+m$ nodes easily. Let $\mathbf{M} = \{N_i \mid i \in \{m+1, \ldots, n+m\}\}$. Then, determining if

$$((\otimes\mathbf{N}) \otimes (\oplus\mathbf{M})) \vdash fire\_seen$$

is true asks if fire is seen in all of the n nodes and in any of the new m nodes.

We can add a fire inference engine to each node F_i, denoted by $(\emptyset, \emptyset, \Theta_i)$, where Θ_i has only one rule that aggregates the context to infer the situation that a fire occurs:

$$\Theta_i = \{((\{fire\_seen, hot\} \Rightarrow fire)\}$$

Such a component F_i might correspond to a physical component (later) attached to each of the sensors.

Then, let $NF_i = N_i + F_i$ represent an extended node, $NF_i \vdash fire$ is true when for node i, $fire$ is inferred. A composition such as $\oplus\mathbf{NF} \vdash fire$ denotes if this is true for any node NF_i in the set $\mathbf{NF}$.

Now, we want to add smoke sensor nodes with an inference component, denoted by N_i', where

$$N_i' = (\{smoke\_detector\}, \Pi_i', \Theta_i')\})$$

where

$$\{('reading(smoke\_detector) = yes' \Rightarrow smoke\_present), ('reading(smoke\_detector) = no' \Rightarrow smoke\_absent)\} \subseteq \Pi_i'$$

and

$$\Theta_i' = \{(\{smoke\_present\} \Rightarrow fire)\}$$

And a network of smoke sensor nodes comprising a set of k such nodes, denoted by **S**, can be distributed in the same areas as the earlier $n + m$ nodes.

A query on the composition such as $(\oplus\mathbf{S}) \otimes (\oplus\mathbf{NF})) \vdash fire$ is true if both the smoke sensor network and the previous sensor network confirms there is a fire. But the composition does not take into account the locations of nodes explicitly.

Composing a NF_i node with a smoke detector node N_j' can be represented, such as $N_{ij} = NF_i \ominus N_j'$ which denotes a node $N_{ij} = (\Sigma_{ij}, \Pi_{ij}, \Theta_{ij})$, where:

$$\Sigma_{ij} = \{camera, temp\_sensor, smoke\_detector\}$$

and

$$\{('reading(smoke\_detector) = yes' \Rightarrow smoke\_present), ('reading(smoke\_detector) = no' \Rightarrow smoke\_absent), ('reading(camera) = flame' \Rightarrow fire\_seen), ('reading(temp\_sensor) \geq 70' \Rightarrow hot)\} \subseteq \Pi_{ij}$$

and

$$\Theta_{ij} = \{(\{fire\_seen, hot, smoke\_present\} \Rightarrow fire)\}$$

Note that N_{ij} might comprise two physically separated nodes NF_i and $N'j$ but abstractly, it can be represented as (conceptually) one node.

Context Attributes for Context-Aware Systems. If we associate a set of context attributes (e.g., location, power available, etc) to each node, denoted using the notation $< node > . < property >$, and form predicates over these properties, we an define sets of nodes.

For example, the following is the set **P** of extended nodes within distance r of *here* with high power levels:

$$\mathbf{P} = \{NF_i \mid NF_i.power = high \wedge dist(NF_i.location, here) \leq r \wedge i \in \{1, \ldots, n + m\}\}$$

We can then define compositions on these notes such as $\oplus\mathbf{P}$. We can also define a set of high powered smoke sensor nodes in the same area:

$$\mathbf{P}' = \{N_i' \mid N_i'.power = high \wedge dist(N_i'.location, here) \leq r \wedge i \in \{1, \ldots, k\}\}$$

A query on the composition such as $(\oplus\mathbf{P}') \otimes (\oplus\mathbf{P})) \vdash fire$ is true if both P and P' confirms there is a fire in the same area (but only as defined by the distance r from *here*). Finer grained areas can, of course, be defined such as a band given by $r' \leq dist(< node > .location, here) \leq r$, for some distance $r' \leq r$.

Summary. What we have shown here are expressions to represent extensions of a sensor network in different ways as well as to represent abstract sensor nodes (which might be physically composed from two or more nodes). They serve as a notation to formally and precisely represent different sensor networks (or different parts of a sensor network) but yet can provide a basis for operational meaning of queries.

Tiling Cooperative Digital Picture Frames

We consider modelling a collection of digital picture frames on a wall; we assume that each digital picture frame has a unique identifier and can sense if another digital picture frame is near it, in four directions: above it, to the left of it, to the right of it and below it. Then, according to what each picture frame thinks of its relative position, each picture frame then show a particular picture. We assume that it is possible to change the position of the frames - that they are fastened to the wall by the owner and they can be repositioned.

Each digital picture frame F_i (assuming $1 \leq i \leq 9$) is modelled as a context-aware action system with a frame sensor (for sensing another frame in four directions which returns a subset of $\{above, below, left, right\}$ depending on what is sensed) of the form $F_i = (R_i, M_i)$, where $R_i = (\Sigma_i, \Pi_i, \Theta_i)$ with

$$\Sigma_i = \{frame\_sensor_i\}$$

and

$$\{('above \in reading(frame\_sensor_i)' \Rightarrow frame\_above_i), ('below \in reading(frame\_sensor_i)' \Rightarrow frame\_below_i), ('left \in reading(frame\_sensor_i)' \Rightarrow frame\_left_i), ('right \in reading(frame\_sensor_i)' \Rightarrow frame\_right_i), ('above \notin reading(frame\_sensor_i)' \Rightarrow \neg frame\_above_i), ('below \notin reading(frame\_sensor_i)' \Rightarrow \neg frame\_below_i), ('left \notin reading(frame\_sensor_i)' \Rightarrow \neg frame\_left_i), ('right \notin reading(frame\_sensor_i)' \Rightarrow \neg frame\_right_i)\} \in \Pi_i$$

and

$$\Theta_i = \{ (\{frame\_left_i, frame\_right_i\} \Rightarrow hor\_linear_i), (\{\neg frame\_below_i\} \Rightarrow bottom\_end_i), (\{\neg frame\_above_i\} \Rightarrow top\_end_i), (\{\neg frame\_right_i\} \Rightarrow right\_end_i), (\{\neg frame\_left_i\} \Rightarrow left\_end_i), (\{frame\_above_i, frame\_below_i\} \Rightarrow ver\_linear_i) \}$$

and M_i is of the form:

$$\{(\{hor\_linear_i\} \Rightarrow show\_hline_i), (\{ver\_linear_i\} \Rightarrow show\_vline_i), (\{bottom\_end_i\} \Rightarrow show\_bottom\_cap_i), (\{top\_end_i\} \Rightarrow show\_top\_cap_i), (\{right\_end_i\} \Rightarrow show\_right\_cap_i), (\{left\_end_i\} \Rightarrow show\_left\_cap_i)\}$$

Note that a frame might show multiple caps at the same time depending on what frames it (not) detects, e.g., a frame that is right ended and left ended at the same time might show both a right cap and a left cap at the same time. Each frame works independently and what emerges is the overall pattern. The frames do not need to communicate with each other, but merely to sense each other's proximity.

Figure 6 gives the idea of how the rules would result in the different patterns being displayed.

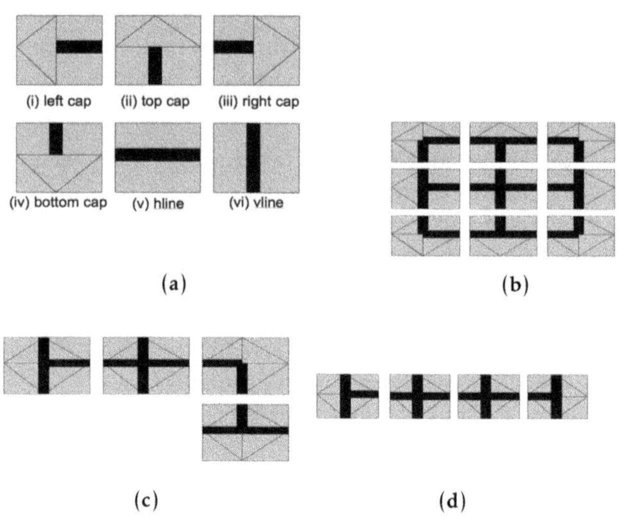

Figure 6. Patterns from rules. (a) shows each of the six display objects.
(b) shows a 3x3 configuration of nine frames and the pattern displayed according to the above rules.
(c) shows the pattern for an L-shaped configuration of four frames according to the rules above.
(d) shows the pattern for a horizontal configuration of four frames.

Next we add a particular frame that can gather information from all other frames (assuming four frames) and so detect the overall arrangement of all the frames (including itself), including horizontally linear, and vertically linear. The frame is denoted by $F' = (R', M')$, where $R' = (\emptyset, \emptyset, \Theta')$, where Θ' contains 12 rules as follows

$$(\{hor\_linear_i, right\_end_j, left\_end_k, hor\_linear_l\} \Rightarrow config\_horizontal) \in \Theta'$$
for $i, j, k, l \in \{1, 2, 3, 4\}, i, j, k, l$ all different

since any of the four frames can be in the situation hor_linear, $right\_end$ or $left\_end$ at a time, and at any time, when the frames are lined up horizontal, two of them will be hor_linear and one $right\_end$ and another $left\_end$, and for vertical configuration, another 12 rules as follows

$$(\{ver\_linear_i, bottom\_end_j, top\_end_k, ver\_linear_l\} \Rightarrow config\_vertical) \in \Theta'$$
for $i, j, k, l \in \{1, 2, 3, 4\}, i, j, k, l$ all different

And M' then voices out which configuration the system detects as follows:

$$\{(\{config\_vertical\} \Rightarrow say\_vertical), (\{config\_horizontal\} \Rightarrow say\_horizontal)\}$$

The combined valid context-aware action system comprising components from F_1, F_2, F_3, F_4 and F' is of the form

$$(R_1 + R_2 + R_3 + R_4 + R', M_1 \cup M_2 \cup M3 \cup M_4 \cup M')$$

Each M_i does its action as before but M' also utilizes "inputs" from the R_is and R' to perform actions. Here, the idea is that F' can be added to the F_is without the F_is needing to change.

For aesthetic reasons, one might want to add new digital frames to the system, that will alter (and subtract from) the behaviour of an existing composite system. For example, suppose we add new "inhibitor" frames that will prevent certain shapes from ever being displayed. Consider a frame $F^i_{inh} = ((\emptyset, \emptyset, \emptyset), M^i_{inh-hline})$ with no context-aware system but has an action system $M^i_{inh-hline}$ which is an inhibitor for frame F_i, preventing the frame from ever showing a horizontal line, of the form:

$$\{(\{\mathtt{false}\} \Rightarrow show\_hline_i)\}$$

So, the system $(R_i, M_i \mid M^i_{inh-hline})$ has a rule of the form

$$\{(\{hor\_linear_i, \mathtt{false}\} \Rightarrow show\_hline_i)\}$$

which will never be satisfied, and so, the $show\_hline_i$ action will never be performed regardless of the what sensor readings are received.

Incidentally, new tailored operators on frames can be defined based on the SIGMA operators. For example, we can define an operator $\odot$ on frames with the semantics:

$$F_1 \odot F_2 = (R_1 \oplus R_2, M_1 \mid M_2)$$

where $F_1 = (R_1, M_1)$ and $F_2 = (R_2, M_2)$.

Then, we have:

$$F_i \odot F^i_{inh} = (R_i \oplus (\emptyset, \emptyset, \emptyset), M_i \mid M^i_{inh-hline})$$

forming the (virtual) frame (or pair of frames) which has restricted display compared to F_i alone.

Smarter Cars and Urban Awareness

There has been interesting developments in using sensors to monitor city and urban environments. The MIT SENSEable Real Time Rome project[11] used cellphones and GPS devices to discover movements patterns of people and transportation systems in Rome, enabling usage of streets and neighbourhoods to be tracked and visualized. The more recent SENSEable CURRENTcity[12] project is working on visualizing the real-time dynamic behaviours of cities including traffic jams and gatherings. The Cityware initiative[13] has a project to monitor the paths of tourists using GPS tracking.

There is also other work towards monitoring, in real-time, traffic behaviour to provide, in advance, information about traffic jams on roads. The notion of crowd sensing and understanding events in the city provides insight into life in the city for its inhabitants and tourists. There is also work towards supporting place-based awareness leading towards what we call "legible places", places that can be "read" and "queried" like a book.

Modern cars will be equipped increasingly with sensors from rain sensors, pedestrian sensing, adaptive cruise control to parking sensors. Pedestrians can carry sensors which cars can pick up to ensure that cars don't hit them. Cars can sense and detect approaching emergency vehicles and react accordingly. Accurate GPS (up to the metre) can help cars in manouvres such as lane change and overtaking. Cars could become aware of nearby parking slots and busy intersections. There is a need for context-aware action systems to work together, the car (as a whole if viewed as a context-aware action system, with its subsystems), and the sensors in the environment of the car.

We illustrate a scenario of car using sensors in the urban environment to become better aware of where vehicular and people crowds are. We imagine a future urban environment which has sensors to detect crowds of people and heavy traffic streets. And the car is able to detect these sensors, connect to them and acquire information about crowds (of people or vehicles) within a preset radius from the current position of the car. Such information might then be displayed on the car's computer on a map indicating nearby crowds. Let this capability of a context-aware car be represented as an action system with a rule that gets the crowd at a position pos and displays that, represented by $M = \{(\{crowd\_at(pos)\}, display\_crowd(pos))\}$. A crowd sensor embedded in the urban environment is represented by context-aware systems of the form

$$\begin{aligned} R =(&\{crowd\_sensor\},\\ &\{('reading(crowd\_sensor) = pos' \Rightarrow crowd\_at(pos))\},\\ &\emptyset) \end{aligned}$$

And the (R, M) represents the overall (distributed) system, comprising a component on the car and a component in the stationary urban environment. When the car detects another crowd sensing system in the infrastructure in the vicinity, say R' in the same form as R, it can employ this system, whether to confirm the crowd position, i.e. we have $(R \otimes R', M)$, or to provide new crowd positions $(R \oplus R', M)$. The car itself might have sensors, represented by the context-aware system R'' which can be combined with infrastructure sensors: $(R \oplus R' \oplus R'', M)$.

Meta-Programming with SIGMA Expressions. It is possible to embed SIGMA expressions into a programming language, in order to form new compositions at run-time, especially when the set of available context-aware systems might change over time, e.g., upon discovery of new systems.

With the smart car example earlier, consider a simple event-driven program that can discover the presence of context-aware systems in the vicinity, and recomputes a composition based on discovered context-aware systems. Let R'' be the car's own context-aware system, and M, the car's action system. We also introduce an operator that takes a context-aware action system E, and executes it (i.e., gets the sensor readings, performs reasoning, and determines if any action should be taken, and executes the action), denoted by **execute**$[E]$. Note that there are two ways that SIGMA expressions can be executed, in the forward-chaining (data-driven) style as in **execute**$[E]$, or using a backward-chaining query style where, given a situation, determine if the situation is occurring of the form $E \vdash S?$.

loop
  initiate discovery of context-aware systems
  $\mathbf{R_d}$:= discovered context-aware systems
  execute$[((\oplus \mathbf{R_d}) \oplus R'', M)]$

[11] http://senseable.mit.edu/realtimerome/
[12] http://www.currentcity.org/
[13] http://www.cityware.org.uk; see Digital Footprints project.

end

Note that the above algorithm works with the just-discovered context-aware systems in the car's vicinity; as the car moves, a different set $\mathbf{R}_d$ of such systems will be discovered.

Modular Robotics

Modular robotics involves building a larger robot from the cooperation of a set of individual robots. The Modular Robotics company[14] have building blocks which are classified into *sense blocks*, *think/utility blocks* and *action blocks*, where sense blocks can sense light, temperature, and distance from other objects, think/utility blocks can map sensor inputs to actions according to specific relationships (e.g., inverse, maximum, minimum, etc), and action blocks performs actions based on sensor inputs. For example, with an *Inverse* thin/utility block between the Light Sensor and the Drive sensor blocks, the robot drives slower when the light gets brighter. Infrared ports are used by robots to connect to neighbouring robots and sense distance from to nearby objects.

Inspired by this approach, we represent a simple system of modular robots, comprising context-aware systems and action systems. Let sense blocks be denoted by context-aware systems of the form $R_s = (\Sigma, \emptyset, \emptyset)$. Let think blocks be denoted by context-aware systems of the form $R_t = (\emptyset, \Pi, \Theta)$. Let action blocks be denoted by action systems M.

Given the above blocks, we can form a configuration of robots, represented by the composition $(R_s + R_t, M)$. x sensor blocks, y think blocks and z action blocks can be represented by compositions of the form

$$((R_s^1 + \ldots + R_s^x) + (R_t^1 + \ldots + R_t^y), M_1 \cup \ldots \cup M_z)$$

But if each block has up to six interfaces, with $z = 1$ and other blocks attached to it, we have $x + y \leq 6$.

The work in [44] mentions the notion of roles which map to sets of behaviours. A robot/block decides what role to take up by sensing its context. Depending on its role assumed, the robot then behaves in a certain predefined way. Such roles can be encoded as a set of rules in an action system.

Related Work

The Internet of Things has caught the attention of the world, and involves the development of suitable hardware, software and paradigms where everyday objects can be linked to the Internet, can be sensed so that data about them can be obtained, and where the things can be composed in useful ways [41, 47], e.g. sensors used in building smart cities [14, 58]. However, recent IoT visions aim to go further, including linking smart things with social networks and embedding their behaviours within the context of social interactions [40], and adding a layer of data processing so that intelligent perception and behaviours can be harnessed from the Internet of Things [55]. This paper focuses on a different, yet complementary, vision for the IoT, where IoT systems or thing-ensembles can be built in an incremental and compositional manner, e.g.,

- sensing capability is increased compared to the previous system with the addition of new sensors, in that what can be sensed previously can still be sensed, and also new aspects can be sensed due to the new sensors;
- action capability is increased with the addition of new actuators so that, what can be done previously is either enhanced or that more actions can be taken, or better actions can now be taken, apart from preservation of some previous actions; and
- "intelligence" or reasoning capability can be increased so that what was previously possible to reason about is still retained but new rules added allow further inferences to be made (even from the same sensor system).

Hence, while IoT work is expanding rapidly, our work focuses on the above aspects for future IoT systems. We focus on reasoning with the components of a system rather than with the data from IoT systems as in other work such as [37, 51].

There are numerous middleware, frameworks and toolkits for building context-aware systems in ubiquitous computing environments, as surveyed in [6, 19, 27, 33]. However, our formalism aims at a programming language based approach, combined with an operator-based formalism; the operators theoretically define how the constituents of two context-aware systems (or action systems) should work together, but does not prescribe how such integration can be carried out; the operators could be implemented in different ways: Web services could be used if the systems are distributed or proprietary protocols might be employed in the case of digital appliances in the home. We aim towards a high level specification language. In [39] is described a framework where integration of heterogeneous legacy systems and reusing components are helpful for incrementally constructing global smart spaces.

Among work on software engineering paradigms and models for building context-aware systems, there has been investigation of suitable abstractions. For example, the sentient object model [8] encapsulates the key features of a context-aware system comprising

[14]See also http://www.modrobotics.com

sensor capture and context-reasoning components. A context-aware system in our model can be mapped to a sentient object in their model, and they do not provide composition operators as we do here. A survey of programming languages supporting the notion of context-oriented programming, as extensions to object-oriented programming, is given in [3]; such languages provide mechanisms to represent context as an explicit concept in the language, and to model context-aware execution of programs (or program fragments) within a software system, especially when they cut across objects. In several of these context-oriented programming languages [15], context is represented in the programs as layers, and during program execution, such layers can be activated or deactivated. Context propagators [5] has been proposed to capture the idea of adaptations having dependencies on each other, and such dependencies are themselves dependent on context. The notions of isotope and context block that represents conditions for execution of isotope elements are proposed in [42]. In [29], a context-aware application is programmed using an abstraction called activity - policies are used to specify context-aware behaviour and resources. The work in [20] uses the notion of context graphs to define the context-aware behaviour of applications, including the transitions among states of the application, and what behaviours are triggered in different situations. Our approach does not represent states of the participating systems but only how they should work together. Model-driven approaches have been investigated for with CAMEL (Context-Awareness Modelling Language) [46], where context-sensing and adaptation to context can be easily represented. In summary, different abstractions have been proposed to represent context-aware behaviours, many embedded within an existing programming language and often also used in designing the applications themselves. However, they have not investigated a compositional operator approach as we do here.

Due to the popularity of context-awareness in pervasive/ubiquitous computing, there has arisen formal systems to model context-aware behaviour. Notable are those based on process calculi approaches, such as the Context-Aware Calculus [59] which has similarities with the ambient calculus (which models processes with spatial notions), the calculus of context-aware ambients which builds on the ambient calculus [45], CONAWA which also builds on ambient calculus ideas with a richer modelling of context with several ambient-like trees [28], a calculi about contextual reactive systems [10, 11], and a bigraphical model of context-aware systems [9]. There has also been work on using the ambient calculus as a basis for a programming paradigm [53]. However, our approach is not based on process calculi and we adopt a software engineering style approach of combining components structurally, rather than combining process descriptions of run-time behaviours as in the context-aware calculi approaches. Then, in our approach, the behaviour of composed components is then determined by following the operational semantics rules and depends on the actual content within the systems. Also, our approach is at a different level of abstraction - details of networking and communication are abstracted away within the semantics of each operator; we only define conceptually how the system should cooperate (a mode of cooperation represented by the meaning of an operator), and leave the protocol details to the implementation. Our approach does not replace but could complement such process calculi models of context-aware behaviour; we use our formalism for structural compositions, and then context-aware calculi could then be used to model what happens at run-time within a composition. Given descriptions of constituent systems in a thing-ensemble, investigating translation of a (a Sigma expression) into an initial context-aware calculi expression and then reasoning about run-time behaviour in the calculi could perhaps be an avenue for future work.

Ideas from component based software engineering (e.g., [23]), service composition (e.g., [18]) and modular logic programming (e.g., [12]), share similarities with our algebraic approach, and in fact, we draw inspiration from operators for combining logic programs in our work; however, the entities we compose are not traditional software components, services, programs but triples representing context-aware systems and relations representing action systems.

The actor model [1] formalizes concurrent computations with the key notion of the actor, and multiagent systems research (e.g., [54]) have formalized representations of software entities that are autonomous, proactive and respond to environmentally sensed inputs. However, the resemblance to our notion of context-aware systems is superficial, given their emphasis on proactive behaviours, and their machinery for that purpose, and our operator language is unique in the way that the constituent contents of systems are composed, rather than multiagent cooperation ideas. While one can consider modelling the context-aware action systems in our examples as agents, and then model their cooperation protocols, and allow cooperation to emerge at run-time (and it would be interesting future work to do so), in our approach, the cooperation among context-aware systems is essentially scripted out (using the Sigma operators), and we adopt a global picture view (from the programmer's perspective) rather than a decentralized view, with the operators abstracting away protocol details.

The work by Beal[15] developed Proto, a functional language where rules can be programmed that describe the behaviour of a collection of nodes spread out across space. Our work is different in that it focuses on a small set of operators for combining and reasoning with thing-ensembles (which may or may not be distributed spatially).

Conclusion and Future Work

We have taken a first step towards providing a language that can serve two purposes, given a set of context-aware things that can communicate with each other, i.e. a thing-ensemble, we can: (i) specify compositions of context-aware things at design-time, and (ii) script ways in which context-aware things should work together at run-time. We built on the design philosophy and operator language first introduced in [36], but here, giving a thorough treatment with new operators and a range of examples to illustrate the versatility of our approach.

Based on the theoretical basis in this paper, future work includes providing design and programming tools based on SIGMA, in order to demonstrate the feasibility of our approach. While demonstrating the applicability of SIGMA in a wide range of applications, we have not yet dealt with examples of greater complexity and of a much larger scale. For example, an entire smart house functionality could be modelled with our approach.

We have not discussed implementation details in depth in this paper. There are two paradigms for implementation when SIGMA expressions are used for scripting behaviours of things:

- *orchestration*: we can use a central coordinator which will interpret SIGMA expressions and when contents of a particular system/thing needs to be used, the coordinator can contact the system (e.g., via Web services if the system provide a Web service interface); essentially, most of the communication is between coordinator and systems, and the coordinator consults the appropriate systems as determined by the operators in the expression.

  As an example, we sketch a strategy to implement a system comprising situation/context-aware recognition and actions via an example given earlier in Figure 5, corresponding to the expression:

  $$((R_p \ominus R_q) \oplus R_r, M \mid M')$$

  To implement a system described by the above expression, the appropriate sensors as defined in R_p, R_q and R_r, are needed, namely, the weight sensor on the chair, the RFID sensor to detect *Seng* and the gaze sensor worn by *Seng*. Computation then proceeds in a (sensor) data-driven fashion by converting the operational rules given earlier into logic programming rules [34]. The readings from the sensors are fed into the rules defining the situation-recognition behaviours in R_p, R_q and R_r to determine if any of the action-triggering situations in $M \mid M'$ are occurring and if occurring, appropriate actions are taken. Backward chaining as illustrated in [34] can be employed, i.e. to determine if any of the situations triggering an action as defined in $M \mid M'$ is occurring, we first do backward chain reasoning on the rules, which then eventually results in interrogating the sensors. The sensors are then queried for their readings and if a situation-triggering action is occurring, the appropriate actions as defined by $M \mid M'$ are taken, and if not, no action is taken. When the sensor readings change beyond a preset threshold, a similar chain of reasoning is triggered to see if any action is now to be taken. Future work will consider efficient implementations of this idea.

- *choreography*: we can use a (mostly) peer-to-peer interaction model where one of the peers/things/systems (or a coordinator, albeit with a diminished role compared to orchestration) receives the SIGMA expression from the user, and then multicasts the expression to all the things/systems in the thing-ensemble; each thing involved in the composition (mentioned in the expression) then communicates with other things, appropriate to the given expressions.

For example, given an expression involving three things, modelled as context-aware systems R_1, R_2 and R_3, and a query $(R_1 \oplus R_2) \otimes R_3 \vdash S$ to determine if a situation S is occurring, then in the orchestration approach, the coordinator interprets this expression, and consults the component systems whenever their sensors or rules needs to be used (according to the operational semantics given earlier), e.g., what we represent as finding a rule whose head matches a situation such as $(\{S_1, \ldots, S_k, C_1, \ldots, C_m\} \Rightarrow S) \in \Theta_i$ results in a service call to system R_i. In the choreography approach, each R_i receives (say, from a coordinator) a copy of the expression and the query about S, and works out which other system(s) to cooperate with and how it should interact with the system(s), e.g., R_1, on receiving the expression sees that it is in $\oplus$ with R_2 and so works with R_2 to get a result, and in a mutual decision determines that it should forward its result to the next level, which is to R_3. R_3 works out its own result, and waits for a result from the composition (or one of) R_1 and R_2, and on receiving

[15] http://proto.bbn.com/commons/?q=user

the result, does an ⊗ with the result and forwards the combined result back to the coordinator. Future work will investigate both paradigms.

While we have detailed a range of scenarios, practical deployments of our approach is still needed, as well as, extensive case studies. Future work will consider case studies in smart city applications [58].[16]

References

[1] Gul Agha, Ian A. Mason, Scott F. Smith, and Carolyn L. Talcott. A foundation for actor computation. *Journal of Functional Programming*, 7:1–72, 1998.

[2] C. B. Anagnostopoulos, Y. Ntarladimas, and S. Hadjiefthymiades. Situational Computing: An Innovative Architecture with Imprecise Reasoning. *Journal of Systems and Software*, 80(12):1993–2014, 2007.

[3] Malte Appeltauer, Robert Hirschfeld, Michael Haupt, Jens Lincke, and Michael Perscheid. A comparison of context-oriented programming languages. In *COP '09: International Workshop on Context-Oriented Programming*, pages 1–6, New York, NY, USA, 2009. ACM.

[4] Luigi Atzori, Antonio Iera, and Giacomo Morabito. The Internet of Things: A survey. *Computer Networks*, 54(15):2787 – 2805, 2010.

[5] Engineer Bainomugisha, Wolfgang De Meuter, and Theo D'Hondt. Towards context-aware propagators: language constructs for context-aware adaptation dependencies. In *COP '09: International Workshop on Context-Oriented Programming*, pages 1–4, New York, NY, USA, 2009. ACM.

[6] Matthias Baldauf, Schahram Dustdar, and Florian Rosenberg. A survey on context-aware systems. *Int. J. Ad Hoc Ubiquitous Comput.*, 2(4):263–277, 2007.

[7] N. Baumgartner and W. Retschitzegger. A Survey of Upper Ontologies for Situation Awareness. In *Proceedings of Knowledge Sharing and Collaborative Engineering*. ACTA Press, 2006.

[8] Gregory Biegel and Vinny Cahill. A framework for developing mobile, context-aware applications. In *PerCom*, pages 361–365, 2004.

[9] L. Birkedal, S. Debois, E. Elsborg, T. Hildebrandt, and H. Niss. Bigraphical models of context-aware systems. In *Foundations of Software Science and Computation Structures (FOSSACS), LNCS 3921*. Springer-Verlag, 2006. http://www.itu.dk/people/birkedal/papers/bigmcs-conf.pdf.

[10] Pietro Braione. *On Calculi for Context-Aware Systems*. PhD thesis, Dipartimento di Elettronica e Informazione, Politecnico di Milano, P.zza Leonardo da Vinci 32, 20133 Milano, Italy, April 2004.

[11] Pietro Braione and Gian Pietro Picco. On calculi for context-aware coordination. In Rocco De Nicola, Gianluigi Ferrari, and Greg Meredith, editors, *Proceedings of the 7th International Conference on Coordination Models and Languages (COORDINATION 2004)*, volume 2949 of *Lecture Notes in Computer Science*, pages 38–54. Springer, 2004.

[12] Antonio Brogi, Paolo Mancarella, Dino Pedreschi, and Franco Turini. Modular logic programming. *ACM Trans. Program. Lang. Syst.*, 16(4):1361–1398, 1994.

[13] Rodney A. Brooks. A robust layered control system for a mobile robot. In *Artificial intelligence at MIT: expanding frontiers*, pages 2–27, Cambridge, MA, USA, 1990. MIT Press.

[14] Shanzhi Chen, Hui Xu, Dake Liu, Bo Hu, and Hucheng Wang. A vision of iot: Applications, challenges, and opportunities with china perspective. *Internet of Things Journal, IEEE*, 1(4):349–359, Aug 2014.

[15] Dave Clarke and Ilya Sergey. A semantics for context-oriented programming with layers. In *COP '09: International Workshop on Context-Oriented Programming*, pages 1–6, New York, NY, USA, 2009. ACM.

[16] P.D. Costa, G. Guizzardi, J.P.A. Almeida, L.F. Pires, and M. van Sinderen. Situations in Conceptual Modeling of Context. In *EDOCW '06: Proceedings of the 10th IEEE on International Enterprise Distributed Object Computing Conference Workshops*, Washington, DC, USA, 2006. IEEE Computer Society. Available at http://www.loa-cnr.it/Guizzardi/DockhornCosta-et-al-VORTE06_final.pdf.

[17] Anind K. Dey. Understanding and using context. *Personal and Ubiquitous Computing*, 5(1):4–7, 2001.

[18] Schahram Dustdar and Wolfgang Schreiner. A survey on web services composition. *Int. J. Web Grid Serv.*, 1(1):1–30, 2005.

[19] Christoph Endres, Andreas Butz, and Asa MacWilliams. A survey of software infrastructures and frameworks for ubiquitous computing. *Mobile Information Systems*, 1(1):41–80, 2005.

[20] Serena Fritsch, Aline Senart, and Siobhan Clarke. Addressing dynamic contextual adaptation with a domain-specific language. In *SEPCASE '07: Proceedings of the 1st International Workshop on Software Engineering for Pervasive Computing Applications, Systems, and Environments*, page 2, Washington, DC, USA, 2007. IEEE Computer Society.

[21] C. Goumopoulos and A. Kameas. Ambient ecologies in smart homes. *Comput. J.*, 52(8):922–937, 2009.

[22] Jayavardhana Gubbi, Rajkumar Buyya, Slaven Marusic, and Marimuthu Palaniswami. Internet of Things (IoT): A vision, architectural elements, and future directions. *Future Generation Computer Systems*, 29(7):1645 – 1660, 2013.

[23] George T. Heineman and William T. Councill, editors. *Component-based software engineering: putting the pieces together*. Addison-Wesley Longman Publishing Co., Inc., Boston, MA, USA, 2001.

[24] Justin Huang and Maya Cakmak. Supporting mental model accuracy in trigger-action programming. In *Proceedings of the 2015 ACM International Joint Conference on Pervasive and Ubiquitous Computing*, UbiComp '15, pages 215–225, New York, NY, USA, 2015. ACM.

[25] Achilles Kameas, Irene Mavrommati, and Panos Markopoulos. Computing in tangible: using artifacts as components of ambient intelligence environments. In *In Ambient Intelligence: The evolution of Technology, Communication and Cognition (G.Riva, F.Vatalaro, F.Davide and M.Alcaniz,eds)*, pages 121–142. IOS Press,

[16] http://smartcities.ieee.org/

2004.

[26] Sung Woo Kim, Sang Hyun Park, JungBong Lee, Young Kyu Jin, Hyun-Mi Park, Amy Chung, SeungEok Choi, and Woo Sik Choi. Sensible appliances: applying context-awareness to appliance design. *Personal Ubiquitous Computing*, 8(3-4):184–191, 2004.

[27] Kristian Ellebaek Kjaer. A survey of context-aware middleware. In *SE'07: Proceedings of the 25th conference on IASTED International Multi-Conference*, pages 148–155, Anaheim, CA, USA, 2007. ACTA Press.

[28] Mikkel Baun Kj¿rgaard and Jonathan Bunde-Pedersen. A formal model for context-awareness. Technical report, BRICS: Basic Research in Computer Science, 2006.

[29] Devdatta Kulkarni. Programming framework for sensor-data driven context-aware applications. *SIGBED Rev.*, 5(1):1–2, 2008.

[30] Mike Kuniavsky. *Smart Things: Ubiquitous Computing User Experience Design*. Morgan Kaufmann, 2010.

[31] Seng W. Loke. Service-oriented device ecology workflows. In *ICSOC*, pages 559–574, 2003.

[32] Seng W. Loke. Representing and reasoning with situations for context-aware pervasive computing: a logic programming perspective. *Knowl. Eng. Rev.*, 19(3):213–233, 2004.

[33] Seng W. Loke. *Context-Aware Pervasive Systems*. Auerbach Publications, Boston, MA, USA, 2006.

[34] Seng W. Loke. *Supporting Real Time Decision-Making: The Role of Context in Decision Support on the Move*, chapter Towards Real-Time Context Awareness for Mobile Users: A Declarative Meta-Programming Approach, pages 89–112. Springer US, Boston, MA, 2011.

[35] Seng W. Loke. Supporting ubiquitous sensor-cloudlets and context-cloudlets: Programming compositions of context-aware systems for mobile users. *Future Generation Computer Systems*, 28(4):619 – 632, 2012.

[36] Seng Wai Loke. Incremental awareness and compositionality: A design philosophy for context-aware pervasive systems. *Pervasive and Mobile Computing*, 6(2):239–253, 2010.

[37] A.I. Maarala, Xiang Su, and J. Riekki. Semantic data provisioning and reasoning for the internet of things. In *Internet of Things (IOT), 2014 International Conference on the*, pages 67–72, Oct 2014.

[38] Nicolai Marquardt, Tom Gross, M. Sheelagh T. Carpendale, and Saul Greenberg. Revealing the invisible: visualizing the location and event flow of distributed physical devices. In Marcelo Coelho, Jamie Zigelbaum, Hiroshi Ishii, Robert J. K. Jacob, Pattie Maes, Thomas Pederson, Orit Shaer, and Ron Wakkary, editors, *Tangible and Embedded Interaction*, pages 41–48. ACM, 2010.

[39] René Meier, Anthony Harrington, Kai Beckmann, and Vinny Cahill. A framework for incremental construction of real global smart space applications. *Pervasive and Mobile Computing*, 5(4):350–368, 2009.

[40] A.M. Ortiz, D. Hussein, Soochang Park, S.N. Han, and N. Crespi. The cluster between internet of things and social networks: Review and research challenges. *Internet of Things Journal, IEEE*, 1(3):206–215, June 2014.

[41] C. Perera, A. Zaslavsky, P. Christen, and D. Georgakopoulos. Context aware computing for the internet of things: A survey. *Communications Surveys Tutorials, IEEE*, 16(1):414–454, First 2014.

[42] Qi Saiyu, Xi Min, and Qi Yong. Isotope programming model for context aware application. *International Journal of Software Engineering and Its Applications*, 1(1), 2007.

[43] Bill Schilit, Norman Adams, and Roy Want. Context-aware computing applications. In *In Proceedings of the Workshop on Mobile Computing Systems and Applications*, pages 85–90. IEEE Computer Society, 1994.

[44] Ulrik Pagh Schultz, Mirko Bordignon, David Johan Christensen, and Kasper Stoy. Spatial Computing with Labels. In *Proceedings of the SASO'08 Spatial Computing Workshop (SCW'08)*, Venice, Italy, October 20 2008.

[45] F. Siewe, A. Cau, and H. Zedan. Cca: A calculus of context-aware ambients. In *Advanced Information Networking and Applications Workshops, 2009. WAINA '09. International Conference on*, pages 972 –977, 26-29 2009.

[46] Andrea Sindico and Vincenzo Grassi. Model driven development of context aware software systems. In *COP '09: International Workshop on Context-Oriented Programming*, pages 1–5, New York, NY, USA, 2009. ACM.

[47] J.A. Stankovic. Research directions for the internet of things. *Internet of Things Journal, IEEE*, 1(1):3–9, Feb 2014.

[48] Norbert A. Streitz, Carsten Rocker, Thorsten Prante, Daniel van Alphen, Richard Stenzel, and Carsten Magerkurth. Designing smart artifacts for smart environments. *Computer*, 38(3):41–49, 2005.

[49] Martin Strohbach, Hans-Werner Gellersen, Gerd Kortuem, and Christian Kray. Cooperative artefacts: Assessing real world situations with embedded technology. In *Ubicomp*, pages 250–267, 2004.

[50] Tsutomu Terada and Masahiko Tsukamoto. Smart object systems by event-driven rules. In *Proc. of the 1st International Workshop on Smart Object Systems*, pages 100–109, 2005.

[51] Sal vatore Gaglio, Giuseppe Lo Re, Gloria Martorella, and Daniele Peri. High-level programming and symbolic reasoning on iot resource constrained devices. *EAI Endorsed Transactions on Cognitive Communications*, 15(2), 5 2015.

[52] Xiao Hang Wang, Da Qing Zhang, Tao Gu, and Hung Keng Pung. Ontology based context modeling and reasoning using owl. *Pervasive Computing and Communications Workshops, IEEE International Conference on*, 0:18, 2004.

[53] T. Weis, C. Becker, and A. Brandle. Towards a programming paradigm for pervasive applications based on the ambient calculus. In *Proc. of the International Workshop on Combining Theory and Systems Building in Pervasive Computing, in conjunction with PERVASIVE 2006*, 2006.

[54] Michael Woolridge and Michael J. Wooldridge. *Introduction to Multiagent Systems*. John Wiley & Sons, Inc., New York, NY, USA, 2001.

[55] Qihui Wu, Guoru Ding, Yuhua Xu, Shuo Feng, Zhiyong Du, Jinlong Wang, and Keping Long. Cognitive internet of things: A new paradigm beyond connection. *IEEE Internet of Things Journal*, 1(2):129–143, 2014.

[56] S.S. Yau and J. Liu. Hierarchical Situation Modeling and Reasoning for Pervasive Computing. In *SEUS-WCCIA '06: Proceedings of the The Fourth IEEE Workshop on Software Technologies for Future Embedded and Ubiquitous Systems, and the Second International Workshop on Collaborative Computing, Integration, and Assurance (SEUS-WCCIA'06)*, pages 5–10, Washington, DC, USA, 2006. IEEE Computer Society.

[57] Juan Ye, Lorcan Coyle, Simon Dobson, and Paddy Nixon. Ontology-based models in pervasive computing systems. *Knowledge Engineering Review*, 22(4):315–347, 2007.

[58] A. Zanella, N. Bui, A. Castellani, L. Vangelista, and M. Zorzi. Internet of things for smart cities. *Internet of Things Journal, IEEE*, 1(1):22–32, Feb 2014.

[59] Pascal Zimmer. A calculus for context-awareness. Technical report, BRICS: Basic Research in Computer Science, 2005.

18

Moving Collaborations: A Critical Inquiry Into Designing Creative Interactive Systems for Choreography

Kristin Carlson[1,*], Thecla Schiphorst[1], Steve DiPaola[1]

[1]The School of Interactive Arts + Technology, Simon Fraser University, Surrey, British Columbia, Canada

Abstract

The use of technology in choreographic process has been encumbered by the richness of data in live human movement and the constraints of computation. While technology is often considered a tool in choreographic process, with developments it can participate as a collaborator by transforming and eliciting creative opportunities. We specifically define 'collaboration' rather than 'tool' to differentiate the nature of collaboration: a dynamic and iterative process with participation from both the user and the technology. This paper presents a contextual inquiry for an interactive system used to provoke creativity in choreographic process. Choreographic process is often distributed, relying on interactions between the choreographer and dancers to develop and evaluate movement material through exploration on different bodies. Based on this interaction model we choreographed and analyzed a dance work in order to design a set of features that support system collaboration in an intelligent choreographic system. Our contribution situates the design and practice of choreographic systems in theory to explore future design of iterative and provocative collaboration.

Keywords: Collaboration, Creativity, Creative Process, Choreography, Choreographic Tools, Creativity Support Tools

1. Introduction

The use of technology in creative practices is often framed as a tool that facilitates human user's creativity. Creativity Support Tools (CSTs) are systems that provide functional support to user creative process, such as the ability to edit video (i.e. using Final Cut Pro or iMovie), visualize data (Spotfire, JMP) or share photos (Flickr and Facebook)[31]. While the term creativity support 'tools' is often synonymously discussed as facilitating 'collaboration' between a user and a creative system, we differentiate the terms here to articulate the distinctive interactive component of creative process. There is a history of designing systems (particularly CSTs) as 'tools' to support user's existing processes, including collaboration (as in the domain of Computer-Supported Cooperative Work and Social Computing, CSCW). However there is less research in designing systems that can 'co-create' with a user, intentionally provoking the user to realize new creative opportunities. CSTs that function as 'tools' are often designed to facilitate task-oriented human user interaction, without adding intelligent features that adapt creativity and cognition theory. As such they lack features that could interject new perspectives into creative process or recommend constraint-based opportunities for creative exploration. Augmenting CSTs with adaptive creative and provocative features requires the design of a set of features that can support system collaboration in an intelligent autonomous system.

While the domain of computational creativity explores generative systems to produce creative results autonomously, there has been far less research into how systems can behave as a creative agent in an interactive collaborative process with a human user [10][29]. To enable such autonomous creativity the system must have the ability to listen and respond to the user with nuanced domain-specific behavior in-situ (described in artificial intelligence as beliefs, desires and intentions).

*Corresponding author. Email: kca59@sfu.ca

The system must also enable iterative development and extensive history-keeping in order to retain a 'memory' of what choices have been made and support editing the history of choices.

We focus on the domain of choreography because movement as a modality is under-researched and its intelligent features are not well understood, while it is also rich source of data in form, function and expressivity [29]. Choreography is developed through embodied decision-making processes and is an established, parameterized exploration of movement composition [1]. We can consider the variables that affect choreographic decision-making as parameters [23][24]. Choreography is necessarily embodied and uses movement as its primary material [22]. Movement phrases are sequenced and developed through interaction with the environment, the conceptual frame of the choreographer and social interaction with dancers and other artifacts such as props, media and computation [10][3][6][20]. From cognitive science, we know that choreography is a distributed practice, devised by the choreographer, executed by the dancers with subtle differences based on their personal movement signatures and iteratively developed [22][2].

This paper presents a pilot study using contextual inquiry to simulate collaboration between a choreographer and a dancer through a set of parameterized choreographic processes. We used a Wizard of Oz technique to design a set of features that support system collaboration for the purpose of designing an intelligent autonomous choreographic system. Based on an analysis of the literature of the feature sets of contemporary choreographic systems, we selected and then blended three individual choreographic systems. Our selection criteria analyzed the systems' ability to generate and edit novel movement and to share data formats. We blended three systems, rather than utilizing a single system, to extend current choreographic models and to provide us with a richer combined feature set to capture, manipulate and playback movement. Using the three blended systems we simulated the iterative development between choreographer and dancer by creating a short solo choreography. We documented this process through contextual inquiry with video and reflective journals. During the choreographic process we extended the limitations of the choreographic feature sets available by focusing on how shifts in choreographic attention exploited a greater range of choreographic choice in decision-making. We found that techniques such as modality shifts and abstraction are useful design strategies for provoking creative compositional choices that can extend current feature sets of choreographic systems. Modality shifts are the translation of movement data from one representation to another. Abstraction refers to the resolution and aesthetics of movement data representation that can modulate between greater specificity and ambiguity as a compositional interpretive strategy. This paper describes our conceptual framework, the systems we reviewed, the contextual study we performed and our findings supporting our design method using simulated computational agency. Our contribution is in identifying design strategies for manipulating complex movement data through compositional concepts illustrating how modality shifts and abstraction of representation can engage a user's attention, interpretation and active engagement towards a collaborative system.

2. Cognitive Framework for Collaboration

Choreography is the art of crafting movement, developed through a long history of techniques [1][3][20]. Like other compositional processes, choreography is a complex creative process that explores a variety of formal procedures that can result in unique artistic creations [6]. The Belgian choreographer Anna De Keersmaeker states that 'what is missing is an interface between mute videos and practical connaissance, or experiential knowledge, that would allow dance to be more than 'a wheel that turns on itself' [15]. Choreographers have been fascinated with the creative possibilities enabled by the use of technology in the compositional process. The use of digital technologies and software programs challenges choreographers to perceive their creative problem space anew through new constraints alongside new possibilities.

There has been a recent rise of research in dance and choreographic cognition to better understand the cognitive decision-making process in collaboration. Kirsh has researched choreographic process by closely observing and discussing the making of a large group work by Wayne McGregor [23]. His findings show that choreographers develop a movement 'idea' and then have dancers 'riff' on it to develop a diverse adaptation of the original idea to collaborative explore movement concepts. He has also explored how dancers 'mark' movement ideas, by performing it half physically and half mentally, in order to better learn movement sequences using embodied forms of cognition [22]. Sawyer and DeZutter investigated improvised theater performance to track how a performance emerged collaboratively [29]. They found that as the dependency of each participant's decisions relied on previous actions, it became harder to predict an individual participant's decisions based on their usual individual traits. The emergent creativity developed from skills and personalities in situ, as well as the actors working with the affordances of the social scenario.

There are strong cognitive overlaps between the processes of movement, decision-making and creativity that can provide unique opportunities for designing

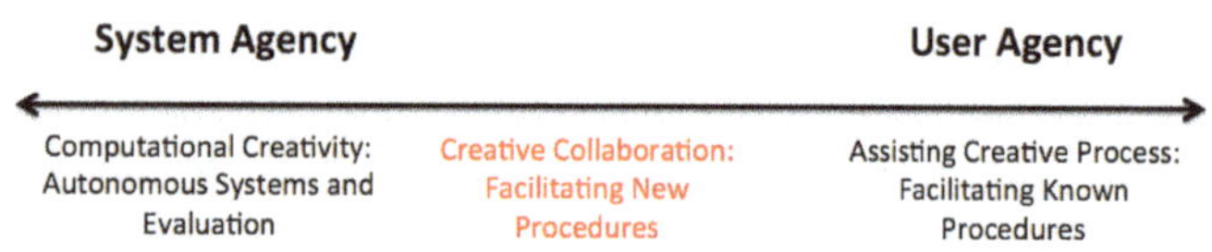

Figure 1. Current Creativity System Design from Autonomous Systems to CSTs

technology [9][10][16][11]. Currently within choreography, computers are used as CSTs to design the presentation of performance (set, lighting, sound, costume). While the limitations of using technology as such will impact choreographic decisions, choreographers often respond to these outside limitations with reinforced use of their own movement habits and styles. We are interested in how computers can be *engaged within the creative choreographic process* to impact habits and style, to shift attention to particular aspects of experience and suggest new choreographic choices.

Creativity Support Tools have historically been designed to support the user's established cognitive processes. This results in momentary novel solutions with a focus on efficient, predictable methods [31]. By continuing to support established creative processes there is minimal engagement with the source of creative patterns. To 'design for designers' we need to explore the iterative creative cognitive process more deeply. Gathering information on process within creativity will develop methods to support and iteratively sustain the 'unknown', situated creative process [13]. While the designer cannot directly control a user's creative cognitive process, they can create tools to indirectly affect the way decisions are made to result in new perspectives to a creative problem.

Defamiliarization, or 'making strange', is a pragmatic tactic used to design for indirectly interacting with a user cognitively. By making the familiar unfamiliar in order to bring new awareness to known practices, user's choices are destabilized while simultaneously engaging agency to re-orient their experience. This form of 'disorientation' incites the user to assert their agency within a creative scenario [4]. Asserting agency utilizes users' own reflections, analysis and assessments of a given situation in order to engage in the task at hand.

Designing for more nuanced collaborative tools that support user experience relies on critical choices around what parameters can be used, how interaction is crafted and how the user is cognitively engaged. DiPaola et al. explored collaborative creative tools by designing a system as a graphic design support tool. This system, titled Evolver, generates design combinations based on the user's seed material using parameters of color, shape, repetition, symmetry and rotation. This system supports exploratory searches through its generative suggestions and provides history-keeping options to bookmark interesting combinations for later use [13]. The use of a genetic algorithm and decision to support the more labor intensive and intuitively-based portion of the design process created a tool that suggests collaboration in the process without fully attempting to re-create collaboration. Improvisational music systems such as Lewis' Voyager and Weinberg and Driscoll's Haile robot have been designed to interact with musicians as players rather than instruments [27][33]. However, they focus on performing compositional decisions in the moment, rather than exploring and structuring a work as a whole. While these systems worked well in well-understood disciplines in visual and sonic mediums, there are many new challenges arising in the design of technology for gesture, full body movement and practice based endeavors. We are interested in designing systems that *provoke* collaborative practices with choreographers through design of existing technology for capturing movement, artificial intelligence techniques for generating movement suggestions and defamiliarization tactics that suggest symbiotic iteration between a user and a system.

3. Computational Systems for Choreography

Though technology continues to develop exponentially quickly, technology continues to have inherent limitations that can be leveraged as creative constraints in artistic processes [9]. These limitations include sensing and capture resolution, in the algorithms to manage large quantities of data, in the circumstances where technologies can be used and in the mediums that technologies can be worked with. These limitations challenge designers and users to develop novel methods for interaction to contribute to the desired outcome, often despite the limitation. However, technology can only fulfill the role of creative collaborator if the interaction provokes the human to devise something divergent from their habitual practice, choices that can extend their creative range away from results they can create themselves, without technology. To address this research agenda, we ask the questions: What specific design features can provoke creative choreographic decision-making in a social and embodied context? How can prototypes through a Wizard of Oz method reveal techniques that can be applied as future design strategies for provoking creative compositional choices?

We reviewed eight systems that support creative process in choreography including: DanceForms, Dancing Genome, Scuddle, Web3D Composer, DANCING, Viewpoints AI, The TKB Creator's Tool and Choreographer's Notebook. DanceForms (previously known

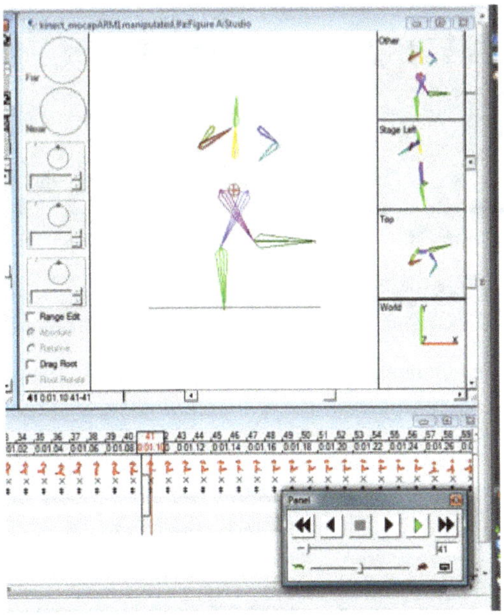

Figure 2. DanceForm's Studio Interface

as LifeForms) [7][8] is a graphical animation compositional tool for designing and visualizing dance movement based on user input or library selection (See Figure 2). The system has three views: space, time and body-position. The space view allows the user to design movement pathways as spatial patterns. The timeline allows the choreographer to design sequences and timings of movement. The body-position view allows the user to design body positions using joint manipulation or to choose codified positions from the libraries. DanceForms supports choreography of multiple figures, spatial patterns and orientation. Merce Cunningham used DanceForms to design movement on avatars, transposing the movement decisions onto live dancers. This process allowed him to explore movement options that he may not have otherwise considered while facilitating his use of chance operations. DanceForms, the most advanced system available for computer-supported choreography, however is not widely used. Perhaps this may be in in part because the form of interaction (low-level detail-oriented, not supportive of whole body interaction) is not aligned to physically exploring movement within choreography. DanceForms does provide multiple levels of abstraction, which encourages the user to view compositions in multiple unique ways.

The Dancing Genome Project [25][26] developed a genetic programming model to explore sequences of movement in performance. The system analyses movement data and re-organizes it to create a new sequence with the same movements. The movement data is created by gathering motion capture data extracted from a dancer performing the movement

Figure 3. Scuddle's Movement Catalysts

sequence, in the studio prior to the performance. The motion capture data is manually segmented, then used as input to the genetic programming model to shift the location of dance movements. The final manipulated sequence is performed by projected avatars on the backdrop as the live dancers performed the original sequence, creating a mixed-reality duet. However, while dancers and avatars performed next to each other, neither are able to make decisions or changes in action during the performance. This genetic crossover and mutation process is similar to exploring themes and variations in the studio choreographic process, where variations of a theme are explored separately by are similar enough to be recompiled together.

Scuddle is a choreographic support tool designed to provoke creative movement decisions using defamiliarization tactics. By constraining the possibilities for movement generation Scuddle simulatenously provokes creative movement opportunities [10](See Figure 3). The system generates unique movement catalysts using a genetic algorithm with a heuristic fitness function, without input from the user. Scuddle is designed to explore and inhibit habitual body positions, levels and effort qualities typically performed by dancers in the studio, in order to bring awareness to new, non-codified movement options. Scuddle is used in the creative process, in the studio, the same way a choreographer would explore creative movement options without technology. However, the addition of technology helps to guide movement decisions towards unusual choices, that may not be addressed otherwise, that both create unique movements and make the decision-making process more apparent to the choreographer. These design choices support Scuddle as both a creativity support tool as well as a research tool.

Web3D Composer creates sequences of ballet movements based on a predefined library of movement material [32]. The system allows the user to select movements from a pool of possibilities, which shift based on structural ballet syntax. This interactive process allows the choreographer to select movements based on the possibilities presented through Markovian selection while presenting nearly complete graphic movement information The Web3D Composer system was designed with the intention of providing a tutoring system for students and generates syntactically correct movement phrases for up to a second year ballet level.

DANCING used a series of music-related parameters, spatial pathway rules and a predefined library of traditional movements to generate Waltz choreography using a Genetic Algorithm [28]. By connecting the correct, predefined 'steps' in a domain- specific sequence that provides stage directions and orientations, this system generates syntactically correct movements in a complete choreography that are represented as ASCII symbols on a bird's eye view of the stage. DANCING illustrates the relationship of individual movements to the whole composition as generative output, however it is not a useful tool contemporary choreographic exploration as a creative problem in the studio since there are no opportunities to intervene in the output.

The Viewpoints AI project looked to the Viewpoints compositional framework to create a real-time interactive system exploring dance improvisation strategies [21]. The system used kinect data and the SOAR reasoning framework to create a repository of short and long-term memory of the choreographer's movements that select and apply different response modes and improvisational strategies. The reasoning framework can respond by: doing nothing, mimicking the user's movement, transforming the user's movement and then performing it, repeating a movement it has learned during its lifetime of experience, or executing various kinds of interaction patterns.

All these systems engage in an interactive creative process with the choreographer yet (outside of Danceforms which has a small following) have rarely been used in creating dance. Tools for facilitating creativity and collaboration include video annotation tools for choreography such as the Creator's Tool and Choreographer's Notebook [11][14]. These are both opportunities for choreographers to objectively view their work, note ideas temporally, and see opportunities for editing their work. The tools afford the choreographer perspective of a whole work within the context of an annotation tool by presenting editing opportunities right in the video format rather than a focus on parts in the studio environment.

Figure 4. Choreographer Working with Kinect and Integrate Systems

Choreographic Process	Pilot Study Process
1. Generate movement	Scuddle System -> Choreography
2. Develop movement into phrases	Choreographer
3. Capture into DanceForms	Choreographer -> Integrate System -> DanceForms
4. Manipulate avatar in DanceForms	Choreographer/ DanceForms
5. Re-learn movement phrase	DanceForms -> Choreographer
6. Develop/ Capture/ Manipulate/ Re-learn	Iteratively between technology and Choreographer
7. Structure Sequences	Choreographer
8. Create spatial pathway plan	Choreographer
9. Reflective journaling	Choreographer

Figure 5. Comparing Choreographic Process Between Systems and the human choreographer

4. Pilot Study

We performed a pilot study to explore what kinds of features could support collaboration in an intelligent autonomous choreographic system. To test our process we devised a critical inquiry by developing a choreography in the studio with the aid of a Wizard of Oz exploration of the interaction between a simulated software tool and a choreographer. Critical inquiry comes from Human Computer Interaction and uses ethnographic methods to collect data in the field, or in situ [18]. Data is collected about how a subject interacts with a system within their normal environment.

We composed a short choreography for a solo performer based on an iterative development process using 3 computational systems in the studio (See Figure 4). This process was documented with video and journal reflections in every stage. The first system is titled Scuddle, to initiate and develop a movement vocabulary [10]. The second system is Integrate, to capture movement data with a Kinect camera and translate the data to bvh format [17]. The third system is DanceForms, an animation platform specifically designed for choreography, which we used to playing back and manipulating movement files [8]. Choreography is often composed in 3 stages: generating movement material, sequencing and developing movement material and structuring movement compositionally. For this study we identified the interactions used and mapped them to the part of the process being explored (Figure 5). This section articulates what was explored in each stage, how technical limitations influenced the process and how the choreographer's choices influenced the process.

4.1. Steps of the Choreographic Process

The goal of this type of compositional process was A) to create a choreography (requiring attention to the choreographic task, not distractions in technology limitations) and B) to identify methods for designing technological support or mediation to engage in the choreographic process. Within this compositional process we were focused on creative and novel movement choices that develop into full compositions. As seen in the above table, the process began by using the Scuddle system to generate movement catalysts, which the choreographer used to create a movement vocabulary (1). The choreographer sequenced movements into phrases herself (2), then used a Kinect camera to capture the phrases (3). The movement files of phrases were imported into DanceForms, and manipulated by adjusting pelvis and limb positions over groups of frames (4). The choreographer then re-learned the new movement phase from the avatar in DanceForms, interpreting the new information (5)(See Figure 6). This exploration between choreographer and mediating technology is performed iteratively to develop movement material (6). Three parameters were imposed by the choreographer as having manageable options for 'chunking' movement information: types of movements (gestural, angular movements, slouched torso), sizes of movement (small, medium, large) and spatial trajectory options (in place, linear, curved)(7). The choreographer then used the parameters to structure sequences spatially and temporally to construct the choreographic progression, by drawing a map on paper (after exploring pathway creation in DanceForms)(8). Finally, the choreographer documented each stage of the process while journaling about her thoughts in each stage.

4.2. Reflections in Composition: Choreographic Process

During the choreographic process we found it easy to develop initial movement 'ideas' from the Scuddle system that were continually re-integrated into the final composition. Because capturing the movement was using the Kinect, our focus while developing movement was on the technology rather than on our explorations. We found that improvising, capturing, manipulating and re-learning our movement was mediating the movement a lot from its original state. The technology did include intelligent interventions to translate our own movement habits when re-learning the movement. The 'information' that was represented to us lost performance quality because it was expressivity filtered 'down' through noise from the camera capture which eliminated individual nuances or complexity in movements themselves. This was a useful step to understand how defamiliarization is a tool for simulating system agency in interaction.

This 'lossy' result of movement information meant that we were required to re- interpret the animated movement rather than re-learn phrases in the performers learn from video. This process became difficult and time consuming without providing additional compositional benefits. Once we had developed sections of movement we noticed the need for high-level compositional tools to structure it. Beyond exploring motion we did not have a metaphorical theme to help guide choreographic decisions. While DanceForms enabled playback and detailed editing it does not yet incorporate high level structural features that can be applied to choreographic explorations. We designed a linear structure with pen and paper that we could be used to manipulate the animation in DanceForms. Ultimately, we focused on our experience of performing the movement and sketched out salient features using contextual inquiry, which provided a much more dense and nuanced map of potential choreographic structure. This map provided many elements to work with because we could easily notate spatial location, speed, transition movements and dynamics in a way that we could remember. Our experiential explorations illustrated the need for shifting modalities (between detail and higher level choreographic context) and using abstraction to move between detailed features such as limb positions and higher level concepts of phrasing, spatial trajectories and structure.

4.3. Technology-Oriented Decisions: Limitations and Creative Opportunities

The inherent limitations of the existing feature sets effected both the realistic applications to choreography as well as manipulating the available creative search spaces. Choreography has traditionally found opportunities to be creatively opportunistic when working with the inherent limitations of the technology by focusing the movement vocabulary, structure and content on solving the new constraints that technology brings to a creative process. The Scuddle system brought a focus to generating unique movements through physical exploration of complex inspirations, however it does not support idea development or iteration. This supported initial movement ideas but did not facilitate interactive dialogue.

Capturing movement with the Kinect made the instantiation of movement in DanceForms a much easier and familiar embodied process than clicking with a computer mouse. However this development alone warrants higher-level controls for manipulating and transforming movement phrases. The automatic capture feature to start capturing with the correct pose was efficient, but needed to be edited out in the animation process. There was a high level of noise in the Kinect data that often presented very jittery data. There

Figure 6. Choreographer Working with Kinect, Integrate and DanceForms Systems

was also a narrow spectrum of capture opportunities for a dancer. The available floor space for capture is small (oriented towards gamers fixated on a screen), the camera still needs to see recognizable limbs even with the depth camera (limbs are easily lost and replaced as standing still), sudden changes of movement or tempo are often lost. Many movements were not able to be captured including curved spine, swinging limbs, legs extended above 90 degrees and any sense of weightedness in the movement. This lack of specificity in the movement data constrained the opportunities for manipulating choreographic process computationally. The choreographer made choices towards basic, more easily recognized movements, attempted to move more slowly, not perform work on the ground which had the effect of reducing complexity and novelty. The most prominent creative understanding in the process with the Kinect was that it brings a strong focus to the body's positions, rather than orientation, spatial relationships and quality or expressivity of movement. There are many qualities that lend themselves towards defamiliarization practices, augmenting the original movement in unique positions and jittery interpolations.

DanceForms brings a strong focus on the details of movement data that allow the user to be very specific with every joint movement in each or any specific frame, however there are no options for higher level manipulation. Once we captured our movement into DanceForms we could not find any simple ways to manipulate the whole body or multiple movements, so we copied and pasted sections of movement to edit the sequences or adjusted single limbs. Spatial pathways could be manipulated most easily, but were difficult to select and move as spatial trajectories. The ability to playback manipulated avatars enabled the choreographer to re-learn and re-interpret the movement in ways that could not have been afforded without the technology. However the re-interpretation was affected by noise in the original captured data and the strong focus and need for detail in the editing interface. In the end of this process we found that the choreographer was continuing to focus on individual movements or very short sequences through the Kinect and DanceForms interaction that were not supporting exploration or development on longer compilations of movement data. Future developments that focus on high level compositional functions would be useful to maneuver in embodied exploration with less articulation of details.

We often find that technology is useful for creating a more objective perspective in which to explore movement, away from the guiding physical sensations of the body's position in relation to itself. However this requires bringing attention to more subtle or compositional aspects of movement that current systems do. While one solution is to wait for technology to 'catch up' and be able to sense and manipulate a higher and more refined resolution of human movement information, we are more interested in current design opportunities that leverage embodied creativity from a perspective of bringing technological innovation. Existing practices in defamiliarization, cognition and creativity theory can be used to engage in creative collaboration. We outline our findings in reflections on the remainder of our discussion on creative choreographic process.

5. Conclusions and Future Work

This paper explored opportunities for designing new feature sets for collaborative creativity into tools used for choreographic process. We developed a contextual inquiry to address this question using three existing systems to inspire, mediate and support choreographic development. Through this iterative process we suggest a variety of opportunities for further investigation. Functional software developments for choreographic systems could focus on furthering current feature sets in body position and movement data. The addition of physics engines to DanceForms would provide new opportunities for exploring qualitative aspects of functional movement. Generative and learning techniques to develop greater autonomous creativity would provide agency in the DanceForms system (or newly designed system), such as using style machines to extract stylistic data from existing movement and manipulating it to become new movement [5], integrating Scuddle to generate unique positions for interpolated movement [10] or using unsupervised algorithms for large datasets such as deep-learning [12].

Alongside the greater creative autonomy of an interactive system, new collaborative systems would benefit from enhanced functional history-keeping and accessibility. Maintaining connections with the history of the creative process enables users to reflect on prior actions and decisions while reverting to prior options if desired. This transparency of the history of choices also supports new creative developments that use and build upon prior actions without losing the central focus. Designing for easy access to history and parallel working processes would also be a useful and important feature when supporting the non-linear and iterative creative process.

Additional methods for manipulating movement data in collaborative systems would be highly useful to support cognitive perspectives of movement and compositional processes. Shifting modalities has also been found to be an important component of choreographic cognition, as studied by David Kirsh [23]. Choreographers often highlight a particular movement feature and develop a creative idea by re-mapping this feature to another part of the body. For example, highlighting the movement of the arm in both position, trajectory and quality and re-mapping it to movement of the hips. This re-mapping strategy could also be more abstract, such as taking the timing of a movement through space and performing the timing by simply walking in time.

Designing a 'modality' system that could be shifted between different data sets would need to consist of parameterizing movement data to enable the user's focus on a specific quality of subtle movement data. Movement parameters could consist of the body part, body position, spatial trajectory, tempo, level of body (on floor, middle, jumping), movement qualities, exertion levels, etc. Patterns in the specific quality of data could then be extracted, abstracted and applied to other another quality of data, hence shifting the same data between modalities. An example of this would be the re-mapping of movement data (such as trajectory and acceleration) from a leg to a subject's head. Various modalities that would be useful in choreography include:

- Qualities of expressive movement such as Effort Qualities from the Laban Movement Analysis framework. Effort Qualities are deconstructed into elements of time, flow, weight and space and combine to create effort actions such as punch, press, dab, flick, wring, float, slash and glide. [24]

- Qualities of tempo and rhythm that would enable the manipulation of an individual and sequence of movements beginning, middle and ending.

- Spatial intention that affects how a subject engages with themselves, their environment, other subjects or objects. This is often depicted as intending to interact through distances, such as presenting oneself strongly to a distant subject or exploring ideas of movements internally.

- Explorations of weighted-ness that provide visuals and control over the subject's center of gravity, ability to perform certain movements and resistance or giving in to gravitational forces.

One aspect of shifting modalities aside from the data itself, is the abstracted representation. The visual presentation of information affects how the choreographer kinaesthetically interprets it. Shifting modalities is a form of changing how movement information is represented, creating new search spaces for creative exploration. While the data could be abstracted by its modality, the level of specificity is also an important factor. The main motivation for a choreographer to use technology is to be able to observe structural choices, creating a more objective stance in which to perceive movement information that can augment the choreographer's own kinesthetic feedback. Presenting information as a 3D avatar with facial features is very different from a stick figure, different from a point cloud or unique form or spatial or geometric temporal or analytics representation. Technology is able to highlight specific features separate from the rest of the movement data, which we cannot do on a person. Being able to highlight the exertion of a choreographer over time or the movement quality of an arm compared to a leg is highly useful information in choreography and many non-creative movement domains.

Our future work is to implement some of the features articulated in this paper in the DanceForms platform. We plan to design a context-specific, adaptable interface for DanceForms that enables high-level editing features for choreography on an iPad device. By leveraging opportunities for mobile platforms we can move a bit closer to embodied interaction in situ, bridging with new opportunities for personal meaning-making between movement and technology. We plan to design editing features based on both the low-level animation keyframe functions as well as higher level features where movement data can be adapted using a combination of modality and abstraction levels. This work is happening concurrently with the MovingStories Research Partnerhip (www.movingstories.ca) in researching methods for capturing, manipulating and representing movement data from a somatic perspective to broaden our understanding of movement in language and computation.

6. Copyright statement

Please be aware that the use of this LaTeX 2_ε class file is governed by the following conditions.

6.1. Copyright

References

[1] Adshead J. (1986). Choreography: principles and practice: report of the Fourth Study of Dance Conference. The University of Surrey: Dance Research Unit by the National Resource Centre for Dance.

[2] Banes S. (1987). Terpsichore in Sneakers: Post-Modern Dance. Wesleyan.

[3] Blom L. A. (1982). The Intimate Act of Choreography. Pittsburgh, Pa: University of Pittsburgh Press.

[4] Boden M. A. (2003). The Creative Mind: Myths and Mechanisms. Routledge.

[5] Brand M and Hertzmann A. (2000). Style Machines. Proceedings of the 27th annual conference on Computer graphics and interactive techniques (ACM Press/Addison-Wesley Publishing Co.), (pgs.183–192).

[6] Butterworth J. and Wildschut, L. (Eds.). (2009). Contemporary Choreography: A Critical Reader. London: Routledge.

[7] Calvert T. and Bruderlin A. and Mah S. and Schiphorst T. and Welman C. (1993). The evolution of an interface for choreographers. Proceedings of the INTERCHI '93 conference on Human factors in computing systems. (pgs. 115–122).

[8] Calvert T. and Welman C. and Gaudet, S. and Schiphorst T. and Lee C. and Goossens M (1991) Composition of multiple figure sequences for dance and animation. The Visual Computer. 7(2), (pgs. 114–121).

[9] Candy L. (2007) Constraints and Creativity in the Digital Arts. Leonardo. 40(4), (pgs. 366–367).

[10] Carlson K.Schiphorst T.Pasquier P. (2011). Scuddle: Generating Movement Catalysts for Computer-Aided Choreography. Presented at the The Second International Conference on Computational Creativity, Mexico City, Mexico: ACM Press.

[11] Carroll E. A., Lottridge, D., Latulipe, C., Singh, V., Word, M. (2012). Bodies in critique: a technological intervention in the dance production process. In Proceedings of the ACM 2012 conference on Computer Supported Cooperative Work (pp. 705–714). New York, NY, USA: ACM.

[12] Coates A., Ng A. Y., and Lee H. (2011). An analysis of single-layer networks in unsupervised feature learning. In International Conference on Artificial Intelligence and Statistics (pp. 215–223).

[13] DiPaola S., McCaig G., Carlson K., Salevati S., and Sorenson N. (2013). Adaptation of an Autonomous Creative Evolutionary System for Real-World Design Application Based on Creative Cognition. In Proceedings of the Fourth International Conference on Computational Creativity (pp. 40–48). Sydney.

[14] Fernandes C. (2013). The TKB Project: Creative Technologies for Performance Composition,Analysis and Documentation. In P. Nesi and R. Santucci (Eds.), Information Technologies for Performing Arts, Media Access, and Entertainment (pp. 205–217). Springer Berlin Heidelberg.

[15] Keersmaeker, A. T. D., and Cvejic, B. A Choreographer's Score: Fase, Rosas danst Rosas, Elena's Aria, Bartok. Mercatorfonds, 2012.

[16] DiPaola S. R., and Gabora L. (2007). Incorporating characteristics of human creativity into an evolutionary art algorithm. In Proceedings of the 2007 GECCO conference companion on Genetic and evolutionary computation (pp. 2450–2456). New York, NY, USA: ACM.

[17] Gaskell A. (n.d.). Kinect BVH Motion Capture. Integrate Systems. Retrieved October 6, 2014 from http://tech.integrate.biz

[18] Holtzblatt K., and Jones S.. (1993) "Contextual inquiry: a participatory technique for system design." Participatory design: principles and practice. Hillsdale, NJ: Erlbaum.

[19] Hollan J., Hutchins E., and Kirsh D. (2000). Distributed cognition: toward a new foundation for human-computer interaction research. ACM Trans. Computer-Human Interaction, 7(2), (pp. 174–196).

[20] Humphrey D. (1959). The Art of Making Dances. Princeton Book Company.

[21] Jacob M., Zook A., and Magerko B. (2013). Viewpoints AI: Procedurally Representing and Reasoning about Gestures. In Proceedings of the 6th Digital Games Research Association Conference, Atlanta, Georgia.

[22] Kirsh D. (2010). Thinking With the Body. The Annual Meeting of the Cognitive Science Society.

[23] Kirsh D. (2011). Creative Cognition in Choreography. Proceedings of the 2nd International Conference on Computational Creativity.

[24] Laban, R., and Ullmann, L. Modern educational dance. MacDonald and Evans, 1963.

[25] Lapointe F.J. (2005). Choreogenetics: the generation of choreographic variants through genetic mutations and selection. In Proceedings of the 2005 workshops on Genetic and evolutionary computation (pp. 366–369). Washington, D.C.: ACM.

[26] Lapointe F.J., and Epoque M. (2005). The dancing genome project: generation of a human-computer choreography using a genetic algorithm. In Proceedings of the 13th annual ACM international conference on Multimedia (pp. 555–558). Hilton, Singapore: ACM.

[27] Lewis G. E. (1999). Interacting with latter-day musical automata. Contemporary Music Review, 18(3), (pp. 99–112).

[28] Nakazawa M., and Paezold-Ruehl A. (2009). DANCING, Dance ANd Choreography: an Intelligent Nondeterministic Generator. In The Fifth Richard Tapia Celebration of Diversity in Computing Conference: Intellect, Initiatives, Insight, and Innovations (pp. 30–34). Portland, Oregon: ACM.

[29] Sawyer R. K., and DeZutter S. (2009). Distributed creativity: How collective creations emerge from collaboration. Psychology of Aesthetics, Creativity, and the Arts, 3(2), (pp. 81–92).

[30] Schiphorst T. (2011). Self-evidence: applying somatic connoisseurship to experience design. In CHI'11 Extended Abstracts on Human Factors in Computing Systems (pp. 145–160). New York, NY, USA: ACM.

[31] Shneiderman B. (2007). Creativity support tools: accelerating discovery and innovation. Communications of the ACM, 50(12), (pp. 20–32).

[32] Soga A., Umino B., Yasuda T., and Yokoi S. (2006). Web3D dance composer: automatic composition of ballet sequences. In ACM SIGGRAPH 2006 Research posters (pp. 5). Boston, Massachusetts: ACM.

[33] Weinberg G., and Driscoll S. (2007). The interactive robotic percussionist: new developments in form, mechanics, perception and interaction design. In Proceedings of the ACM/IEEE international conference on Human-robot interaction (pp. 97–104). New York, NY, USA: ACM.

19

Synchronous networks for bio-environmental surveillance based on cellular automata

Bao Hoai Lam[1,*], Hiep Xuan Huynh[2], Bernard Pottier[1]

[1]Université de Bretagne Occidentale, LAB-STICC, UMR CNRS 6285, France

[2]DREAM Team/UMI 209 UMMISCO-IRD, Cantho University, Vietnam

Abstract

The paper proposes a new approach to model a bio-environmental surveillance network as synchronous network systems, systems consist of components running simultaneously. In the network, bio-environmental factors compose a physical system of which executions proceed concurrently in synchronous rounds. This system is synchronized with a synchronous wireless sensor network, the observation network. Topology of the surveillance network is based on cellular automata to depict its concurrent characteristic. Several aspects of the above model is simulated by using the case study *Brown Planthoppers surveillance network in the Mekong Delta.*

Keywords: Bio-environmental surveillance network, Brown Planthopper, synchronous network, cellular automata.

1. Introduction

Bio-environmental phenomena occur continuously and concurrently. Continuous occurrence means that they compose an unbroken whole, without interruption while concurrency allows them to happen at the same time. For example, some factors such as temperature and wind influencing Brown Planthoppers (BPHs) invasion [6][7] from one place to another, are continual. Besides, they are concurrent since the motivation for propagating from a source to a destination comes from surrounding conditions of the source and its neighbors. Conditions from such different places must be executed simultaneously. Similarly, floods which cause overflows of water are also uninterrupted and concurrent.

To confront with disasters caused by bio-environmental phenomena, several experienced as well as academic solutions are proposed and a solution with wireless sensor network (WSN) emerges as a suitable capable choice [4][8]. This kind of solution uses sensors to measure environments and other factors. The sensed values from sensors will be sent via a wireless network to a data center periodically. Next, a back-end system will manipulate these values and propose solutions relating to collected situations of surrounding conditions. Such application is called bio-environmental surveillance network.

There exists a relation between a bio-environmental phenomenon and its observation WSN in a surveillance network. Indeed, the WSN provides sampling inputs for the phenomenon and the phenomenon occurs in accordance with its rules based on these inputs. On the other hand, damage levels of bioenvironments caused by the phenomenon are sampling measured by sensor nodes of the WSN.

The paper proposes a new approach to model a bio-environmental surveillance network as synchronous networks [14] in order to emerge the relation between bio-environmental phenomena and their WSNs. In this work, the topology of the surveillance network is based on cellular automata [24], a parallel structure.

The structure of this paper is as follows. Section 2 depicts some previous work relating to wireless sensor network as well as bio-environmental surveillance modeling. Next section is about modeling a bio-environmental surveillance network as synchronous networks. Definitions of a synchronous network and a synchronous wireless sensor network are also depicted in this section. To be an example for this model, case study Brown Planthopper surveillance network in the Mekong Delta of Vietnam is introduced in section 4. A data model for a bio-environmental surveillance network is described in section 5. Section 6 depicts

*Bao Hoai Lam. Email: lhbao@cit.edu.vn

implementations of a bio-environmental surveillance network. Next section illustrates some simulation results of the surveillance network described in the case study. The last section is our conclusion and future plans.

2. Related work

Examples of bio-environmental surveillance network is variant such as light trap networks, flood surveillance, fire forest surveillance. Light trap network uses light traps to measure environmental factors and insect densities. Thanks to it, people can make suitable decisions to get better protection for their crops from insect attacks.

Light trap method is one of solutions to prevent high densities of spruce budworm in Canadian forests [9][10]. It allows people to participate insect trapping by giving light traps to them and track their traps from June to end of August every year. Periodically, people only report estimated densities of insects via a website, an application or even with a paper and a pen. Finally, trap samples are collected and counted in a lab environment. Applications of data collections from these light traps are variant, for example, thanks to wing wear and body size measurements of adult spruce budworms captured at light traps in some previous years, some useful inference on seasonal patterns related to reproduction can be archived [11]. However, these light traps seem not to compose a network, instead, they create a combination of traps to collect data for post processing. Therefore, there are few information about the model of the light trap network.

A forest fire detection system can be modeled as a k-coverage problem in WSNs [36]. In this work, Fire Weather Index (FWI) System [37] is used in designing an efficient fire detection system in order to optimize the communication and sensing modules of the observation WSN. The WSN life circle is prolonged due to a data aggregation schema based on the FWI since the schema only delivers the data that is of interested by the application.

An insect surveillance network is modeled as an interaction between an insect physical system and a WSN system, both can be modeled as synchronous networks [1]. In this model, the physical system is considered as a synchronous network based on cellular automata topology by being dividing into cellular cells. This physical system is synchronized with a synchronous WSN system. Nevertheless, data communication aspect of the WSN system seems not to be focused in this work.

3. Bio-environmental surveillance network

3.1. Synchronousnetwork

Synchronous network [14] is a network describing synchronized rounds of message exchange and computation. It consists of pieces of processes which may send and receive messages simultaneously.

Mathematically, a synchronous network can be considered as a graph G where processes are located at its nodes and these processes communicate together via their edges using message sending.

Each node in a synchronous network is termed as a process which consists of the following components:

- $states_i$: a collection of states at process i.
- $msgs_i$: a message-generation function specifies that the process i sends to an indicated neighbor, starting from the given state.
- $trans_i$: a state-transition function specifies a new state to which the process i moves from the current state and messages from incomming neighbors.

Wireless Sensor Network A Wireless Sensor Network (WSN) [17] consists of n wireless sensor nodes distributed in a two dimensional planes (figure 1). It can be considered as a graph G=(V,E) where each sensor node is a node in V and an edge of E is established between 2 nodes if the distance between them is at most a transmission range r_t, the maximum distance that the single transmission of a node can be received by all nodes in its vicinity.

Besides, each sensor node can measure its surroundings within a sensing range r_s. Normally, the sensing range of a node is much smaller than the communication range.

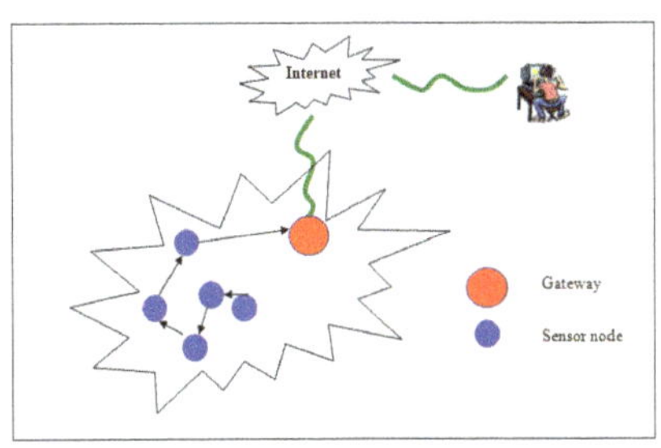

Figure 1. Sensor nodesin WSN.

Gateway node is a special node in which data sensed from other sensor nodes is integrated. Commonly, there is an application connecting to this node to process these pieces of data for decision making.

A WSN can be considered as a synchronous network [19] since the WSN shares a reference clock and allows sensor nodes periodically listening and emitting interleaved with silences. The WSN behaviors also require time division, frequency or channels division in operating the data exchange among sensor nodes.

In practice, both message-generation and state-transition functions can be shortly called as "transition rules", rules allowing the process i to send messages to neighbors in order to compose its new state.

3.2. Description of bio-environmental surveillance network

A bio-environmental surveillance network is a network to monitor bio-environmental phenomena due to environmental factors based on WSN approach [4][5][8]. In this network, there are 2 distinct systems: a physical system and a network system which interact and exchange data together in order to compose a whole one: a *bio-environmental surveillance network* (figure 2).

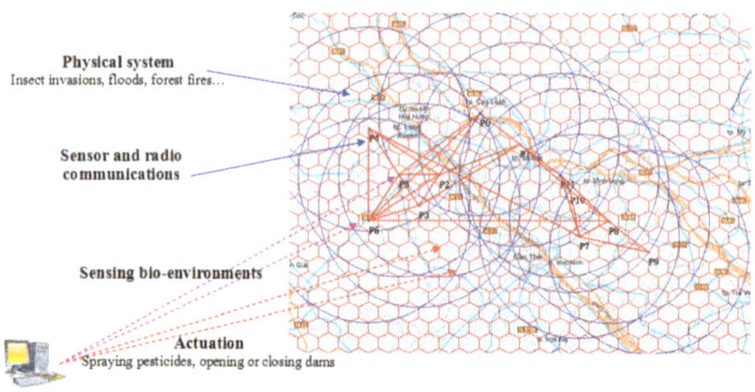

Figure 2. A bio-environmental surveillance network is composed of 2 systems. A physical system is a phenomena' working space which is divided as cells. These cells are monitored by a WSN system (*sensor and radio communications*).

A physical system is a system that the phenomena occur, in other words, it is a working space or working environment of the phenomena. This space consists of environmental factors and other conditions. For example, brown planthoppers desire to suck rice plants under some environmental factors, therefore, rice fields and these factors compose a system termed: *physical system*.

The above physical system is monitored by computing components which connect together to constitute a wireless network. It is a WSN in which sensor nodes can periodically sense factors in the physical system. Next, sensed values are transmitted through the wireless network to a data center for post processing.

In short, the integration of the physical system and wireless network system composes the bio-environmental surveillance network. Both systems can be modeled as synchronous networks.

3.3. Physical system

Description. The *physical system can be considered as a synchronous network* since components of this network can be found in the physical system.

The physical system of a bio-environmental surveillance network is the working space (or environment) in which the phenomena occur. This space is divided as units called *cells* (figure 2). Each cell has 4 neighbors (Von Neumann neighborhoods), 8 neighbors (Moore neighborhood), or 6 neighbors (hexagonal cell) [25] (figure 3).

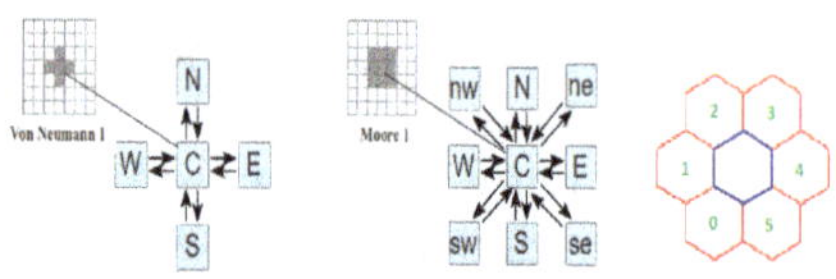

Figure 3. Neighbors of a cell.

The space in figure 2 represents a cellular automaton which is depicted by a triple (S, n, f) where:

1. A finite state set S. A state of a cell describes the status of that cell.
2. Distance n identifying neighbor cells, *normally* $n=1$. When $n=1$, a cell has at most 4, 6, or 8 surrounding cells.
3. Transition rule f: $S^n \rightarrow S$ depicts the change of a cell's state at a specific time based on the current state of the cell and its neighbors.

   At the time t, the state of a cell depends on the state at time t - 1 of its neighbors. The cell itself can be integrated in its neighborhoods. Updating cells are done by a transition rule. All cells have the same transition rule and the transition rule is applied to all cells at the same time. Whenever the rule are applied to the entire system, they could change the entire system synchronously.

Mathematically, the above cellular automaton of the physical system is a topology of **a synchronous network** which is modeled as a directed graph $\mathbf{G_1}=(V_1, E_1)$ where V_1 is the collection of nodes and E_1 is the set of edges.

Nodes. Each cell in figure 2 represents a node in the graph $\mathbf{G_1}$. By the time passing, each node i (also called process i) holds a collection of states $states_i$ of which a state describes a status of the cell at a time t.

Edges. Edges in the graph $\mathbf{G_1}$ are composed by links between a node and its neighbors (4, 6, 8 neighbors in figure 3). Because $\mathbf{G_1}$ is a directed graph, there are 2 edges between the node and a neighbor of it (2 directions). Thus, these 2 edges can become an undirected edge and $\mathbf{G_1}$ can be seen as an undirected graph.

Behaviors. Behaviors at node i are expressed by transition rules of states $transitions_i$. Normally, these rules are functions which specify a set of conditions causing bio-environmental phenomena.

3.4. Synchronous Wireless Sensor Network

Sensor nodes are distributed in the some cells of the physical system (circles in figure 2) to measure some factors in the physical one. These sensor nodes can compose a synchronous WSN (section 3.1) in which each node is a node of a graph $\mathbf{G}_2$ and each edge between 2 nodes is identified thanks to their communication ranges.

In this case, behaviors of the WSN can be operated as followed:

1. Firstly, states for all sensor nodes are initialized to prepare output messages. These states is the initial environmental conditions and insect behaviors sensed from the physical insect system. They are also messages sent through the network.
2. Next, a loop will be executed forever throughout each node.
   - Each node sends messages to its output neighbors.
   - States of the node and neighbors are changed according to their current states and input messages
   - It prepares next output messages.

3.5. Cyber Physical System

A Cyber Physical System (CPS) [13] is a system of collaborating computation and physical processes. Embedded computers and networks monitor and control the physical processes, usually with feedback loops where physical processes affect computations and vice versa. In physical world, the passage of time and concurrency are two core characteristics.

The bio-environmental surveillance network fits into a CPS framework. Environmental factors become physical entities while the observation network is the computation. Sensor nodes in the surveillance network can sense the surrounding conditions and thanks to monitoring the data collected in sensors, people can have some decisions relating to their situations. These build a physical loop between physical entities and computation with timed characteristics.

3.6. BIOSYN

Both physical system and network system are modeled as synchronous networks, therefore, the insect surveillance network can be called as **BIO**environmental **S**urveillance s**Y**nchronous **N**etworks (BIOSYN).

It is necessary to have some synchronizations between these two systems in operating such BIOSYN. The point is that the physical system is a continuous system while the WSN is a discrete one with an interval time between two adjacent sensing times. For example, it is clear that the cricket invasion is continuous, however, the WSN, if applicable, does its jobs discretely. Thus, both physical system and network system need to have some synchronous points to synchronize the data exchange between them (figure 4).

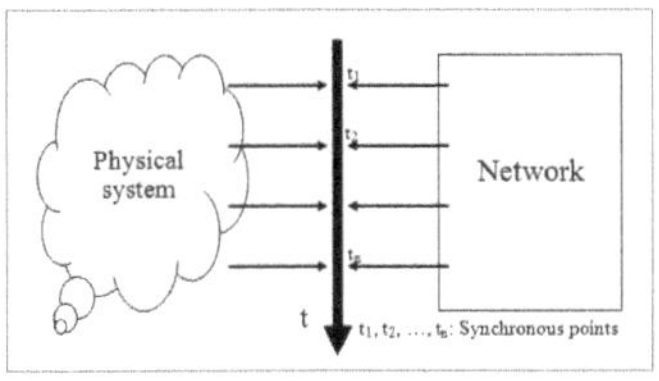

Figure 4. Synchronous points in BIOSYN.

4. Case study: The Brown Planthopper surveillance network in the Mekong Delta

4.1. Problematic

Light trap network is one of solutions for insect management in Vietnam. The network uses light traps to attract insects (due to their phototactics) in order to classify and count insect densities. Thanks to these density values, people can know situations of their fields better and make decision if necessary.

The Wireless Sensor Network approach applied to a light trap network (as proposed solutions in [4][5]), may help calculating Brown Planthopper (BPH) densities and measure environmental factors automatically. This kind of solution uses sensors, new automatic light traps, to measure environments and hopper behaviors. These sensed values from sensors will be sent via a wireless network to a data center periodically. Next, a back-end system will manipulate these values and propose solutions relating to situations of collected data. Such application is called BPH surveillance network.

4.2. Synchronous networks for BPH surveillance

A BPH surveillance network is a network to monitor BPH behaviors due to environmental factors based on WSN approach [4][5][8]. The network consists of 2 systems: insect physical system and network system [1]. These systems can be considered as synchronous networks which use cellular automata as their topologies.

In this network, working space or working environment of hoppers is divided as a grid of cells (a cellular automaton) (figure 5). Some cells of the grid contain automatic light trap sensor nodes to sampling measure surrounding conditions and hopper densities. These sensor nodes compose a massive coordinated sensing machine which its topology is similar to mesh connected WSN.

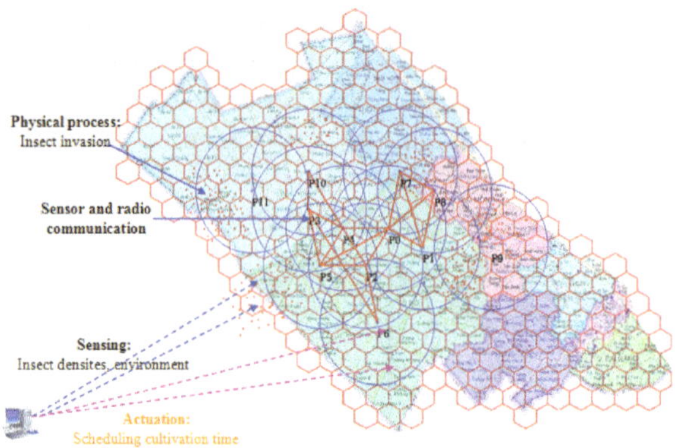

Figure 5. A BPH surveillancenetworkis composedof 2 systems. A insect physical system is the insect's workingspace whichis dividedas cells. These cells are monitoredby a WSN system.

4.3. Insect physical system

The insect physical system represents the insect working space which is divided as cells. Each cell has 4 neighbors (Von Neumann), 6 neighbors (hexagonal) or 8 neighbors (Moore).

The space in figure 5 represents a cellular automaton (**S**, **n**, **f**), where:

1. A finite state set S. A state of a cell describes the hopper status of that cell. This status is calculated thanks to the density of hoppers at that cell. In practice, people use following table [6] (figure 6) to depict hopper statuses in their fields:

BPH density (BPH/m²)	Color	Meaning
< 500	Background	Normal
500 -< 1500	rgb[0,255,0]	Light infection
1500 -< 3000	rgb[255,255,0]	Medium infection
3000 - ≤ 10000	rgb[251,153,234]	Heavy infection
> 10000	rgb[255,0,0]	Hopper burn

Figure 6. Ascendinglevels of infested BPHs in rice fields

Therefore, at the time t, a cell in the hexagonal CA can be valued as an element of the set *{**Normal**, **Light**, **Medium**,**Heavy**, **Burn**}*.

In addition, a cell may locate an automatic light trap sensor node. This trap can catch hoppers and the density of hoppers in the trap may indicate the real infected situation at that cell. The following table [30] (figure 7) is used for describing hopper statuses in a sensor node:

BPH density	Color	Level of hoppers
< 1000	Background	1
1000 -< 2500	rgb[0,255,0]	2
2500 -< 5000	rgb[255,255,0]	3
5000 -<10000	rgb[251,153,234]	4
≥ 10000	rgb[255,0,0]	5

Figure 7. Ascendinglevels of infested BPHs in light traps.

2. Distance *n* identifying neighbor cells, *normally n=1*. When *n=1*, a cell has at most 4, 6, 8 surrounding cells (depends on neighbor type choice).

3. Transition rule f: $S^n \rightarrow S$ depict the change of a cell's state at a specific time based on the current state of the cell and its neighbors. For example, if the center cell and its neighbors have the state **N**ormal at the time t, then the state of that cell at the time t+1 is **N**ormal: f(NNN**N**NNN) = N.

   In fact, the transition rule f is a function depending on some variables such as: density of hoppers in a cell as well as its neighbors, rice age, wind, hopper velocity and other environmental factors.

- Rice age: The young rice is a very good food for hoppers, therefore, they tend to locate at the young rice fields [6][7][15]. On the other hands, hoppers can not suck ripe rice so they will propagate to other fields due to wind if their rice in their current fields become mature or ripe. In addition, young rice is the first condition for hoppers landing. The green color of young rices mapped into the water is a very attractive color source for hoppers, therefore, they tend to take landing to the young ones. On the other hands, ripe rice color does not attract hoppers because they are not sensitive with this color.
- Wind: the wind velocity, calculated in cells/time step. It illustrates the maximum distance that adult hoppers can propagate in a time step. For example, 5 cells/t means that hoppers can propagate to another cell with the distance 5 from the current cell under the wind direction. If there is no wind, hopper can transmit to its neighbor cells. Only a part of adult one can propagate to other fields. In this paper, it is an predetermined constant.
- Hopper velocity. Without wind, hoppers can propagate to near rice fields by this their velocities, approximate 0.4m/s [16].
- Hopper age. Totally, the life circle of BPHs is 26-30 days [15] depending on environmental factors and it spreads in 3 phases: eggs, nymphs and adults. The growth time lapse of each phase is as followed: eggs 6-8 days, nymph 12-15 days, adults 19 days. Some experiments show that a female adult BPH can lay 100-300 eggs during its life circle [15].
- Density of hoppers in a cell and its surroundings. The relation between the

density of a cell and its state depicted in figure 6.

Mathematically, the above cellular automaton is a topology of **a synchronous network** which is modeled as a graph $\mathbf{G_1}=(V_1, E_1)$ where V is the collection of nodes and E is the set of edges.

Nodes. Each cell in figure 5 represents a node in the graph $\mathbf{G_1}=(V_1, E_1)$. By the time passing, each node (or process) *i* composes a collection of states $states_i$ of which a state holds values of rice age, wind, and BPH density at the time t.

Each node may consist a sensor node to sense above factors of a state. When the sensor node senses environment, it transmits the collected data to a gateway for storing and post processing.

Edges. Edges in the graph G_1 are composed by links between a node and its neighbors (4, 6, 8 neighbors). Because G is a directed graph, there are 2 edges between the node and a neighbor of it (2 directions).

Example. Figure 8 is an example of a graph of the BPH surveillance network in Phongdien district, Cantho, Vietnam. In this example, the map of this district is divided as hexagonal cells, each cell is almost a commune of the district. For example, if Phongdien and Nhonai communes are considered as a hexagonal center (**Center cell** in the figure), following communes such as Truonglong, Tanthoi, Giaixuan, Mykhanh, Nhannghia (hexagons 1, 2, 3, 4, 5) become neighbors of the hexagonal center approximately. The hexagon 0 is another neighbor of the center, however, it seems to occupy few area of Phongdien district.

Figure 8. The graph of Phong Dien district, Canthowhenit is dividedas hexagonalcells.

The graph of the BPH surveillance network of this district is illustrated in the right of figure 8. In this graph, each hexagonal cell becomes a node and 2 cells in a neighborhood compose an edge. The **Center cell** is illustrated as a node A which has 6 neighbors named B, C, D, E, G, F (corresponding to cells 0, 1, 2, 3, 4, 5 respectively). However, each of B, C, D, E, G, F only has 3 neighbors.

Behaviors. Behaviors at node *i* are expressed by transition rules of states $transitions_i$. Normally, these rules are functions mapping a collection of states at a cell and its neighbors at the time t to create the new state of that cell at the time t+1. These transition rules are applied simultaneously at every cell.

The following pseudo code depicts the transition rule of a node n:

ALGORITHM: Transition rule (behaviors) of node n at the time t

---

INPUT:

d[n,v]: density of v-days old BPH at the time t-1 (v=1..hopperLifeCircle)

environment: surrounding conditions including wind, temperature, humidity, etc

OUTPUT: state of the node n

```
//At the time t, BPHs increase 1 day old
for(i=2; i<= 28; i++)
  d[n,v] = d[n,v-1];

// Egg: 1..6 days, nymph: 7-14 days;
// adult: 15-28 days
eggDays = 6;
nymphDays = 14;
adultDays=hopperLifeCircle;

//Choose hoppers in day 15-28 to give Eggs,
//if these hoppers were used before,
//choose other ones
//When eggs are created,
//they are accumulated into d[1]
layEggs(15,28,d[n], Environment);

//Calculate death rate with a rate r
calculateHopperDeath(d[n], r, environment);

//density of node n
density = calculateDensity(d);
if (density >= THRESHOLD or
      environment.riceAge IS NOT young)
  if (environment.windVelocity==0) //no wind
      for (j in neighbors of n){
        //Calculate the number of
        // migrating adults from n to j
        propagateHopperNoWind(d,n,j);
      }
      else // has wind velocity
        for (j in neighbors of n)
          if (j is in leeward of n){
            //Calculate the number of migrating
            // adults from n to j  according to
            // the wind velocity
            propagateHopperWind(d,n,j,
                    environment.windVelocity);
          }
      }
```

```
  }

  //calculate incoming hoppers from neighbors
  for(j in neighbors of n){
    calculateIncomminghoppers(d,n,j);
  }

  //calculate state of n at the time t
  state = calculateState(d,n);
}
```

Hopper propagation under wind direction In the above algorithm, the method **propagateHopperWind**(*d, n, j, environment.windVelocity*) calculate the number of adult hoppers migrating from n to j. This hopper migration number can be calculated as followed:

Let node *n* is a source cell which is able to propagate *d adult hoppers* under the wind velocity *v (m cells/t)*. Thus, the wind velocity causes adult hoppers distribute to *at most m cells* in a period of time *t*. In this case, the number of hoppers propagating to a cell n_k which has *distance* k from the *source cell n* under the wind direction, is estimated as $\frac{d}{2^k}$ (figure 9).

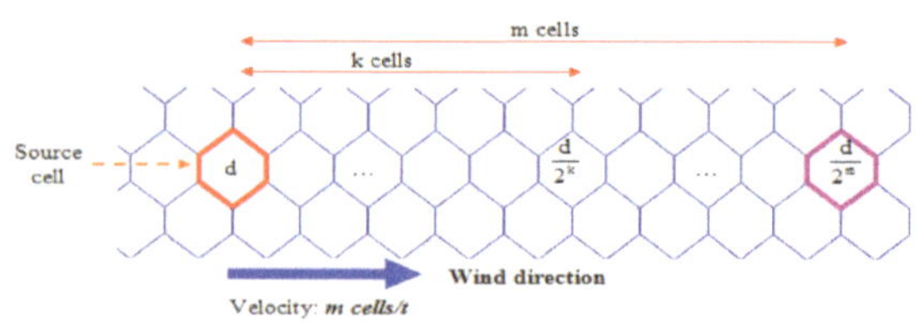

Figure 9. Estimationof propagatedadult hoppersdue to wind.

4.4. HopperobservationWireles Sensor Network

A WSN is created by distributing sensor nodes in some cells to sampling measure environmental factors and hopper density (sensor P1, P2,... in figure 5). Data collections from the WSN can be considered as indications for the operating of the whole BPH surveillance network as well as for post processing. The WSN can be considered as a synchronous network.

The WSN represents a mesh network by being considering as a graph $\mathbf{G}_2=(V_2, E_2)$ where V_2 is a collection of nodes and E_2 is a collection of edges. A node in V_2 represents a sensor node while an edge is created between 2 sensor nodes if their distance is at most the transmission range (figure 5). In addition, each sensor node in the WSN has its own behaviors.

Behaviors. Behaviors of the WSN depict the data dissemination inside the network. This data dissemination includes data sensing, packaging, and transmitting via the network topology.

Sensor nodes of the WSN can sense environmental factors and hopper density. Sensor nodes are able to measure temperature, humidity, wind, light intensity, which are factors influencing hopper behaviors. Besides, nodes can also support in situ hopper density counting or estimation.

Data packet Before transmitting, data is packed as packets which each of them has the structure as in figure 10. In this structure, ID is the unique ID of the packet and location ID represents the ID of the sensor node that senses environmental factors. The time that the sensor node senses data is described by a time stamp. Source ID depicts the sensor node that sends the packet while destination ID is the ID of next node that the packet is received. Normally, the destination ID is identified thanks to a routing table [31], a table that routes packets to a sink (or a gateway). These attributes compose a header of the packet. In addition to the header, the packet contains a data part which stores surrounding conditions such as light intensity, temperature, humidity, wind velocity, wind direction and hopper density.

HEADER						DATA					
ID	time Stamp	source ID	destID	location ID	data Size	density Hoppers	light Intensity	temperature	humidity	wind Velocity	wind Direction

Figure 10. Packet structurein the hopperobservationWSN.

A packet here depicts a structure to maintain a piece of spatial-temporal data. Indeed, time stamp illustrates the temporal aspect while location id describes the spatial one and the data part of the packet becomes data aspect. Figure 11 is an example of this piece of data. It can be translated as: at the time *01/02/2016 08:07:56 AM*, the sensor node *10* has *500 hoppers caught* at the temperature 29^oC and *5.5 km/h* wind velocity.

HEADER						DATA					
ID	01/02/2016 08:07:56 AM	...	...	10	...	500	...	29	...	5.5	...

Figure 11. Exampleof a piece of spatial-tempoal data.

Consequently, a packet can be declared as the following C based pseudo code:

```
typedef struct{
  int ID;
  int sourceID, desID;
  // local ID of sensor node that senses data
  int location;
  //time stamp of sensing
  int timeStamp;
  int dataSize;
  float densityHoppers;
  float lightIntensity;
```

```
    float temperature, humidity;
    float velocityWind;
    float directionWind;
}LocalPacket;
```

Packet dissemination These behaviors compose of sensing surrounding conditions, packing into a packet and sending the packet to a gateway node (figure 12).

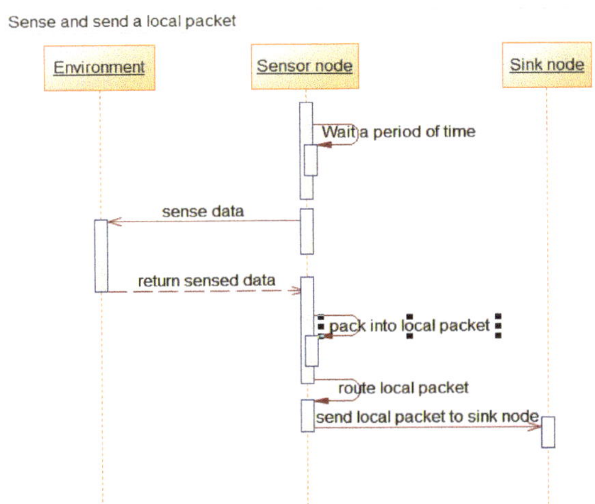

Figure 12. Procedureof sensing and sendinga packet.

Figure 12 depicts behaviors of a sensor node in the hopper observation WSN. First, next execution time of a sensor node is identified by a timer. Next, the sensor node measures surrounding conditions and receives sensed data from environment, then these pieces of data are packed as a packet (as in figure 10). A pre-calculated routing table is used to route the local packet to its gateway where data is concentrated.

Timer In the WSN, timer is a mechanism calculating the time when nodes (sensor, sink and gateway) execute their tasks. This time can be figured out based on sensing times of sensor nodes and communication times to transmit packets to the gateway.

Assume that each sensor node contains n sensors, t_i is a duration of time to read sensor i. Therefore, $T_{read} = \{t_1, t_2, ..., t_m\}$ is a collections of times to read sensors in a sensor node.

Let $T_{sensing}(s)$ is the sensing time, a duration of time that a sensor node finish reading all its sensors. This time is calculated as:

$$\Rightarrow T_{sensing} = Max\{t_1, t_2, ..., t_m\} = Max\{T\} \quad (1)$$

To transmit data to the gateway, following factors are considered:

- f: the frequency used to transmit data (Hz).
- DataSize: sizes of a packet(bit).
- DataRate: the rate (speed) to transmit data ($Kbps$). This rate depends on the frequency used to transmit data.

The communication time $T_{local\_transmit}$ from one node to another node in the WSN is shown as below:

$$T_{transmit} = \frac{DataSize}{DataRate} \quad (2)$$

Therefore, the duration of time that a sensor finish executing its tasks (Figure 8) is calculated as:

$$T_{node} = T_{sensing} + T_{transmit} \quad (3)$$

Let $T_{to\_gateway}$ be a duration of time that the gateway finishes collecting data from all other nodes (figure 8), then $T_{to\_gateway}$ is estimated as:

$$T_{node} \leq T_{to\_gateway} \leq nT_{node} \quad (4)$$

where n is the number of sensor nodes of the WSN. Indeed, in the best case, all sensor nodes transmit their data directly to the gateway at the same time, then $T_{to\_gateway}$ is approximately T_{node}. However, in the worst case, sensor nodes perform sequentially, then $T_{to\_gateway}$ is around nT_{node}.

The interval time T between 2 next actions of a sensor node in the BPHSUN can be calculated as the maximum of $T_{to\_gateway}$ after adding an error-time Δt. Therefore, T is estimated as:

$$T = nT_{node} + \Delta t \quad (5)$$

$$\Rightarrow T = n(T_{sensing} + T_{transmit}) + \Delta t \quad (6)$$

Routing table Routing table [31] is a structure to store shortest paths from a node to other nodes in a WSN. In the BWSN, because automatic light trap sensor nodes located at fixed positions, routing table is pre-calculated and stored to disseminate data to the gateway.

Distance vector algorithm [31] is used to calculate routing tables for a WSN $G(V, E)$. The algorithm is shown as below:

ALGORITHM: Calculate a routing table for node v to other nodes in $G(V, E)$

INPUT: c[v, w]: the direct cost from v to w ($w \in V$).
OUTPUT: D[v, w]: distance between v and w ($w \in V$).
Next(v, w): next node to reach to w from v. ($w \in V$)

```
createRoutingtable(){
  for each (w in V)
        D[v, w] = 0;
  D[v, v] = _INFINITY;
  // Find route from v to others
  for each (w in V){
```

```
            if (w == v) continue;
            /* Select the shortest distance
               from v to its neighbors
                  after adding direct costs */
            minCost = D[v, w]; // old distance
            neighbor = -1;
            nextNode = Next(v, w);

            for each (j in v.getNeighbors()){
                    neighbor = j;
                    newCost = c[v,j] + D[j, w];
                    if (newCost < minCost){
                            minCost=newCost;
                            nextNode=neighbor;
                    }
            }

            // Update routing table
            D[v,w]=minCost;
            Next[v,w]=nextNode;
    }
}
```

The above algorithm is executed parallely for all nodes $v \in V$, then paths from all node $v \in V$ to other nodes are found after 1 execution step. This procedure of calculating routing tables for all nodes is looped until distances are unchanged.

Example Let a WSN given by a graph $G_2(V_2, E_2)$ (figure 13). Assume that the direct cost from a node to its neighbors is 1.

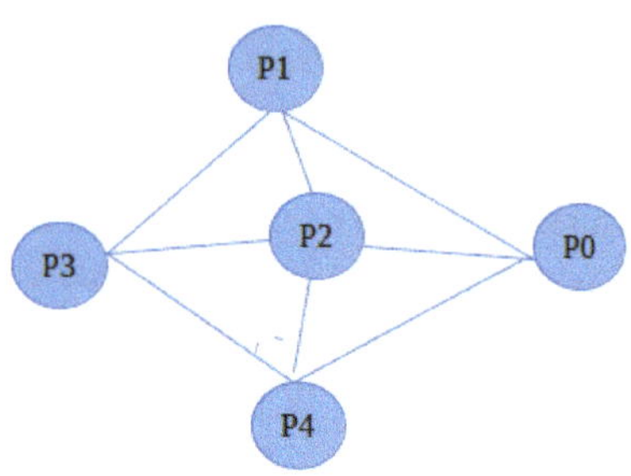

Figure 13. Exampleof a graph to calculate routingtables.

The following result are routing tables of all nodes in the graph G_2 after 2 execution steps. After 2 steps, routing tables of all nodes are unchanged, thus, they become final routing tables.

Routing packets It is a process of sending and receiving packets at each sensor node based on a routing table so that all packets are concentrated at the gateway.

Before sensed data is transmitted, it is packed as a packet (as figure 10). The packet consists of an ID, source ID, destination ID, location, time stamp and other measured values of environmental factors. The

	P0		P1		P2		P3		P4	
	D	*Next*	*D*	*Next*	*D*	*Next*	*D*	*Next*	*D*	*Next*
P0	0	-1	1	0	1	0	2	1	1	0
P1	1	1	0	-1	1	1	1	1	1	1
P2	1	2	1	2	0	-1	1	2	1	2
P3	2	1	1	3	1	3	0	-1	1	3
P4	1	4	1	4	1	4	1	4	0	-1

Figure 14. Routingtables after 2 steps.

attribute location depicts the local ID of the sensor node that senses these surrounding conditions.

Example: Assume that P1,P2,P3,P4 are sensor nodes and P0 is the sink node of the graph in figure **??**, according to routing tables in figure 14, collected data from P1, P2, P4 can send directly to P0 while P3 needs 2 hops to reach to P0 (P3 - P1 - P0). Figure 15 depicts a round how sensed data from sensor nodes reaches the sink node P0.

HEADER						DATA					
1	...	P1	P0	P1	...	500	...	30	...	...	...
2	...	P2	P0	P2	...	550	...	29	...	...	...
3	...	P3	P1	P3	...	600	...	30	...	...	...
4	...	P4	P0	P4	...	650	...	29.5	...	...	...
5	...	P1	P0	P3	...	600	...	30	...	...	...

Figure 15. Exampleof routinglocal packets.

To implement this mechanism, a sending buffer and a receiving buffer are maintained at each sensor node. These buffers are used in 2 following methods: sendLocalPackets() and receiveLocalPackets() to send and receive local packets at a node v. All nodes in a sub network at level 1 execute these methods concurrently.

Sending packets Sending packets takes place at sensor nodes of the WSN after they are granted execution times from the timer. First, environmental factors are sensed by sensors of a node and these pieces of sensed data are packed as a packet. Next, the packet is added to sending buffer of the node. The next step is to move all packets from the node's receiving buffer to its sending buffer in order to send to the gateway thanks to the routing table.

The algorithm is described as followed:

ALGORITHM: send local packets at a sensor node

INPUT: Node *v*, routing table *t*, receiving buffer *v.receiveBuff*

OUTPUT: sending buffer *v.sendBuf*

```
sendLocalPacket(){
 if (isExecutionTime()){
   if (!isLeader(v)){
     senseData();
//Sense data from environment
     createPacket(p); //Create a packet
```

```
      l = Leader(v);    //Find leader of v
      //Next hop of v is destination ID of p
      p.destinationID = t.Next[v][l];
      v.sendBuff.Add(p);
      for each (packet p in v.receiveBuff)
        v.sendBuff.Add(p);
      v.receiveBuff.Clear();
    }
  }
}
```

Receiving local packets When sensor nodes are granted execution times, they start receiving packets from their neighbors. Main idea of receiving packets at a node v is to locate in the sending buffer of each neighbor of v in order to find packets considering v as their destination IDs, then these packets are added to the receiving buffer of v.

ALGORITHM: receive local packets at a node

INPUT: Node v, routing table t
OUTPUT: receiving buffer *v.receiveBuf*

```
receiveLocalPacket(){
 if (isExecutionTime()){
   l = Leader(v); //Find leader of v
   for each (Neighbor j of v){
     for each (packet p in j.sendBuf){
       desID = p.destinationID;
       if (desID == v.ID){
         //Next hop of v is destination ID of p
         NextID =  t.Next[v][l];
         p.destinationID = NextID;
         v.receiveBuff.Add(p);
       }
     }
   }
 }
}
```

4.5. Cyber Physical System synchronization

Sensor nodes are deployed into some cells to sense meteorological conditions and to count the BPH density. In this problem, sensor nodes are automatic light traps which are able to sense these above factors and transmit these sensed values to a gateway. Circles in figure 5 depict communication ranges of sensor nodes while centers of these circles are sensor node localizations.

Normally, manual light traps are turned on at night to attract insects and insect densities are counted in the next morning. Therefore, automatic light trap sensor nodes almost sleep during day time. In addition, during night time, environmental conditions and hopper densities do not need to be sensed frequently; instead, it is suggested that they are calculated every interval time calculated in formula (6).

Synchronous points $t_0, t_1, ..., t_n$ (as in figure 4, section 3.6) between physical system and WSN system are calculated thanks to the interval time in formula (6). These synchronous points are necessary because the physical system is continuous while the WSN operates discretely. In addition, light traps are turned on at night to attract insects and insect densities are counted in the next morning; during night time, environmental conditions and hopper densities do not need to be sensed frequently because they do not change so much in a short period of time.

5. Bio-environmental surveillance network data model

A bio-environmental surveillance network plays important role in collecting data. Thanks to the current data collection and historical data, people can know better their solutions and can make suitable decisions. For example, in comparison with historical light trap data, the current trap data may indicate the peak season of BPH and people can propose a suitable cultivation time.

The data model of the bio-environmental surveillance network consists of 2 parts: meta-data model and data collection model. Meta-data model depicts the meta structure of a bio-environmental surveillance network, includes of the physical system and WSN system. On the other hand, data collection model is the model of collected data; it depends on what types of data people want to collect or sense.

5.1. Meta-data model

Figure 16 illustrates the meta-data model of a bio-environmental surveillance network. In this model, TCellularSystem, TCell, TSensorNode are important entities to depict 2 systems in a surveillance network: physical and network system. However, the main focus of the model is the entity TCell and its relationships.

Cell is a unit of the surveillance network. Each cell maintains a position in the image (xPos, yPos) and its geographical location (longitude, latitude, elevation). Each cell can be a Von Neumman type (4 neighbors), hexagon type (6 neighbors) or Moore type (8 neighbors).

The entity TCellSystem maintains the cellular system of the surveillance network. It can show the width and the height of a Von Neumman and Moore cell, or show the radius of a hexagonal cell. Besides, when a cellular system is created, a time stamp and a name is assigned to it. Each cellular system has an own ID.

Each pixel of the map of the physical system is stored in the entity TPixel. RGB color space is used to process image data. Besides, each pixel is in a fixed position (x,y)

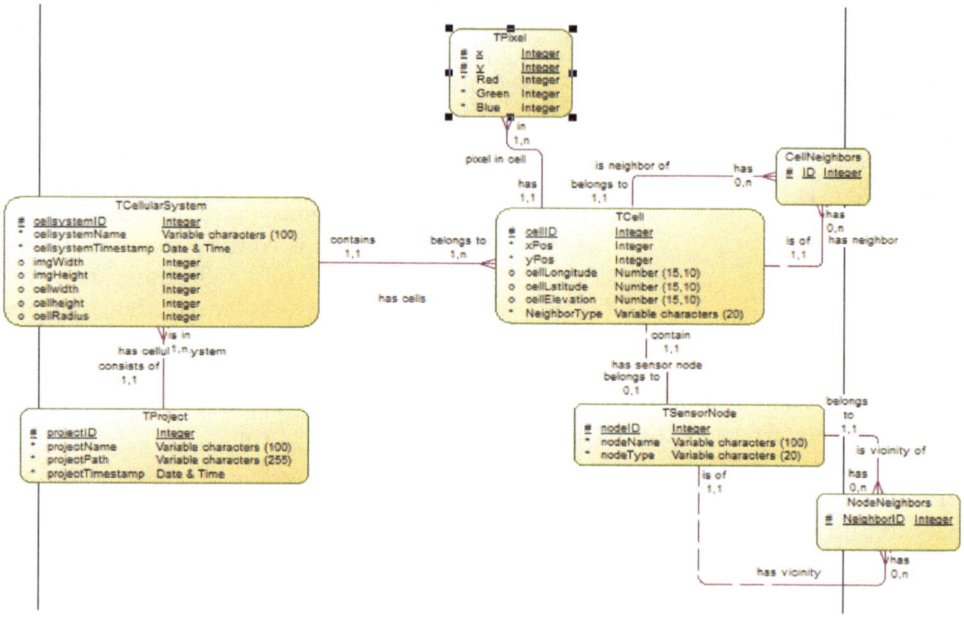

Figure 16. Data modelfor bio-environmentalsurveillancenetwork.

in the image. A pixel belongs to a cell, but a cell can consist of many pixels.

The entity CellNeighbors maintains neighbors of cells. Each cell can have 4, 6, 8 neighbors. Each neighbor is assign an order from 0..3 (4 neighbors), 0..5 (6 neighbors), 0..7 (8 neighbors).

These above entities illustrate the physical system of a BIOSYN, however, entity TSensorNode depicts the WSN system in the BIOSYN. Each sensor node is assigned an ID, a name and type (sensor node, sink node, gateway). Each sensor node is in a cell while a cell can contain at most a sensor node.

In a more general case, people can have many cellular systems interacting each other, the entity TProject is introduced to solve it. A project can have many cellular system while a cellular system belongs to only one project.

5.2. Data collectionmodel

Different surveillance applications have different data collection model. For instance, water level may be taken into account in the data collection model of a flooding surveillance. However, temperature, humidity, light intensity, wind, hopper density are considered as factors in a BPH surveillance data model.

To be simple, the data model consists of 2 entities TData, TSensingData which has relations with TCell and TSensor, respectively. Fields in these 2 entities are almost the same such as: time stamp and other environmental factors. The difference is the TData depicts the data generated from the physical system while TSensingData contains sampling sensing data from surrounding conditions.

For example, the data collection model of a BPH surveillance network is as followed (figure 17):

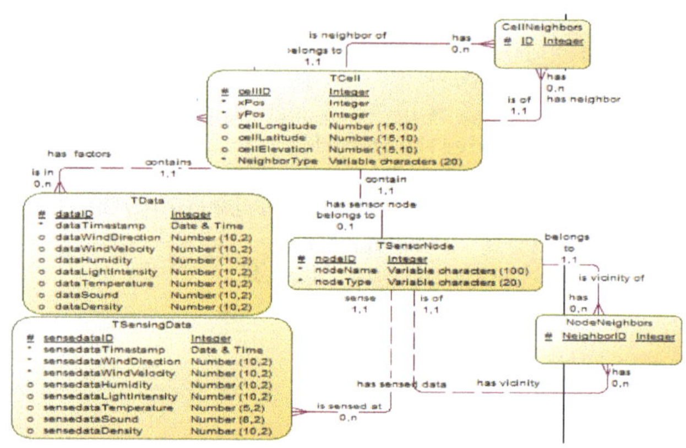

Figure 17. Data modelfor BPH surveillance.

6. Implementation

6.1. Workflo

The implementation of a BIOSYN contains 3 important parts: data structure, states and behaviors. Firstly, data structure (cells in figure 2) is generated from geographic data. Next, states and behaviors are implemented in CUDA [20] to illustrate synchronous characteristics in the model.

CUDA is chosen for implementing BPHSYN since the parallel programming paradigm of CUDA is well-suited for the model's concurrency.

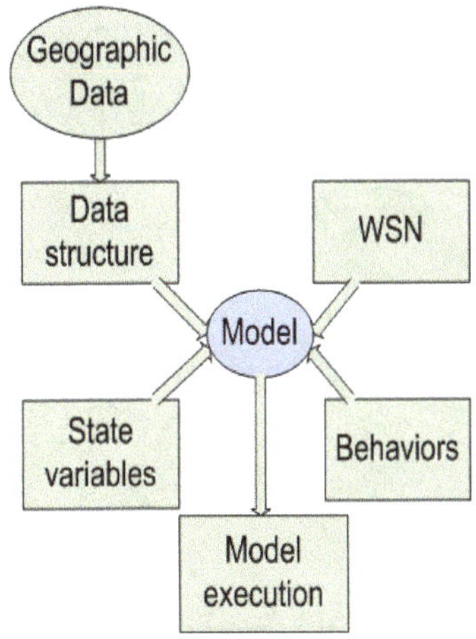

Figure 18. Workflo for modelingphysical and WSN system.

6.2. Physical system data structure

Graph. The following CUDA code is the definition of the physical system graph. Supposes that the physical system uses hexagonal cellular automata.

```
typedef struct{
  int x, y;
}Point2D;
typedef struct{
  int xPos, yPos; //offset coordinates
  Point2D Corners[6]; // 6 corners
  int Neighbors[6]; // 6 neighbors
}Hexagon;
// Honeycomb structure
typedef Hexagrid Hexagon[N];
```

State variables. Each node in the graph represents a process in a synchronous network. This process maintains a structure to store its state variables by the time passing. The following code is the definition of a state variable in the case study BPH surveillance network:

```
typedef struct{
  int xPos, yPos; //position
  float         riceAges;    // rice ages
  InsectDensity densityBPH;
  float         windVelocity;
  Direction     windDirection;
}State;
```

6.3. Wireles Sensor Networkdata structure

Sensor nodes are distributed in some cells of the graph. The following code describes the definition of the WSN in a bio-environmental surveillance network:

```
typedef struct{
  int node;
  float distance;
}Neighbor;
typedef struct{
  //offset coordinates of the hexa cell
  int xPos, yPos;
  // neighbors of sensor node
  Neighbor Neighbors[MAXFLOW];
}Channel;
typedef Network Channel[N]; // WSN
```

6.4. Behaviorsimplementation

Behaviors of the physical system and WSN system depends on different surveillance networks. For example, these behaviors of a BPH surveillance network are implemented as descriptions in 4.3 and 4.4.

7. Experiment

7.1. Data used

The simulation of the BPH synchronous network uses data collections in Cantho city (figure 5), a typical rice city in the Mekong Delta. Current light traps (8 traps till 2015) are considered as sensor nodes, automatic light trap sensor nodes, in the simulation (circles in figure 5). Hexagonal cellular automata is used by dividing the area of Cantho into hexagonal cells with approximately $0.18km^2$ each.

7.2. Tools used

Some tools are used in this experiment. A self developed tool HexGen is used to generate codes of hexagonal cellular automata synchronous networks in Cuda. Besides, the map of Cantho city is processed by the tool PickCell in the framework NetGen [18]. Behaviors of the BPHSYN are implemented in CUDA to run the simulation on the NVIDIA card GeForce GTX 680 1.15GHz with 1536 CUDA Cores (8 Multiprocessors x 192 CUDA Cores/MP).

7.3. Sensor nodeintervaltimecalculation

According to Formula 6 the sensor node interval time of the BPHSUN depends on sensing times and communication times.

Actually, sensing time of a sensor is the response time of that sensor. This time relies on type of sensors as well as concrete surrounding conditions. A sensor node here is an automatic light trap [35] consists of following sensors shown in figure 19. This figure also depicts the response time of each sensor [32][33][34].

Therefore, the sensing time is $T_{sensing}$ = 20s. There are 8 sensor nodes ⇒ n = 8.

LORA technology [28] is used to transmit data in the BPH surveillance network in Mekong Delta[35]. The board Semtech SX1276 [28] is used since it is suitable for allowance frequencies in Vietnam. The specification of this board shows that it has 0.018-37.5 kbps data

Sensor	Response time (s)
Wind sensor WindSonic M	0.25
Temperature and humidity sensor DT22	20
Light sensor BH1750FVI	Approximately human eye response

Figure 19. Responsetimesof sensorsin a light trap sensornode

rate, the frequency 433Mhz. Assume that the board is configured to work with 11kbps data rate.

According to 10, the size of each packet is 48 bytes ($DataSize = 48$).

$$T_{transmit} = \frac{DataSize}{DataRate} = \frac{48 * 8}{11 * 1000} = 0.384s$$

According to formula 6, the interval time T is calculated as:

$$\Rightarrow T = 163.072 + \Delta t(s)$$

Because environmental factors do not change so much during a short period of time, $\Delta t = 1636.928$s is chosen $\Rightarrow T$=1800s = 30 minutes. Thus, the interval time between 2 adjacent actions of a sensor node is 30 minutes. If a light trap works 4 hours every night, the sensor node senses environment and transmits data 9 times.

7.4. Scenario: physical system and networksystem synchronization

This simulation depicts the interaction between the insect physical system and WSN system in the BPH surveillance network. Sampling data from sensor nodes play as inputs for the insect physical system. On the other hands, when the insect physical system occurs, the collected data from sensor nodes illustrates the damage level caused by BPHs.

In the beginning, the WSN gives environmental conditions at 8 locations of sensor nodes. The temperature is 30^oC (the common temperature in Mekong Delta). The wind direction is 2 (the direction from the **center cell** to **cell 5** in figure 8 with the velocity is 5km/h (approximately 10 hexagonal cells/h). That means BPHs can transmit to the cell distance 10 from the source cell in one hour. Besides, almost rice fields in Cantho are almost young and suffer lightly from hoppers (light infection color of rice fields and sensor nodes as in figure 20).

The density of hoppers tends to increase in the next days. Due to wind, hoppers spread in other rice fields of Cantho and their densities it rises gradually until the 5th day. On this day, hopper burns appear in almost communes in Cantho while few places are light or medium infection (figure 22). The infected area reaches maximum value on the 13th day, then it begins

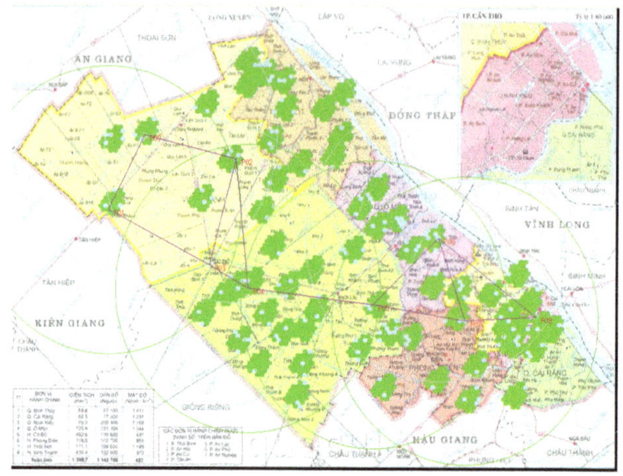

Figure 20. BPHs in rice field and sensor nodes in the beginning.

decreasing. On the 30th day, it almost becomes normal on the rice fields and in sensor nodes.

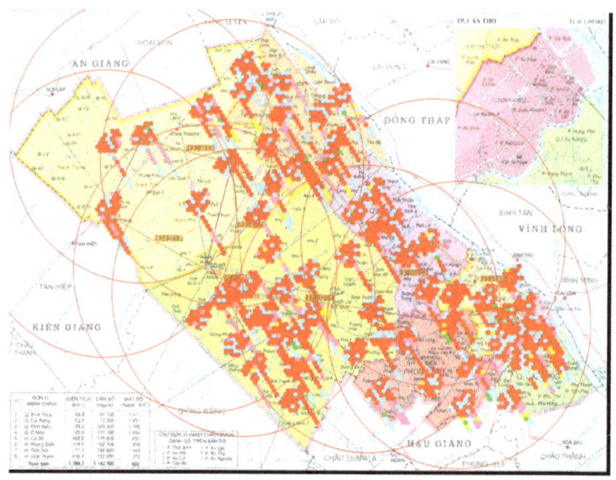

Figure 21. Hopperburnson the 13th day.

Figure 22 illustrates tendencies of BPHs in rice fields and automatic light trap sensor nodes during this simulation. Starting with light infections in the beginning, densities of hoppers increase gradually and reach the peak point at the $13^{th}, 14^{th}, 15^{th}$ day. From that day, they decrease slowly and they almost become normal densities at the 30th day.

Figure 22 also describes that the density of hoppers caught in automatic light trap sensor nodes is ratio with that of hoppers in rice fields. In other words, data collection in sensor nodes can be used as an indication for the hopper infection in rice field. However, there is not many investigation about the relation between them.

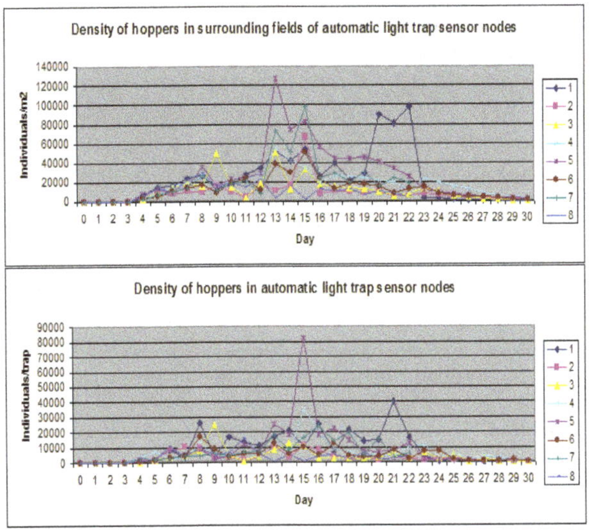

Figure 22. Rice field and automatidight trap sensornodedata collectionafter 30 days of simulation.

8. Conclusion

We have described the a bio-environmental surveillance network BIOSYN as synchronous networks. In our work, a BIOSYN is an interaction of a physical process, modeled as a synchronous network, and a synchronous WSN. Environmental factors sensed in the WSN provide inputs for the execution of the bio-environmental physical system. On the other hands, the effect of the bio-environmental phenomena is sampling measured by sensor nodes. These composes a feed back loop between the physical system and the network system.

Cellular automata are considered of the bio-environmental surveillance network topology, especially in the physical system. Indeed, bio-environmental phenomena behave in an environment which is divided as cellular cells of which each has 4, 6, 8 neighbors. These cells compose a synchronous network which is modeled as a graph by considering each cell is as a node and two vicinity cells compose an edge. In addition, concurrent characteristics of the network are depicted as node behaviors.

The mesh network is used to disseminate data via the WSN network. The WSN is considered as a graph of which nodes are sensor nodes and edges are composed thanks to sensor node communication ranges. A timer is used to schedule sensor node behaviors in synchronous rounds to disseminate data to a gateway. A database schema is also introduced to maintain historical sensor node sampling data for post processing.

9. Acknowledgements

The work of this paper was in the scope of the "OBSNET - Optimizing the Brown Planthopper surveillance network using automatic light traps in the Mekong Delta region" project of the Ministry of Education and Training, Vietnam. The first author was funded by the program 911 of Ministry of Education and Training, Vietnam.

References

[1] Bao Hoai Lam, Hoang Van Tran, Hiep Xuan Huynh and Bernard Pottier (2015) *Synchronous network for insect surveillance* (ACM), in Proceedings of the Sixth International Symposium on Information and Communication Technology, Hue City, Vietnam, December 3-4, 2015, ISBN 978-1-4503-3843-1, pp. 27–34.

[2] Kelvin F. Conrad, Richard Fox and Ian P. Woiwod (2007) *Monitoring biodiversity: measuring long-term changes in insect abundance* (Cabi Publisher), Insect conservation biology, ISBN 9781845932541, pp. 203–225.

[3] Woiwod, I. P. and Harrington, R. (1994) *Flying in the face of change: The Rothamsted Insect Survey* (Wallingford: CAB International), in Long-term Experiments in Agricultural and Ecological Sciences (ed. R. A. Leigh & A. E. Johnston), pp. 321–342.

[4] Hoai Bao Lam, Tai Tan Phan, Long Huynh Vuong, Hiep Xuan Huynh, and Bernard Pottier (2013) *Designing a brown planthoppers surveillance network based on wireless sensor network approach*, in: ISCRAM (Information Systems for Crisis Response and Management) Vietnam 2013 Conference.

[5] Bao Hoai Lam, Hiep Xuan Huynh, Mahamadou Traoré, Pierre Yves Lucas, and Bernard Pottier (2014) *Monitoring environmental factors in Mekong Delta of Vietnam using Wireless Sensor Network approach*, in: 8th International conference on Simulation and Modelling in the Food and Bio-Industry FoodSim'2014, ISBN 978-90-77381-84-7, pp. 71–78.

[6] Cuong H. Phan, Hiep Xuan Huynh, and Alexis Drogoul (2010) *An Agent-Based Approach to the Simulation of Brown Plant Hopper (BPH) Invasions in the Mekong Delta* (IEEE), in: 2010 IEEE RIVF International Conference on Computing & Communication Technologies, Research, Innovation, and Vision for the Future (RIVF), Hanoi, Vietnam, 1–6.

[7] Nguyen, Vinh Gia Nhi and Huynh, Hiep Xuan, Vo, Tung Thanh, and Drogoul, Alexis (2011) *On Weather Affecting to Brown Plant Hopper Invasion Using an Agent-based Model* (ACM, New York, NY, USA), in: Proceedings of the International Conference on Management of Emergent Digital EcoSystems, MEDES '11, ISBN 978-1-4503-1047-5, pp. 150–157.

[8] Viet Xuan Truong, Hiep Xuan Huynh, Minh Ngoc Le, and Alexis Drogoul (2012) *Modeling a Surveillance Network Based on Unit Disk Graph Technique - Application for Monitoring the Invasion of Insects in Mekong Delta Region* (Springer), in: PRIMA 2012: Principles and Practice of Multi-Agent Systems - 15th International Conference, Kuching, Sarawak, Malaysia, pp. 228–242.

[9] http://www.healthyforestpartnership.ca/

[10] Marc Rhainds, and Stephen B. Heard (2014) *Sampling procedures and adult sex ratios in spruce budworm*, Entomologia Experimentalis et Applicata.

[11] Marc Rhainds (2015) *Wing wear and body size measurements of adult spruce budworms captured at light traps: inference on seasonal patterns related to reproduction* (Springer Japan), in Applied Entomology and Zoolog, Volumn 50, no. 4, pp. 477–485, ISBN 0003-6862.
[12] K. L. Heong, and B. Hardy (2009) *Planthoppers: new threats to the sustainability of intensive rice production systems in Asia* (International Rice Research Institute), ISBN 978-90-77381-84-7.
[13] Edward A. Lee (2010) *CPS Foundations* (ACM), in Proc. of the 47th Design Automation Conference (DAC), pp. 737–742.
[14] Lynch, Nancy A (1996) *Distributed Algorithms* (Morgan Kaufmann Publishers Inc), ISBN 155860348.
[15] WH. Reissig, E.A Heinrichs, J.A. Litsinger, K. Moody, L. Fiedler, W.Mew, and A.T. Barrion (1986) *Illustrated guide to integrated pest management in rice in tropical Asia*, IRRI, Philippines.
[16] S. Cheng, J. Chen, H. Si, L. Yan, T. Chu, C. Wu, J. Chien, and C. Yan (1979) *Studies on the migrations of brown planthopp ers nilaparvata lugens*, Std. Acta Entomol. Sinica, 22: pp. 1–21 (Chinese, English summary).
[17] Yingshu Li, My T. Thai, and Weili Wu (2008) *Wireless Sensor Networks and Applications* (Springer).
[18] Bernard Pottier, and Pierre-Yves Lucas (2015) *Dynamic networks "NetGen: objectives, installation, use, and programming"*, Université de Bretagne Occidentale, https://github.com/NetGenProject.
[19] Guillaume Kremer, Jimmy Osmont and Bernard Pottier (2009) *A Process Oriented Development Flow for Wireless Sensor Networks*, International Workshop on Smalltalk Technologies - ESUG'09 Brest.
[20] NVDIA, https://developer.nvidia.com/cuda-zone.
[21] GPU, http://www.nvidia.com/object/what-is-gpu-computing.html.
[22] IEEE Standard for Modeling and Simulation (M&S) High Level Architecture (HLA)– Federate Interface Specification, IEEE Std 1516.1-2010, pp. 1–378.
[23] Dufay, C (1964) *Contribution a l'Étude du phototropisme des Lépidopteres noctuides*, Annales des Sciences Naturelles Zoologie, Paris, vol 12. no 6, pp. 81–406.
[24] Stephen Wolfram (1994) *Cellular Automata and Complexity* (Westwiew Press, Perseus Books Group), ISBN: 0-201-62716-7.
[25] Stephen Wolfram *Statistical mechanics of cellular automata* (American Physical Society), Reviews of Modern Physics vol. 55, pp. 601–644.
[26] Ho Van Chien, Nguyen Huu Huan, and Le Quoc Cuong (2012) *Escape Strategy can successfully manage BPH and virus disease in the Mekong*, http://ricehoppers.net/2012/09/escape-strategy-can-successfully-manage-bph-and-virus-disease-in-the-mekong
[27] José Cecílio, and Furtado Pedro (2014) *Wireless Sensor Networks: Concepts and Components* (Springer International Publishing), in Wireless Sensors in Heterogeneous Networked System, Computer Communications and Networks, pp. 5—25.
[28] LORA technology, http://www.semtech.com/wireless-rf/rf-transceivers/sx1276/.
[29] G. Lasnier, J. Cardoso, P. Siron, C. Pagetti, and P. Derler (2013) *Distributed simulation of heterogeneous and real-time systems*, in Distributed Simulation and Real Time Applications (DS-RT), 2013 IEEE/ACM 17th International Symposium on, pp. 55—62.
[30] Truong Chi Quang, Vo Quang Minh, Nguyen Thi Hong Chau, and Ho Van Chien (2013) *Managing the Brown Planthopper caught by light traps for supporting of rice cultivation in the Mekong Delta*, in National GIS application conference (in Vietnamese).
[31] Vutukury, S.; Garcia-Luna-Aceves, J.J. (2001) *MDVA: a distance-vector multipath routing protocol*, in INFOCOM 2001, Twentieth Annual Joint Conference of the IEEE Computer and Communications Societies. Proceedings. IEEE , vol.1, no., pp.557–564 vol.1, doi: 10.1109/INFCOM.2001.916780.
[32] WindSonic M datasheet, http://gillinstruments.com/data/datasheets/windsonic-m.pdf?v=01.2014
[33] DT22, http://www.dfrobot.com
[34] BH1750FVI, http://www.electronic-circuits-diagrams.com
[35] Ky Minh Nguyen, Bao Hoai Lam, Tuyen Phong Truong, Hieu Minh Thi Pham, Chien Van Ho, Bernard Pottier and Hiep Xuan Huynh (2015) *Automatic hopper light trap*, in Cantho University Journal of Science, pp. 168–178, Cantho University.
[36] Hefeeda M., Bagheri M. (2007) *Wireless Sensor Networks for Early Detection of Forest Fire*, in Mobile Adhoc and Sensor Systems, MASS 2007, pp. 1-6.
[37] Government of Canada http://cwfis.cfs.nrcan.gc.ca/

20

A federation of simulations based on cellular automata in cyber-physical systems

Hoang Van Tran[1,*], Hiep Xuan Huynh[1], Vinh Cong Phan[2], Bernard Pottier[3]

[1]Department of Software Engineering, College of Information & Communication Technology, Cantho University, Cantho city, Vietnam
[2]Department of Computer Engineering, Faculty of Information Technology, Nguyen Tat Thanh University, Hochiminh city, Vietnam
[3]Département Informatique, Université de Bretagne Occidentale, Brest, France

Abstract

In cyber-physical system (CPS), cooperation between a variety of computational and physical elements usually poses difficulties to current modelling and simulation tools. Although much research has proposed to address those challenges, most solutions do not completely cover uncertain interactions in CPS. In this paper, we present a new approach to federate simulations for CPS. A federation is a combination of, and coordination between simulations upon a standard of communication. In addition, a mixed simulation is defined as several parallel simulations federated in a common time progress. Such simulations run on the models of physical systems, which are built based on cellular automata theory. The experimental results are performed on a federation of three simulations of forest fire spread, river pollution diffusion and wireless sensor network. The obtained results can be utilized to observe and predict the behaviours of physical systems in their interactions.

Keywords: Cyber-physical system (CPS), cellular automata (CA), federation, high-level architecture (HLA), mixed simulations.

1. Introduction

In recent years, more and more research focus on the cyber-physical systems (CPSs) [1], which are defined as integrations of computation, networking, and the physical systems. Taking advantages of wireless sensor network (WSN) [12], the sensing ability of CPSs is widely considered over the last years. This ability allows CPSs to be able to sense the physical world and transmit sensed data to base station for performing studies, analysis, and decision making.

Simulating the sensing ability before real implementations highly reduces cost and effort of the development of CPSs. However, due to uncertain interactions of complex physical systems, simulating such type of system is much more complicated compared to the traditional computing systems. One of the critical challenges is involving of interoperability in the models.

Recently, several approaches have been suggested to confront with that issue [5][6][20]. But, they do not so far consider on federating physical systems instead of tightly combining of existing tools and languages. Furthermore, those solutions are targeting embedded systems. Natural systems and phenomena have not still been involved and examined in literature. Thus, modelling and federating such systems and phenomena are taken into account in the context. For the rest of this paper, physical system and natural system are interchangeably used.

At present, cellular automata (CA) [2][3] model has emerged as a very promising technique for solving complex physical systems [4]. It has been used to address complex problems in many fields of science, engineering, computer science, and economy. In particular, parallel cellular automata models are effectively applied in fluid dynamics, molecular dynamics, biology, genetic, chemistry, road traffic flow, image processing, and environment modelling. Hence, we propose to use CA to facilitate modelling complex and large natural systems. A CA typically consists of two main components:

- *Cellular space:* This presents a lattice of cells. For each cell, we define a neighborhood that locally determines the evolution of the cell. All cells have

*Corresponding author. Email: tvhoang@ctu.edu.vn

the same size in the lattice. Each cell can be in a certain state.

- *Transition rule:* This rule acts upon a cell and its neighborhood. The cell's state changes from one *discrete time step* to another. The CA evolves in time and space as the rule is subsequently applied to all the cells in parallel.

In this article, we present a new approach for federating simulations. This solution enables several CA-based simulations working together as a distributed simulation system, so-called mixed simulation. To achieve it, we at first detail a method to model physical systems in accordance with the CA [2][3] model. Then, the parallelism is employed to accelerate large simulations. A federation of several simulations is conducted based on the high-level architecture (HLA) [19] standard.
The remainder of this paper is organized as follows. In Section 2, we introduce related work. Section 3 describes the process of modelling physical systems under the terms of CA. Section 4 presents the definition of the mixed simulation. And then, a federation of mixed simulations and related discussions are presented in Section 5. Section 6 gives some experiments of running a federation. The last section concludes the paper.

2. Related work

In the last decade, several studies focus on handling challenges of interoperability of various components in CPSs [1]. A few commonly interesting trends are briefly described as follows.
First of all, an approach about coordinating data communication and time synchronization between simulation frameworks is considered. Some practical works following that track are presented in [5][23]. These solutions allow to federate several simulations, however, the issues related to synchronization time are not taken into account.
Likewise, [22] proposes a framework for exchanging data and time synchronization by integrating two available tools. Its aim is to facilitate design and evaluation of networked control and cyber-physical system (NCCPS) [22] in CPSs.
To sum up, no work considers on using of the CA model for representing natural phenomena in the context of CPSs. Modelling such systems as well as their interactions in space and time are not regarded in most cases.

3. Modelling physical systems based on cellular automata

A physical system is viewed as a portion of the universe and has interactions with other ones outside it. For instance, water flows in a river may impact on riverbanks or even a road system located near that river. As mentioned earlier, since complex behaviours and large scale of such systems, modelling this type of system confronts with many issues. Cellular automata (CA) [2][3] is considered as a good alternative supposed to deal with those issues [4].
a) Cell system: At first, we propose a definition of cell system constructed based on the CA model. Such system consists of a collection of cells. Each cell holds its local state (interested parameters), which may be values of pollution density, insect population, humidity, or wind speed. A cell has connections and directions to adjacent cells (neighbouring cells). Simply, directions can be organized as pairs of number, shown in Table 1.

Table 1. An organization of directions in a cell system.

Direction	Value
East (E)	(1,0)
West (W)	(-1,0)
North (N)	(0,-1)
South (S)	(0,1)

In this study, we suggest using two common types of two dimensional neighborhoods Von neumann 1 (4 neighbours: E, W, S, N) and Moore 1 (8 neighbours: E, W, S, N, NW, NE, SW, SE) [2], as presented in Figure 1.

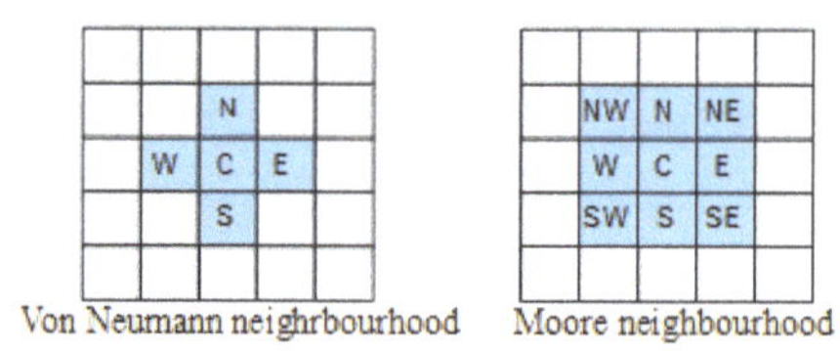

Figure 1. Two commonly used CA neighbourhoods: The Von Neumann neighbourhood (left) consists of the central cell itself plus 4 adjacent cells, and Moore neighbourhood (right) in which there are 8 adjacent cells.

Table 2 formally shows an example of a cell system, which is generated from geographic data except for its local state being *"pollution density"*.

Table 2. An example about a cell system of 601 cells (Von Neumann 1).

Cell	Pollution density	Neighbour	Neighbour directions
1	100	591, 26, 2, 601	(-1,0), (1,0), (0,-1), (0,1)
2	50	590, 1	(-1,0), (0,1)
...	...	...	...
26	10	1, 27	(-1,0), (0,1)
...	...	...	...
601	78	1	(1,0)

Notably, the number of neighbours of each cell is determined by the chosen CA and they are typically equal for all cells. But, in this case, cells' neighbourhood can differ from cell to cell since their original positions in the physical system. For example, Figure 2 visualises a river cell system in which cells are close to the riverbanks have less neighbour than other cells.

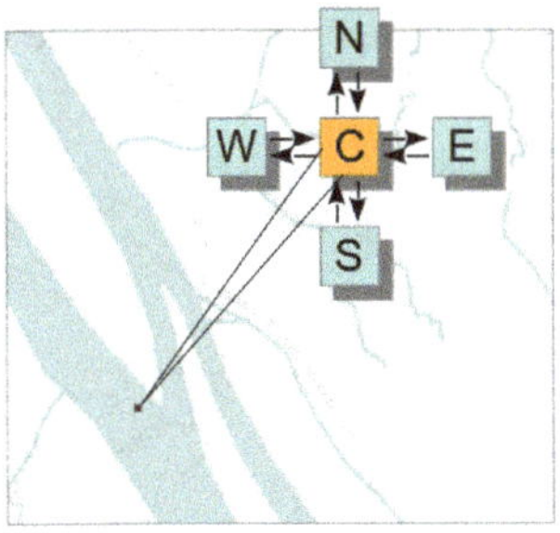

Figure 2. A cell system of a river system with Von Neumann 1.

b) The process of modelling physical systems: This is a set of ordered activities to achieve physical simulations from geographic data. Figure 3 simply depicts the process under the terms of the cell system definition. Geographic data are initially processed to generate cell systems, which are associated with definitions of states and transition rules to make up complete models. Physical simulations are obtained by just running the models.

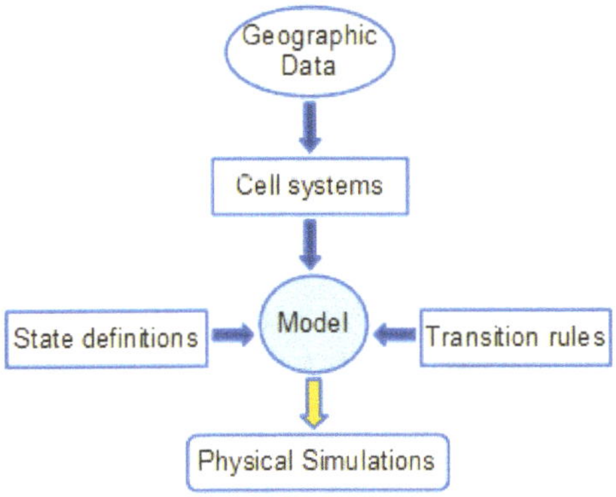

Figure 3. A process of developing physical simulations based on cell systems.

3.1. Eliminating useless calculations according to cell system model

Raster data are widely used as inputs in many natural phenomenon models, however, this often leads to useless time because of computations on the uninterested regions. In other words, they are regions outside the target systems, as depicted in Figure 4. Actually, several works have done to deal with that issue and one representative is presented in [26], in which cells are not belonging to any target areas are marked a label *"NoData"* in the preprocessing phase. During the execution of the models, those cells are omitted. It effectively works in most cases, however, time cost for data checking has not yet eliminated completely.
Cell system model helps to avoid meaningless processing by default. For example, in Figure 4, a cell system, which is extracted from geographic data, represents the river system. During executions, computations only occur on the river cell system. This approach is apparently more productive than the previous solutions.

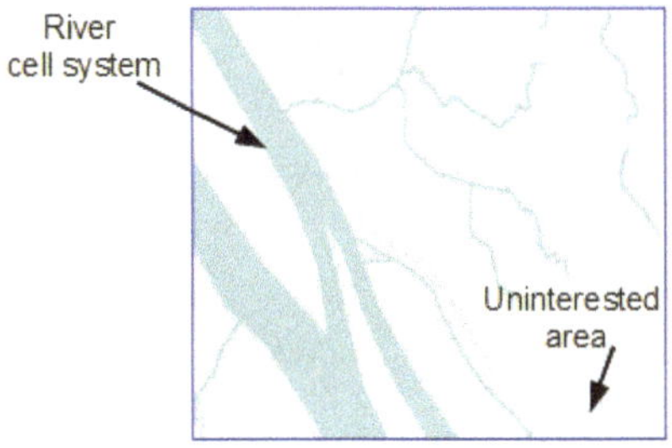

Figure 4. The river is considered as a target system.

In the next sections, we are going to present the models of two physical systems: forest fire spread model and river pollution diffusion model. In addition to those models, we suppose that a WSN [12] is deployed for monitoring fire in the forest, and its model is thus presented as well. Since the generations of cell systems are automatic by facilitating of an open source tool, the next considerations are definitions of local states and transition rules.

3.2. River pollution diffusion model

River pollution diffusion model is built based on the method presented in the previous section. In the context of pollution, it is possible to think of various potential situations such as chemical, oil, or contaminant. The diffusion of those generally depends on density. Thus, pollution density is chosen as cell state for this model.

- *state:* The cell state holds a value of the pollution density.
- *transition rules:* Updating pollution density at $cell_i$ at time t+1, termed as S_i^{t+1}.

$S_i^{t+1} \leftarrow S_i^t/2$
for *(j in neighbours of* $cell_i$*)*
  n ← *number of neighbours of* $cell_j$
  $S_i^{t+1} \leftarrow S_i^{t+1} + S_j^t/(2 * n)$
end for

Generally, to make an evolution of the system, pollution density of each cell is subtracted by a half. This portion is equally delivered to its neighbours. A reversing transfer also takes place at the same time.

3.3. Forest fire spread model

The fire spread model is defined for simulating the fire spread in the forest. In reality, temperature and humidity are common parameters to be monitored to decide whether there exist fire or not in the forest. However, for the sake of simplicity, the cell state is abstractly represented by one of the four values: tree, fire, ash, and empty.

- *state:* tree, fire, ash, and empty.
- *transition rule:* Updating the new state of $cell_i$ at time t+1, termed as S_i^{t+1}.

if (S_i^t *is TREE* ***and***
    at least one of its neighbor is FIRE)
  $S_i^{t+1} \leftarrow FIRE$
else if (S_i^t *is FIRE*)
  $S_i^{t+1} \leftarrow ASH$
else if (S_i^t *is ASH*)
  $S_i^{t+1} \leftarrow EMPTY$
end if

3.4. Wireless sensor network model

To represent a sensing component of CPS, a wireless sensor network (WSN) [12] is added to the context. It consists of a set of wireless sensor nodes that are scattered on target regions. Its featuring is to collecting environmental data and sending those data to centers (base stations) for processing.

In WSN, connections between nodes are ensured via their waves. The maximum distance waves of one node can cover is determined by its communication range. Hence, nodes lie within communication range one node can connect to it.

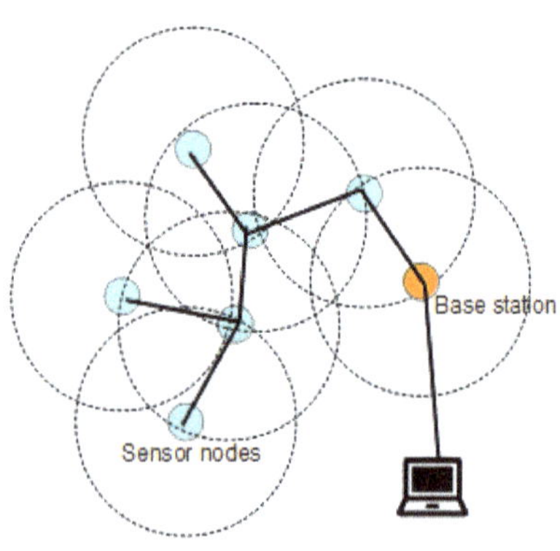

Figure 5. Depicting wireless sensor network. Large dotted circles imply communication ranges of nodes.

We assume that a collection of sensors is deployed in the forest. Each sensor node is represented by one cell in the context of the cell system. Then, the cellular space of the model is formed by links among those cells.

Figure 6 shows an example in which there are three sensor nodes deployed on a field (left figure), n1, n2, and n3. The corresponding cell system is given in the right figure. Cell states of the model are represented by sensing data holding one of two values (fire or normal).

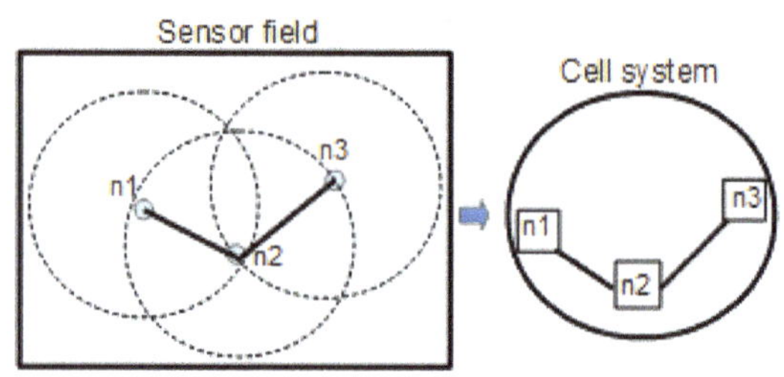

Figure 6. WSN model based on cell system.

- *state:* The cell state holds values of sensing data, which is collected from the forest (fire or normal).
- *transition rule:* Every step, all cells perform two main tasks, sensing and communicating. To prepare communications with its neighbours, cells store sensing data and their identity into packets.

//Actuation
if (S_i^t *is FIRE*)
  Raise signals
end if
//Communication
for (*p in packets held at* $cell_i$)
  Send p to neighbours
end for
for (*j in neighbours of* $cell_i$)
  Receive packets from j
end for
//Sensing
$S_i^t \leftarrow S_i^{t+1}$

4. Mixed simulations

4.1. Parallel simulations

The simplicity of the CA models brings about several advantages in modelling and simulation, especially complex physical systems. However, since huge sizes and complex behaviours of such system, modelling and simulating them pose many challenges to existing simulators. And this becomes more serious in the case of a system including several simulations cooperating together.

In recent years, simulations based on the parallelism

mechanism has been regarded in academic as well as industrial works. Several new architectures have been proposed to make possible designing and development of high performance environment based on the CA theory. Significant instances of these environments are CAM [28], CAMEL [29], StarLogo [30], and NEMO [31]. These environments allow the exploitation of the inherent parallelism of the cellular automata model for the efficient simulation of complex systems that can be modeled by a very large number of cells with local interaction only.

To carry it out, we propose to use the graphic processing unit (GPU) [7][8] in hopes of accelerating very time-consuming simulations. The GPU provides hundreds of threads running in parallel. In this case, a cellular automaton is instanced as a SIMD (Single Instruction, Multiple Data) program. In fact, CA implemented as a number of processes mapped on threads that execute the same code on different data simultaneously. According to this approach, the transition function of a single cell of the system must be specified. As a result, the computations on cells are thus to be executed at the same time.

The details of the implementations of parallel simulations in accordance with the Cuda [10][9] model are formally presented in Figure 7.

(1) Initialize cells' states
(2) Copy data from CPU to GPU
    and launch the kernel on GPU
        All processes run in parallel
        (3) Compute new states for current cells
        (4) Update: current states ← new states
(6) Copy data from GPU to CPU for visualising

Figure 7. An implementation model of parallel simulations based on Cuda programming.

4.2. Mixed simulations

A mixed simulation is defined as a collection of parallel simulations organized as a distributed system. In such systems, many parallel simulations are concurrently run on different hosts connected by a network infrastructure. However, many important aspects of this type of system have to be taken into account the context.

For such systems, the major consideration is how to synchronize their activities among hosts having their own time, which may be different. Obviously, it is impossible to achieve a global time for all hosts. Therefore, a new mechanism is required for synchronizing time between simulations in the context.

To do that, we propose a central component for the proposed architecture. It is not only responsible for connecting parallel simulations, but also for coordinating their activities in time, as depicted in Figure 8. In other word, this component plays a role as a coordinator in the system.

Along with time synchronization, data exchanging among simulations is one of main focuses of this study. Due to interacting via a network, it is necessary to find out a way to effectively transfer data among hosts, but still ensure loose coupling and scalability characteristic of the system. This will clearly be described in the next section.

By achieving interoperability among distributed simulations, mixed simulations are expected to be able to imitate not only behaviours of real systems, but also interactions between them.

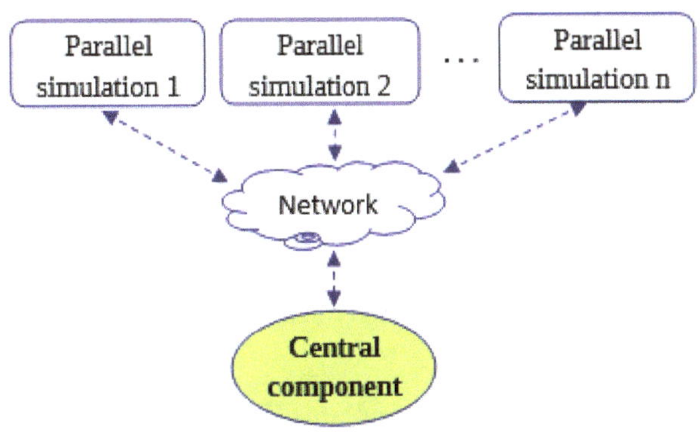

Figure 8. A general architecture of mixed simulations.

For instance, considering the interactions between the river and the forest system, as shown in Figure 9. Ashes produced by fire in the forest can pollute the river at the frontier between them. Otherwise, evaporation will also affects fired spread in the forest. Thus, those physical interactions will be put into models in this type of simulation.

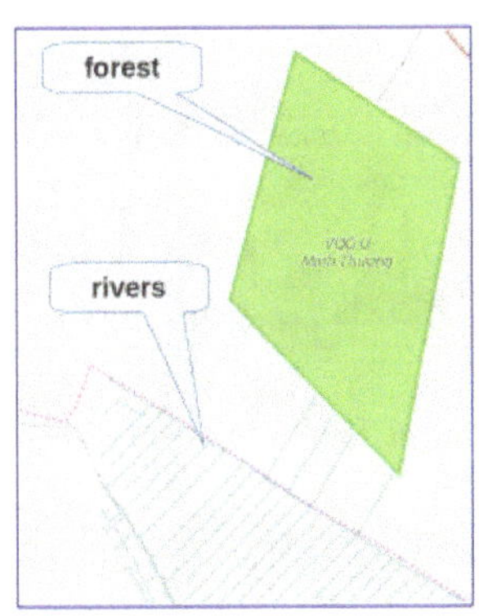

Figure 9. An example about interactions being happened between river and forest system. (OpenStreetMap [16]).

5. Federation of simulations

In this section, we will answer the question: *"How can several simulations work together?"*. As mentioned earlier, a mixed simulation consists of several parallel simulations concurrently running on different networked hosts. Time synchronization is required to make data exchanged between simulations worthwhile.

Towards that goal, an architecture is designed in accordance with the high-level architecture (HLA) [13][14][19]. In which, the entire system is viewed as a *federation* containing several *federates* linked via the central component *run-time infrastructure (RTI)* [19]. *federates* are . The The HLA is formally defined by three components.

1. A set of rules describes the responsibilities of federates and their relationship with RTI. An example is that *all exchange of data among federates should occur via the RTI during a federation execution* [13].
2. An interface specification provides services for managing federates and interactions. For example, it indicates how a federate join or leave a federation [14].
3. An Object Model Template [15] defines how information is communicated between federates, and how the federates and federation have to be documented (using Federation Object Model FOM) [15]. FOM defines the shared objects, attributes, and interactions for a whole federation.

As shown in Figure 10, we present the federation of three main federates representing the three parallel simulations being forest fire spread, river pollution diffusion, and WSN. An observer federate is also designed to visualise the federation.

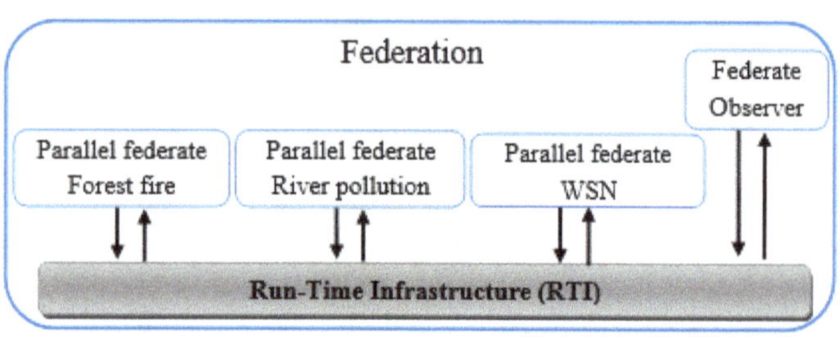

Figure 10. A federation of four federates: River, forest, WSN, and observer.

To prepare for the next explanations, we assume that there exist interactions between the three main simulations. Firstly, as fires of the forest approach to the river, ashes produced by the fires will pollute the river. Secondly, the evolution of the physical environment results in the changing of WSN's behaviour, particularly emitting signals at sensor nodes as soon as fire is detected in the forest.

5.1. Exchanging data

In the proposed model, exchanging data between federates is performed via the central component. However, sending and receiving a large amount of data among networked systems are not very feasible. Thus, a publish-subscribe mechanism is employed to serve that goal. This mechanism enables subscribers and publishers to determine what data they will send or receive without knowledge about each other. This obviously provides the network scalability for the system.

To achieve it, federates have to pre-declare what data they publish and subscribe to the central component before execution of the federation. As briefly presented in Table 3, a FOM file contains such declarations of sharing data that are encapsulated as object classes of attributes. It is noted that along with sharing status data, positions of objects are also considered as well.

Table 3. Describing publishers and subscribers of shared data between the four federates: forest, river, WSN, and observer.

Object Class	Attributes	Publishers	Subscribers
ForestNode	FState, Position	Forest	River, WSN, Observer
RiverNode	RState, Position	River	Observer
WSNNode	WState, Position	WSN	Obsever

Technically, two exchanges exist in the entire system. One is internal transferring between GPUs and CPUs after new states simultaneously computed on the federates' GPU, and the one externally occurs as data are exchanged among federates. Therefore, the states of publishers at time t may be involved in the evolution of other subscribers at time t+1, as depicted in Figure 11.

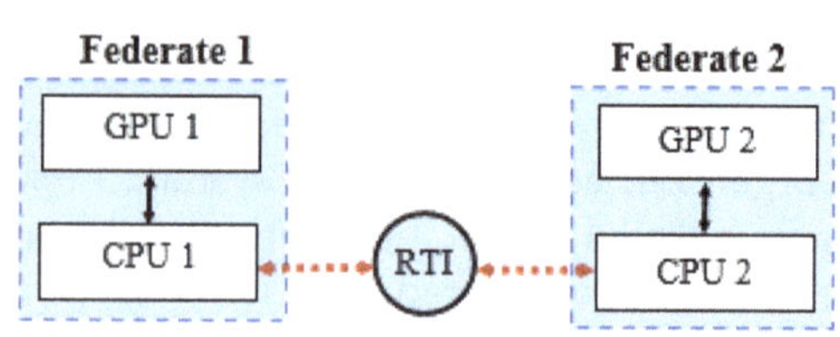

Figure 11. The technical view of exchanging data in the federation of two federates.

5.2. Time management

Time management is one of the important components of the recommended architecture. It controls the advancement of each federate in simulation time according to its time. This component allows federates in one federation to be able to synchronize their activities based on local logical time together to ensure the causal relationships. It is understandable that two types of time are considered in the context, one is local time maintained by federates, the other is global time controlled by the central component.

This mechanism comes up with two properties, constrained and regulating. The former ensures the federates to be able to send updates. Meanwhile, the latter allows the federates to receive updates from the central component. Therefore, both of them are often enabled for all federates, and only the constrained property is assigned to the observer federate since it is designed without any sending. Table 4 shows time policies proposed for the proposed federation.

Table 4. Time management of the four federates.

Federate	Constrained	Regulating	Time advance
Forest	Yes	Yes	Time stepped
River	Yes	Yes	Time stepped
WSN	Yes	Yes	Time stepped
Observer	Yes	Yes/No	Time stepped

In order to synchronize time, each federate associates its logical time with sending data, so-called time-stamp. Thanks to this information, the central component is capable to calculate the next time step for the federation as well as to coordinate federates as a synchronous system.

Time stepped federates will calculate values based on a point in time and process all data being sent up to the next point in time (current time + time step). Thus, to advance logical time for the time-stepped simulation, each federate has to send its request to the central component. Then, all receive data, which have been sent from federates, with the time-stamp less than or equal to the time requested will be released from the central component. After those data have been received by federates, a time grant is returned to the requesting federate. And then, the federate is able to advance its logical time.

The time advancement of a Federate F is posed after sending UpdateAttributeValues(UAV) service, this phase has three steps:

Step 1: X sends a request using TimeAdvanceRequest (TAR) service.

Step 2: X can receive reflectAttributeValue (RAV) callbacks. Then, X may update its system with received data, for example.

Step 3: X waits for the granted time t2, TimeAdvancedGrant(TAG). At the TAG(t2) reception, the local time of the federate will be advanced to t2.

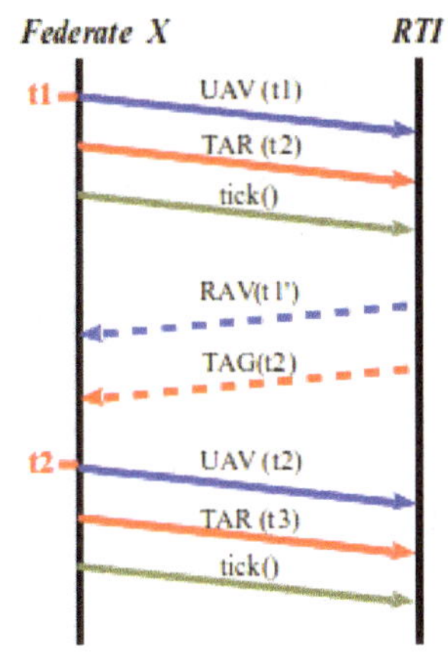

Figure 12. Time advancement process.

At the beginning of a federation execution, a synchronizing point is basically required. Figure 13 illustrates how to initialize the synchronizing point for all federates.

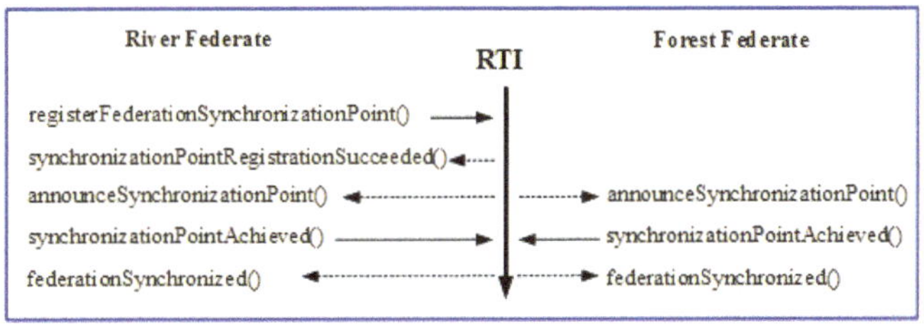

Figure 13. Federate synchronization for the river and the forest federate.

First of all, the river federate sends a synchronizing request to the RTI. The RTI will then responses to it and send an announce to the forest federate to achieve a synchronization point. Next, services will be used by both federates to confirm the synchronized point achieved.

Therefore, for the proposed federation, the time synchronization of the four federates can be obtained via the following steps.

Step 1: All federates connect to the federation initialized earlier.

Step 2: The federates are put into a common time progress by requesting synchronization point services, Figure 13.

Step 3: The federates update the new states, and then send the updates to the central component if they are publishers.

Step 4: Each federate sends an advance time request to the central component.

Step 5: For each federate, it needs to wait until all data on the central component with the time stamps are less than or equal to its requested time are received by subscribers, then it is able to advance its logical time.

Obviously, the proposed solution enables to create an environment in which several parallel simulations synchronously working together. Although it costs a little of time due to data transportation, it is significantly advantageous in the case of several complex and large systems.

6. Experiment

6.1. Data used

Data used in this study was taken from OpenStreetMap [16]. In which, we considered on a small area located in the Mekong Delta of Vietnam as a study region 9. Those data were used to generate cell systems of the river, forest, and WSN. The CA patterns were 8, 4, and 4 neighborhood, respectively. Since our current focus is the federation of mixed simulation, input data were randomly created for simulations.

6.2. FESI tool

We have developed the FESI (FEderation of SImulation) tool by using C/C++ language. It enables to develop parallel simulations and federate them as distributed simulations. Cell systems generated by PickCell open source software [11] are used as inputs. The framework CERTI [17][18] was used in this project as the role of RTI. Meanwhile, the X Window System [24] was employed to provide a GUI (Graphical User Interface) environment for displaying simulation results and interacting with users.
The computations of the parallel simulations run on the NVIDIA card GeForce GTX 680 1.15GHz with 1536 CUDA Cores (8 Multiprocessors x 192 CUDA Cores/MP). This board was connected to the Intel CPU 3.4GHz (8 CPUs x 4 cores/CPU).

6.3. Scenario 1 - The parallel simulation of pollution diffusion in the river

We run the river model for pollution diffusion presented in Section 3.2. Initially, two points were randomly polluted in the river. After 4 steps, the diffusion of pollution is shown in Figure 14. The darker regions implied the larger density of pollution, and vice versa.

Figure 14. The parallel simulation of diffusing pollution in a river. This is initialized with two polluted points (dark points).

6.4. Scenario 2 - Accelerating simulations based on GPU

We still used the pollution diffusion model in this case, its aim is to measure how faster model running on GPU compared to running on CPU. Besides, we used five target rivers with different sizes. Their cell systems generated from PickCell with the different number of cells ranging from 1,220 to 83,661 cells.

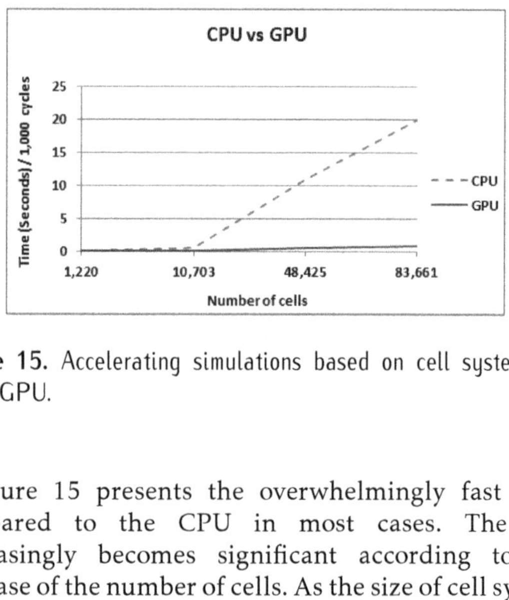

Figure 15. Accelerating simulations based on cell systems by using GPU.

Figure 15 presents the overwhelmingly fast GPU compared to the CPU in most cases. The gap increasingly becomes significant according to the increase of the number of cells. As the size of cell system is **83,661**, the GPU was approximately 22 times faster than the CPU.
The number of cells influences the performance for both the CPU and the GPU. On the CPU, the upward trend is very noticeable. The great increase starts from the size of **10,703** to **83,661** at a rate of **0.26(seconds)/1,000 cells**. It is projected that the trend anticipation will be maintained with bigger sizes. Whereas, the rising on the GPU is not dramatic. It gradually rises between **1,220** and **83,661** at a rate of **0.01(seconds)/1,000 cells**.

6.5. Scenario 3 - The interactions between the forest and the river federate

Regarding simulate the interactions of physical systems in CPSs. We launched a federation of a mixed simulation with three federates: river, forest, and observer. In which, the two first ones run on the GPUs. In this case, the river federate created and joined the federation on the RTI-CERTI. It waited for the forest federates to enter. The river federate sent a request to others to achieve a synchronization point in the federation. And then the synchronization point was achieved.

Figure 16 showed that the ashes (brown points) formed by the fire (red points) polluted the river from the step 4 as they spread close to the river. This also shows that models based on the CA can work together in the common time progress.

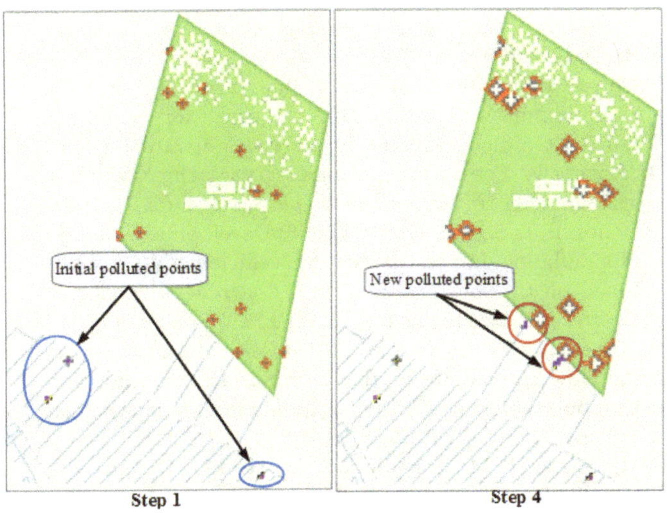

Figure 16. The screen shoot was taken from the observer federate. It shows the data exchange between the the river and forest federate in the federation. Two regions marked the small red circles represent the new polluted points created by the ashes, which are formed from the forest fire after 4 steps.

6.6. Scenario 4 - The federation of mixed simulation with the four federates: river, forest, WSN, and observer

The models were presented in Section 4.2. The sensors nodes were represented by black points in the forest. They appeared with the sensing ranges (small circles) and communication ranges (large circles).

As the previous case, the four federates first need to achieve a synchronous point. At each step, these federates exchange data together via the RTI.

Figure 17 presents the results captured from the observer federate. In this scenarios, due to no fire close to the river, until step 4, there were no new polluted points created in the river. Meanwhile, since a sensor recognizes that the fire appeared within its sensing range (smaller circle), it will change its color, sensing and communication range (larger circle) to red color.

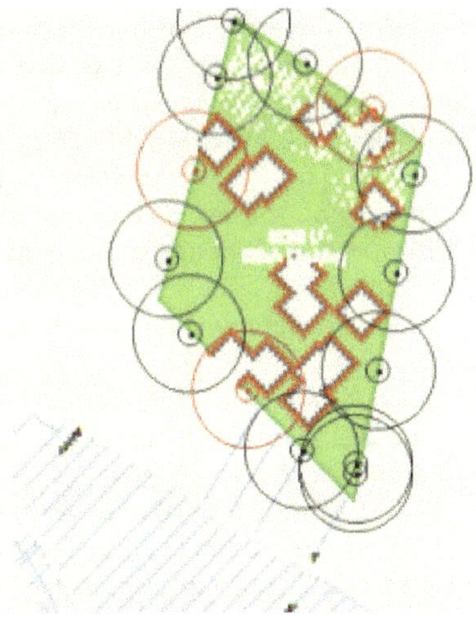

Figure 17. The interoperability of four federates under the context of the mixed simulation: river, forest, WSN, and observer. The sensors changed to red color since the fire was detected close to them.

7. Conclusion

In context of modelling and simulating for cyber-physical systems, we have described a new approach on the federation for simulations. The models of physical systems are based on cellular automata. In this method, they must have at least two components: cell system and transition rules. The FEMIS tool has developed in order to simulate those models in parallel and perform the federations of those simulations. The parallel computations on the GPU aim to reduce the simulating time for large and complex models. The experimental results were obtained by federating the three parallel simulations for forest fire spread, river pollution diffusion and wireless sensor network. By using federated simulations, the behaviour of physical systems with their interactions could be observed in simulation progress.

References

[1] Edward A. Lee (2010) CPS Foundations. *Proc. of the 47th Design Automation Conference (DAC). ACM.* 737-742

[2] Stephen W. (1984) Computation theory of cellular automata *Communications in Mathematical Physics, vol.96, n.1. Springer-Verlag.* 15-57

[3] Alfons G.H. Hoekstra, Jiří K., Peter M.A. Sloot (2010) Simulating Complex Systems by Cellular Automata *Berlin and Heidelberg, Springer-Verlag, Chapter 1.* 1-16

[4] Wolfram, S. (1984) Cellular automata: a model of complexity. *Nature 31*, 419-424

[5] Riley D., Eyisi E., Bai J., Koutsoukos X., Xue Y., Sztipanovits J. (2011) Networked Control System Wind Tunnel (NCSWT)- An Evaluation Tool for Networked Multi-agent Systems. *The fourth International Conference on Simulation Tools and Techniques (SIMUTools), Barcelona, Spain,* 9-18

[6] Al-Hammouri, A.T., Branicky, M.S., Liberatore, V. (2008) Co-simulation Tools for Networked Control Systems. *Hybrid Systems: Computation and Control, Springer and Heidelberg, vol. 4981.* 16-29

[7] Li, Dan and Li, Xia and Liu, XiaoPing and Chen, YiMin and Li, ShaoYing and Liu, Kai and Qiao, JiGang and Zheng, YiZhong and Zhang, YiHan and Lao, C. (2012) GPU-CA model for large-scale land-use change simulation. *Chinese Science Bulletin, vol.57, n.19. SP Science China Press.* 2442-2452

[8] Du, Jun and Liang, Qiang and Xia, Yongchun (2012) Parallel Simulation Based on GPU-Acceleration. *AsiaSim 2012. Springer Berlin Heidelberg.* 355-362

[9] Blecic, Ivan and Cecchini, Arnaldo and Trunfio, GiuseppeA (2013) Cellular automata simulation of urban dynamics through GPGPU. *The Journal of Supercomputing, Vol.6, N.2. Springer US.* 614-629

[10] Gulati, Kanupriya and Khatri, SunilP (2010) GPU Architecture and the CUDA Programming Model. *Hardware Acceleration of EDA Algorithms. Springer US, Chapter 3.* 22-30

[11] Bernard P., Pierre-Yves L. (2014) Dynamic networks âĂŞ NetGen: objectives, installation, use, and programming. *Université de Bretagne Occidentale*

[12] LuÃŋs M.L. Oliveira and Joel J.P.C. Rodrigues (2011) Wireless Sensor Networks: a Survey on Environmental Monitoring. *Journal of Communications. Vol. 6, N.2.* 143-151

[13] IEEE (2010) IEEE Standard for Modeling and Simulation (M&S) High Level Architecture (HLA)– Framework and Rules. IEEE Std 1516TM-2010, 1-38

[14] IEEE (2010) IEEE Standard for Modeling and Simulation (M&S) High Level Architecture (HLA)– Federate Interface Specification. IEEE Std 1516TM-2010, 1-378

[15] IEEE (2010) IEEE Standard for Modeling and Simulation (M&S) High Level Architecture (HLA)– Object Model Template (omt) Specification. IEEE Std 1516.2TM-2010. 1-110

[16] OpenStreetMap Mekong Delta Region, South of Vietnam. `https://www.openstreetmap.org` (accessed on 18 October 2015)

[17] Noulard E., Rousselot J.-Y., and Siron P. (2009) CERTI, an open source RTI, why and how. *Spring Simulation Interoperability Workshop*

[18] Bruno A., Siron P., Eric N. (2008) Running Real Time Distributed Simulations under Linux and CERTI. *2008 Euro Simulation Interoperability Workshop Proceedings, 08E-SIW-061*

[19] Alvarado, J.Rodríguez and Osuna, R.Vélez and Tuokko, R. (2008) Distributed Simulation in Manufacturing Using High Level Architecture. *Micro-Assembly Technologies and Applications, vol.260. Springer US.* 121-126

[20] Lasnier G., Cardoso J., Siron P., Pagetti C., and Derler P. (2003) Distributed simulation of heterogeneous and real-time systems. *Proceedings of IEEE International Symposium on Distributed Simulation and Real-Time Applications, 2013.* 55-62

[21] Kyoung-Soo W., Jong-Chan K., Chang-Gun L. (2011) A novel simulation framework for supporting real-time cyber-physical interactions. *The Fifth IEEE International Conference on Service-Oriented Computing and Applications (SOCA).*1-3

[22] Lin, Jinzhi and Wu, Ying and Wu, Gongyi and Xu, Jingdong (2012) NCCPIS: A Co-simulation Tool for Networked Control and Cyber-Physical System Evaluation. *Network and Parallel Computing, Springer Berlin Heidelberg, vol.7513*

[23] ForwardSim Inc. Simulation and Technologies: HLA Toolbox for MATLAB - HLA Blockset for Simulink. (2013). [Online]. Available: http://www.forwardsim.com

[24] Robert W. Scheifler and Jim G. (1986) The X window system. *ACM Transactions on Graphics. vol.5, n.2.* 79-109

[25] Zhou F., Li S., and Hou X. (2008) Development method of simulation and test system for vehicle body CAN bus based on CANoe. *The 7th World Congress on Intelligent Control and Automation (WCICA). IEEE.* 7515-7519

[26] Cirbus J., Podhoranyi M. (2013) Cellular Automata for the Flow Simulations on the Earth Surface, Optimization Computation Process. *Applied Mathematics & Information Sciences 7, No.6.* 2149-2158

[27] Quartieri J., Mastorakis N.E., Iannone G., Guarnaccia C. (2010) A Cellular Automata Model for Fire Spreading Prediction. *The 3rd WSEAS International Conference on URBAN PLANNING AND TRANSPORTATION (UPT '10), Greece, Volume: Latest Trends on Urban Planning and Transportation.* 173-179

[28] Toffoli T. and Margolus N. (1986) Cellular Automata Machines A New Environment for Modeling. *The MIT Press*

[29] Spezzano G., Talia D., Di Gregorio S., Rongo R., Spataro W. (1996) A Parallel Cellular Tool for Interactive Modeling and Simulation. *IEEE Computational Science & Engineering 3.* 33

[30] Resnick M. (1994) Turtles, Termites, and Traffic Jams. *The MIT Press, Cambridge.*

[31] Hutchinson D., Kuttner L., Lanthier M., Maheshwari A., Nussbaum D., Roytenberg D., Sack J.-R. (1996) Parallel Neighbourhood Modeling: Research Summary. *Proc. SPAA '96, Padua, Italy.* 204

21

A Mathematical Model of Game Refinement and Its Applications to Sports Games

Arie Pratama Sutiono[1,*], Rido Ramadan [1], Peetikorn Jarukasetporn [3], Junki Takeuchi [2], Ayu Purwarianti[1], Hroyuki Iida[2]

[1]Bandung Institute of Technology, Bandung, Indonesia
[2]Japan Advanced Insitute of Science and Technology, Ishikawa, Japan
[3]Chulalongkorn University, Bangkok, Thailand

Abstract

This paper explores a mathematical model of game progress. We claim that a realistic model of the game progress during the in-game period is not linear but exponential. The second derivative value, i.e., acceleration in the sense of dynamics, is derived from the model and we propose to use the value as a measure of game refinement. This is because acceleration of game progress should be related to the emotional impact in our minds, as thrill or engagement in games. We also evaluate well known games using our theory like sport games, that would further be classified by the rule to finish the game. It is expected that the game refinement theory will be widely used as a tool to assess the quality of various types of games as a new game theory.

Keywords: game refinement theory, engagement assessment, sports games, board games, video games

1. Introduction

Many efforts have been devoted to the study of strategic decision making in the framework of game theory with focus on mathematical models of conflict and cooperation between intelligent rational decision-makers or game-players. Game theory originated in the idea regarding the existence of mixed-strategy equilibria in two-person zero-sum games [10], which has been widely recognized as a useful tool in many fields such as economics, political science, psychology, logic and biology.

However, little is known about mathematical theory from the game creator's point of view. It is interesting to know theoretical aspects of increasing attractiveness of games and its sophistication. An early work in this direction has been done by Iida *et al.* [8], in which a measure of game refinement was proposed based on the concept of game outcome uncertainty. A logistic model was constructed in the framework of game refinement theory and applied to many board games including chess variants and Mah Jong [7]. The proposed measure of refinement concerns the information gained on the average branching factor and the average game length for a given game.

It is natural but challenging to apply the game refinement theory to various types of games such as sports games and video games. This encourages us to study a general model of game refinement in the domain of sports games. Although there are some rule changes in the history of every sport games, the core mechanism is still unchanged from the original idea. That is why sports is an interesting subject to game refinement theory.

2. Game Refinement Theory

In this section we construct a game progress model which can be used for various types of games. A measure of game refinement will be derived from the model. Then we consider the gap between board games and other games such as sports games and video games. Moreover, we apply some data from well-known games to confirm the effectiveness of the proposed measure.

2.1. Game progress model

In this study "game progress" has twofold. One is game speed or scoring rate, while another one is game information progress with focus on the game outcome.

* Corresponding Author. E-mail: arie.pratama.s@gmail.com

In sports games such as Soccer and Basketball, the scoring rate will be calculated by two factors: (1) goal, i.e., total score and (2) time or steps to achieve the goal. For example, in Basketball the total score is given by the average number of successful shoots, whereas the steps to achieve the goal is estimated by the average number of shoots attempted. Then the game speed of Basketball is given by

$$\frac{average\_number\_of\_successful\_shoots}{average\_number\_of\_shoots}$$

We need to consider a reasonable way to obtain the game speed for various type of games. For some sports games such as Basketball and Soccer, we can obtain statistics of average number of shoots and goals per one game.

For other sports games such as Volleyball and Tennis in which the goal (i.e., score to win) is set in advance, the average number of total points per game may correspond to the time or steps to achieve the goal. For video games such as Pac Man, the steps can be estimated by the average number of movements of Pac Man and the goal is estimated by the scores gained at each stage. In the domain of board games, the steps can be estimated by the average depth of game tree, (i.e., game length) and the goal is estimated by the average branching factor. For the board game case we discuss more detail later on.

Now we consider a model of game information progress. Let G and T be the average number of successful shoots and the average number of shoots per game, respectively. If one knows the game information progress, for example after the game, the game progress $x(t)$ will be given as a linear function of time t with $0 \leq t \leq T$ and $0 \leq x(t) \leq G$, as shown in Equation (1).

$$x(t) = \frac{G}{T}\,t \tag{1}$$

However, the game information progress given by Equation (1) is usually not known during the in-game period. The game outcome is uncertain until the very end of game in many games which we call *balanced game* or *seesaw game*. It means that the change of game information becomes greater near the end of game. Thus, the game information progress should be not linear but something like exponential.

Hence, we propose a realistic model of game information progress in Equation (2).

$$x(t) = G\left(\frac{t}{T}\right)^n \tag{2}$$

Here n stands for a constant parameter which will depend on the perspective of an observer in the game considered. [1]

If one knows the game outcome, for example after the game, or if one can exactly predict in advance the game outcome and its progress, then we have $n = 1$, where $x(t)$ is a linear function of time t. During the in-game period, various values of the parameter n for different observers including players and supporters will be determined. For example, some observers might be optimistic with $0 \leq n < 1$. However, when one feels any challenges to win or achieve the goal, the parameter would be $n > 1$.

Meanwhile, we reasonably assume that the parameter would be $n \geq 2$ in many cases like balanced or seesaw games. Thus, we have the second derivative of $x(t)$ in Equation (2), as shown in Equation (3).

$$x''(t) = \frac{Gn(n-1)}{T^n}t^{n-2} \tag{3}$$

Solving the formula at $t = T$, the equation becomes

$$x''(T) = \frac{G}{T^2}n(n-1)$$

Newton mechanics indicates that the force in the physical world is derived from mass and acceleration. It is assumed in this study that the game information progress in any type of games is happening in our minds. We do not know yet about the physics in our minds, but it is likely that the acceleration of information progress is related to the force in mind. Hence, it is reasonably expected that the larger the value $\frac{G}{T^2}$ is, the more the game becomes exciting due to the uncertainty of game outcome. Thus, we propose to use the value $\frac{G}{T^2}$ or its root square, as shown in Equation (4), as a game refinement value for the game considered.

$$R = \frac{\sqrt{G}}{T} \tag{4}$$

We will consider this game refinement value as R value in the rest of this section.

2.2. Game Progress in board games

Here we consider the gap between board games and other games such as sports games. We assume that the game information progress in sports games is related to game progress in board games. To justify the assumption, we begin the explanation for the case of board games. From Equation (1) we know the game

[1] Note that Equation (2) might be a promising and realistic model of game progress, but its uniqueness (to be exponential) should be investigated more in the future.

information value as a function of game progress or depth of game tree t in board games.

We show, in Figure 1, an illustration of game progress. The whole triangle illustrates a game tree and the middle line illustrates a game progress. As players make their moves, the game progress begins to move down.

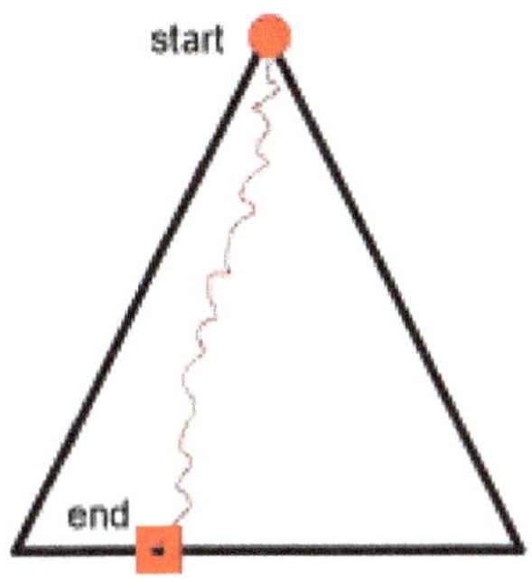

Figure 1. Illustration of game progress through a game tree

At each depth of the game tree, one will choose a move and the game will progress. Calculating the length of middle line in Figure 1 is the key to get the total game progress value. To calculate it, consider Figure 2. In the game tree, we have the horizontal difference between neighboring nodes (say Δl), the difference between levels of tree (say Δt), and distance between them (say d).

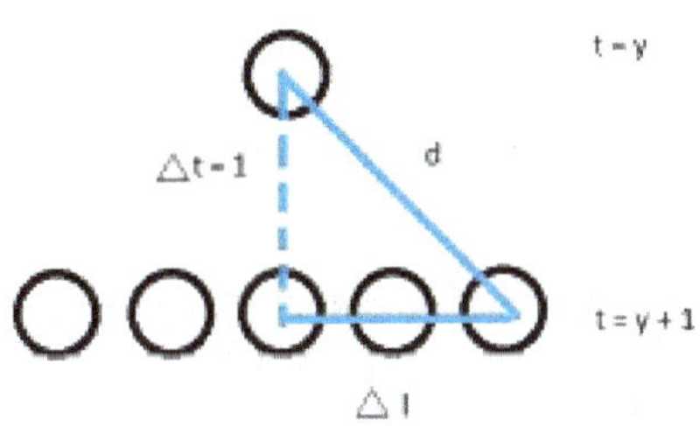

Figure 2. Illustration of distance in game tree

The distance d can be found by using simple Pythagoras theorem, thus resulting in

$$d = \sqrt{\Delta l^2 + 1} \tag{5}$$

B is the branching factor for a decision in a board game. Note that $\Delta l < B$ holds because the maximum horizontal difference between neighboring nodes will be $B - 1$. Assuming that the approximate value of horizontal difference between nodes is $\frac{B}{2}$, then we can make a substitution in Equation (5).

$$d = \sqrt{(\frac{B}{2})^2 + 1}$$

The game progress is the total distance between nodes considered in the game tree. Here, we do not consider Δt^2 because the value ($\Delta t^2 = 1$) is small compared to B. The game length will be normalized by the average game length D, then the game progress $x(t)$ is given by

Table 1. Game progress factors and game refinement value for various games

	Time	Goal to achieve	R
Sports games	N	G	$\frac{\sqrt{G}}{N}$
Board games	D	B	$\frac{\sqrt{B}}{D}$

$$x(t) = \frac{t}{D} \cdot d = \frac{t}{D}\sqrt{(\frac{B}{2})^2} = \frac{Bt}{2D}$$

In general we have

$$x(t) = c\frac{B}{D}t$$

Where c is a different constant which depends on the game considered. [2] However, we manage to explain how to obtain the game progress value itself. The game progress in the domain of board games forms a linear graph with the maximum value $x(t)$ of B. Assuming $c = 1$, then we have a realistic game progress model for board games, which is given by

$$x(t) = B(\frac{t}{D})^n. \tag{6}$$

Equation (6) shows that our present study comes to the same formula described in the early works [7].

3. Application to Time-limited Sports

In this section we will evaluate the time-limited sports, especially for Soccer and Basketball. Then we compare it with previous research for board games to see if the principle could also be applied to this type of sports.

3.1. Soccer and Basketball

Here we show, in Table 1, a comparison of game progress factors and game refinement value for various type of games. In sports games, G is number of goals and N is the number of attacks. While in board games, B is the branching factor and D is the depth of game tree.

To confirm the effectiveness of proposed measures, some data of games such as Chess and Go [8] from board games and Basketball and Soccer from sports are compared. For Basketball the data were obtained from the NBA website [9], while the data for Soccer were obtained from the UEFA championship [11]. We show, in Table 2, a comparison of game refinement value for various type of games. We suspect when the values

[2]Further investigation is needed to obtain the exact value of c.

Table 2. Measures of game refinement for board games and sports games

Game	B or G	D or N	R
Chess	35	80	0.074
Go	250	208	0.076
Basketball	36.38	82.01	0.073
Soccer	2.64	22	0.073

go too high, then observer will get the excitement too fast and when the value goes too low the excitement comes slower thus would not make a game interesting. Sophisticated games should have a common factor (i.e., same degree of acceleration value) to feel engagement or excitement regardless of different type of games as we can see from sophisticated games that we have observed. We understand that Chess, Go, Soccer and Basketball are both sophisticated with long history.

In the previous study with focus on Chess variants [8], we observed that the proposed measure seems promising to explain the evolutionary changes of game rules. For example, in the Chess history most variants were outsourced and only a few variants survived to the present. The surviving variants went through the sophistication of the game rules to maximize the entertainment impact making the depth of lookahead (i.e., intelligent aspect of games) more critical for the outcome of the game. Experienced players often noted that large and complex games were not attractive at all and interesting and enjoyable games are those with more entertainment impacts. The evolutionary process has produced the present version of Chess which seems a well-balanced search-space complexity and entertainment impact. Modern Chess may be considered a highly matured and optimized Chess-like game. Similarly, through this study we make the same observation for sophisticated sports games such as Basketball and Soccer as well as for board games such as Chess and Go.

4. Application to Score-limited Sports

We consider the application of game refinement theory to score-limited sports such as Badminton, Table Tennis and Volleyball in this section. For each case we need to examine an appropriate model of game progress to derive a game refinement value.

4.1. Badminton

Badminton's official rules are described in the official site of Badminton World Federation [1]. There are two sides of player(s), either singles or doubles. There are 5 types of events: Men's singles, Women's singles, Men's doubles, Women's doubles, and Mixed doubles. Under the new current scoring system, the side which wins the rally scores the point regardless of which side serves. The information length of Badminton game is 21 points, while maintaining a minimum 2-point lead. The first side to win 2 games out of 3 wins the match. This is the g3 x 21h rally point scoring system.

In the past, a g3 x 15h side-out scoring system was used (3 x 11 for women's singles). For this side-out scoring system, only the server can score the point. If the service side loses the rally, no point is awarded, and the service passes to the other side. It can be seen that this could, and often did, result in irregularly long match times. The scoring system was changed in December 2005.

In Badminton, the game progress or scoring rate will be calculated by two factors: winner's score (say W) and total score (say T). Then the game progress of Badminton is given by

$$\frac{average\_winner's\_score}{average\_total\_score}$$

We show, in Table 3, the game refinement values for the old scoring system and the current scoring system. Data was collected from BWF world championship [2][4]. Game refinement values for the two variants is significantly different, indicating the change to game refinement values as affected by the change in scoring. The side-out system gives the higher value 0.121. Under the new system, the game refinement value recedes to a more balanced 0.086. The result implies that the change in the scoring system makes the game more interesting and attractive for observation.

Table 3. Measures of game refinement for Badminton

Scoring system	W	T	R
Side-out	30.070	45.154	0.121
Current	46.336	79.344	0.086

4.2. Table Tennis

Table tennis was first played in 1880s in England as an after dinner activity. It became popular with the introduction of the name 'Ping-Pong' by J. Jacques & Son, and has since undergone a few change of rules and equipment [3]. In this part, we will explain the basic rules of table tennis and the change in equipment, and compare the results from applying game refinement theory on table tennis.

The official rules are available on the official site of ITTF [3]. There are two sides of player(s), either singles or doubles. There are 4 types of events: Men's or Women's singles, and Men's or Women's teams. As with the new badminton and volleyball systems, in table tennis the side which wins the rally scores the point regardless of service. The information length of game is 11 points, while maintaining a minimum 2-point lead. The match consists of any odd number of games, usually 7.

At first, table tennis rackets were pure wood. Around 1900, the use of layered foam-core rackets, topped with rubber sheeting, dramatically increase the game's speed and added some spin to the game. Prior to the 2000 Summer Olympics, table tennis used a 38 mm ball. The ball size was changed to 40 mm which effectively reduced the game's speed. The scoring system was also changed in September 2001. Until that year, game lengths were 21 points with a 2 point lead, while matches were usually a best-of-3 or -5. Most recently, the doubles event was integrated as a part of team events for 2008 Olympics. Each team features 3 players, 1 singles player and 2 double players, playing a best-of-5 series of 2 singles, follow by 1 doubles, followed by up to 2 more singles. The first side to win 3 wins the match.

We show, in Table 4, the game refinement values for pre-2001 games, post-2001 games, and the team events. Data of Olympic matches from 1988 to present was collected from the records of the ITTF official site [3]. The result indicates that in the case of table tennis, the changes do not significantly affect the game progress. The average game refinement value of 0.076 holds for table tennis since the 1988 Olympics. As for the result of the team event, the game refinement value is lower than for the others, and lower than expected for a well-balanced sophisticated game. We also notice the higher average score for team events. It can be safely assumed that the change to integrate the doubles event into the team events, while lessen the emphasis on less popular doubles event [6], reduces the value of game sophistication beyond the 0.07 – 0.08 window of good balance.

Table 4. Measures of game refinement for Table Tennis

Scoring system	W	T	R
Pre-2001	57.869	101.530	0.075
Post-2001	54.863	96.465	0.077
Team	131.283	232.123	0.049

4.3. Volleyball

In this study we select three important variants from the history of Volleyball: (1) side-out scoring system with 15 points, (2) rally point system with 30 points, and (3) rally point system with 25 points (see Table 5).

Table 5. Three rule variants of Volleyball

rules	points	set
side-out scoring system	15	best-of-five
rally point system	30	best-of-five
rally point system	25	best-of-five

We first focus on the current rule, i.e., rally point system with 25 points. We show, in Table 6, statistics on the average point per game in Volleyball games from V-league in Japan [5]. The max point and min point are also shown. Since the average point per game is 44, it is expected that the final score on average is 25 – 19. Likewise, the score in max point case would be 37 – 35 due to deuce and 15 – 6 (15 points) in min point case.

Table 6. Statistics on point per game in rally point system Volleyball with 25 points (n=486).

	Points
Max points	72
Min points	21
Ave points	44

The game progress in Volleyball can be given by the average number of goals ($G = 25$) and the average number of all scores ($T = 44$). By applying Equation (4), we obtain as the value as follows:

$$R_{25pts} = \frac{\sqrt{25}}{44} = 0.114$$

We next consider the rally point system with 30 points. Since the data for 30 points rally currently is unavailable, we estimate it based on the statistics of 25 points rally point system Volleyball. For this purpose, we assume the same ratio (25 : 19) of winning points and losing points and obtain 30 : 22.8. It means that the average goals G is 30 while the average total points T is 52.8. By applying Equation (4), we obtain as the value as follows:

$$R_{30pts} = \frac{\sqrt{30}}{52.8} = 0.104$$

We show, in Table 7, some statistics of rally point system which include some other cases.

Table 7. Some statistics for rally point system with various goal points.

G	T	R
25	44	0.114
30	53	0.104
35	62	0.095
40	70	0.090
50	88	0.080
60	106	0.073

For the side-out scoring system, we try to simulate to obtain the data since we have no real data. We assume some different *scoring percentage* γ for the side of serving the ball. From the previous results it is estimated that we have the final score $15 : 11.26$. The game progress values for some different parameters can be calculated and shown in Table 8.

Table 8. Some statistics for side-out system with 15 points (simulation).

scoring γ (%)	total score	R
60	43.77	0.088
50	52.52	0.074
40	65.65	0.059
37	72.0	0.054
33	86.66	0.045
25	105.04	0.037

We suspect based on our experience that the scoring percentage $\gamma = 50$ holds in the side-out system with 15 points Volleyball when both opposing teams are well balanced. Hence, we take the $R_{25pts} = 0.074$ for representing the game refinement value of 15 points side-out scoring.

We show, in Table 9, the comparison between three Volleyball variants based on the game refinement value.

Table 9. game refinement value for three variants of Volleyball.

variants	points	R
side-out scoring system	15	0.074
rally point system	30	0.104
rally point system	25	0.114

The side-out scoring system (with scoring rate roughly $\gamma = 50$) had been played long time (1947-1999). As we can see from the Volleyball's history and rules changes, the γ should be lower than 40. The game refinement value is different with other sophisticated games such as Chess and Soccer. However, the rule was changed in 1999 to improve game understandability. At the same time, the rule change has made rise in excitement, depicted in higher value of game refinement value.

The study using board games suggests that the game refinement value of sophisticated games are somewhere in the range 0.07 – 0.08 [7]. The higher R value in the current volleyball rules might mean the game become more exciting, but it is not alligned well with our previous classification of sophisticated game. Thus, it can be assumed that this rule changes might not the optimum method to improve the game attractiveness in the *comfortable range*.

5. Conclusions

In this paper we presented a mathematical model based on game progress and game information model with focus on game outcome uncertainty. Game refinement value was derived from the game information progress model. Its second derivative, which is the acceleration in the sense of dynamics, was derived from the model to use the value as a game refinement value. This is because the acceleration of game information progress should relate to the emotional impact such as entertainment and engagement which may correspond to the force in physics. We then investigated a model of game progress in various sport game domains. Applying data from different type of games, we see that this model may only fit for the time-limited sports, such as soccer and basketball. Either this model does not fit for the score limited sports or this type of sport has its own refinement value, which is not 0.07 – 0.08.

This research left some further works that need to be done. More investigation in collecting data of many other games to test and elaborate the model we discussed in this paper will be needed. Automated tasks to collect data from media might also help instead of collecting data from simulation and the internet. Moreover, it is expected to establish the game refinement theory, which will be widely used like classical game theory, not to find the optimal strategy to play a given game, but to assess the sophistication of games or quality of life in game playing.

Acknowledgement

The authors wish to thank the anonymous referees for their constructive comments that helped to improve the article considerably. This research is funded by a grant from the Japan Society for the Promotion of Science, in the framework of the Grant-in-Aid for Scientific Research (B) (grant number 23300056).

References

[1] Badminton world federation's official site. http://www.bwfbadminton.org.

[2] Bwf world championships. http://en.wikipedia.org/wiki/BWF_World_Championships.

[3] International table tennis federation's official site. http://www.ittf.com.

[4] Tournament software badminton. http://www.tournamentsoftware.com/ tournamentlist.aspx?id=2.

[5] V-league official site. http://www.vleague.or.jp/.

[6] ETTA. English table tennis association fs press. http://ettapressoffice.com/history.

[7] H. Iida, H. Takahara, J. Nagashima, Y. Kajihara, and Y. Hashimoto. An application of game-refinement theory to mah jong. In *International Conference on Entertainment Computing (ICEC)*, pages 333–338, 2004.

[8] H. Iida, N. Takeshita, and J. Yoshimura. A metric for entertainment of boardgames: Its implication for evolution of chess variants. *Entertainment Computing Technologies and Applications*, pages 65–72, 2003.

[9] NBA. Nba team statistics. url: http://stats.nba.com. Accessed: 30-10-2013.

[10] J. Neumann. Zur theorie der gesellschaftsspiele. *Mathematische Annalen*, 100(1):295–320, 1928.

[11] UEFA. Uefa 2012/13 team statistics. url: http://www.uefa.com/uefachampionsleague/. Accessed: 8-11-2013.

22

Modeling the Connections of Dynamic Sensor Fields Based on BT-Graph

Tuyen Phong Truong[1,*], Huong Hoang Luong[2], Hiep Xuan Huynh[2], Vinh Cong Phan[3] and Bernard Pottier[1]

[1]Lab-STICC, UMR CNRS 6285, Université de Bretagne Occidentale, France
[2]Can Tho University, Vietnam
[3]Nguyen Tat Thanh University, Vietnam

Abstract

In this article, we propose a new approach to model and optimize the dynamic sensor field for both internal network connections and LEO satellite connection based on BT-Graph. Due to the shift of LEO satellite's orbit at each revolution, a dynamic sensor field (DSF), which is able to redetermine its gateways, is suitable to improve successful data communications. It is convenient to present a DSF as a BT-Graph that aims to utilize optimization algorithms. Parallel search algorithms are also deployed for more efficient execution time when DSFs consist of a significant large number of nodes. The simulation experiments are performed on an abstract forest fire surveillance network to validate our proposed approach.

Keywords: associate analysis, BT-Graph, parallel processing, sensor field, LEO satellite.

1. Introduction

Wireless Sensor Network (WSN) [1] is known as a network of sensors cooperatively operating in order to surveillance or collect the environmental parameters. Although the short transmission range can be compensated by applying a mesh topology, it would be economically infeasible to deploy in large geographic areas or behind obstacles (mountains, oceans, ...). To overcome these disadvantages, satellite-based wireless sensor networks have emerged as a promising technology with various applications.

Because the orbit of a LEO satellite shifts in westward direction around the polar axis at each revolution [12], the meeting points of a gateway on the Earth's surface and the LEO satellite will be changed over time. With a static sensor field, it can be occasionally unsuccessful in communication with the LEO satellite because the meeting time is not enough for data exchange [12]. Therefore a dynamic sensor field (DSF) [1], which has the ability to redetermine its gateway to adapt with the shifts of LEO satellite's paths, is proposed to improve the connection time.

It is necessary to choose proper gateways for the longest length of connection time. Consequently, the connections between a LEO satellite and a dynamic sensor field should be presented by a graph-based model because it is convenient to determine number of neighbors in the satellite's communication range and then apply the optimization algorithms. Moreover, BT-Grap model [2] based on balltree structure is efficiently support not only for searching the range nearest neighbors (RNN) but also finding the shortest path to a given target node [4] [18] [20]. In this article, we propose a new approach, namely dynamic sensor field optimization model based on BT-Graph (BT-DYNSEN), to model and optimize the DSF in communication with LEO satellite.

In addition, with the development of Compute Unified Device Architecture (CUDA), general purpose Graphics Processing Unit (GPGPU) is widely used to implement parallel algorithms, which can speed up the large and complex processing tasks [7] [8]. Nonetheless, serial graph algorithms are hard to be parallelized on GPUs due to irregular memory access and huge

*Tuyen Phong Truong. Email: phong-tuyen.truong@univ-brest.fr

memory-intensive operations. For this reason, we also propose parallel algorithms to improve search speed based on BT-Graph model using NVIDIA CUDA GPUs. In our work, the experimental results were deployed on parallelization of RNN algorithm and Dijkstra's shortest path search algorithm.

The rest of this article is organized as follows. In section 2, we overview related works. Section 3 presents the graph-based model for a DSF based on BT-Graph (BT-DYNSEN). How to optimize the DSF for satellite connection is presented in section 4. Implementation of BT-DYNSEN is described in section 5. Section 6 gives the simulation experiments on an abstract network for forest fire surveillance in Vietnam before a conclusion is drawn.

2. Related work

The LEO satellite communication can be used in mobile satellite communications, surveillance the Earth surface, geological surveys, so on [1] [16] [19]. In the last decade, researches on communication services provided by LEO satellites have focused on several main directions such as optimizing the mechanics, interconnections, electric circuits, power supply. On another approach, many surveys show the research topics in design trajectory, handover traffic and constellation, as well as design protocols, radio frequencies, onboard transceivers and antenna designs [13] [14] [1] [19]. However, the direct radio links between sensor fields and LEO satellites are not considered in literature. In recent years, it emerges as an attractive topic because of the current innovation solutions such as LoRa Semtech and solutions from vendors QB50 [11].

Furthermore, graph-based model has emerged as well approach to present the structure and elaborate the performance of wireless sensor networks [3]. For example, random geometric graph [20] was used to determine the probability of the whole network being connected. Secure communications between large number of sensor nodes in WSNs can be elaborated on expander graph [25] and finding transmission path in network was performed based on Pascal graph [5]. Furthermore, hyper-graph [24] was utilized to support for reducing the transmission energy consumption and improving the fault-tolerant ability of the system.

Additionally, in recent years parallelization on GPU platforms is an emerging strategy to improve significantly performance speed [9]. When the graph consisting of a vast number of nodes, it is necessary to process in parallel based on NVIDIA CUDA technology to improve search speed [10]. In the next section, we introduce BT-DYNSEN model for optimizing the DSF for the connection with LEO satellite.

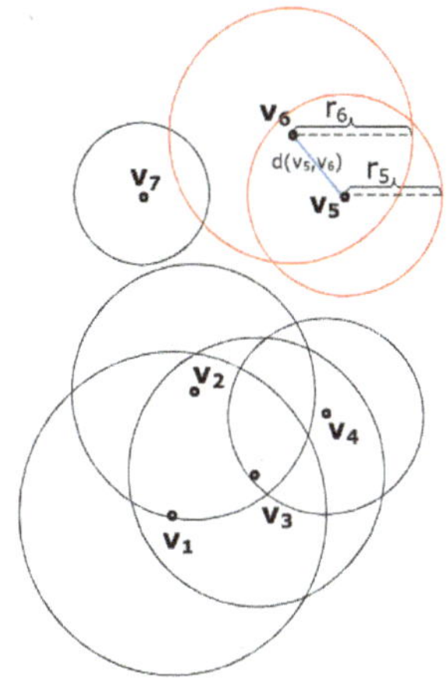

Figure 1. A dynamic sensor fiel in which the communicatiomanges of sensor nodes are indicated by the radii of balls.

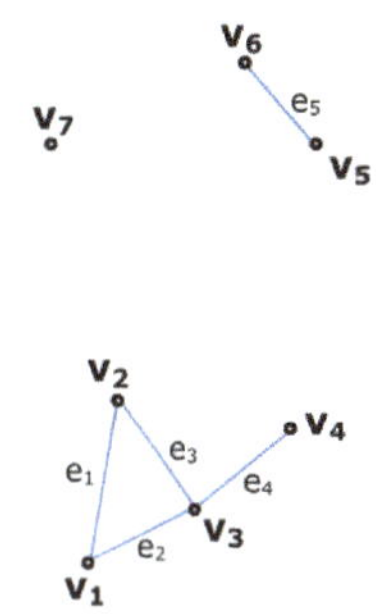

Figure 2. A BT-Graph of the dynamic sensor fiel with 07 vertices $V = \{v_1, v_2, v_3, v_4, v_5, v_6, v_7\}$ and 5 edges $E = \{e_1 = e(v_1, v_2), e_2 = e(v_1, v_3), e_3 = e(v_2, v_3), e_4 = e(v_3, v_4), e_5 = e(v_5, v_6)\}$.

3. BT-DYNSEN

3.1. BT-DYNSEN modelof DSF

BT-DYNSEN model of a dynamic sensor field (DSF) [1] based on BT-Graph [2] is a graph G(V,E). In this graph, set of vertices $V = \{v_i\}$, $i = 1..n$ corresponds to sensor nodes and set of edges, $E = \{e_j\}$, $j = 1..m$ are connections between the nodes with associated weight functions $W = \{w_j\}$, $j = 1..m$. The value of each w_j is given by Euclidean distance $d(v_i, v_j)$, $i \neq j$. Additionally, $R = \{r_i\}$, $i = 1..n$ are the radii of the communication ranges of nodes [2] [4] [6]. An edge is established if and only if the distance between two nodes is less or equal to the minimum value of their communication radii, $d(v_i, v_j) \leq \min(r_i, r_j)$, $i \neq j$. Note that the terms *node* and *vertex* are used interchangeably in this article as a matter of convenience.

In Figure 1, for an example, a pair of vertices (v_5, v_6) has communication ranges r_5 and r_6 respectively. Because the distance between v_5 and v_6 is less than r, $d(v_5, v_6) < r$, so there exists an edge e_5 connecting them as can be seen in Figure 2. Similarly, the others edges

Table 1. An example of satellite connection data with three rows: *Connection number*, *Connection vector* and *Time vector*.

Connection number	1	2	3	...
Connection vector	v_1	v_3	v_4	...
Time vector	t_1	t_2	t_3	...

of this graph namely $e_1 = e(v_1, v_2), e_2 = e(v_1, v_3), e_3 = e(v_2, v_3), e_4 = e(v_3, v_4)$ could be established. There are $2^n - 5$ edges between a pair of nodes of this graph that are not existed due to inadequacy of the condition. The vertex v_7 is isolated because all distance values between it and the other nodes are inadequate to the condition.

3.2. BT-DYNSEN model of LEO satellite connection

The connections between a LEO satellite and a dynamic sensor field are also described in a BT-Graph. Let $V = \{v_i\}$, i=1..n, is a set of sensor node coordinates in a DSF. In this scenario, connection time is defined when any sensorset [1] of the DSF under the satellite coverage. The LEO satellite communication range is considered as a circle whose center is sub-point on the ground (sub-satellite point), s. It is noted that sub-satellite point, s, is where on the ground the straight line connecting the center of the Earth and the satellite meets the Earth's surface [12]. The associated BT-Graph consists of $n+1$ vertices $P = \{V, s\} = \{v_1, v_2, ..., v_n, s\}$. If a node is within the communication range of the satellite, there exists an edge with weight given by the Euclidean distance between them. Otherwise edge weight is set to infinity. Consequently, the number edges of the graph are $m + n$ by adding n new edges $C = \{c_1, c_2, ..., c_n\}$ with $c_i = c(s, v_i)$, $i = 1..n$. The weights of the n new edges are denoted by $Z = \{z_1, z_2, ..., z_n\}$. In this case the set of edges is $R = \{E, C\} = \{e_1, e_2, ..., e_m, c_1, c_2, ..., c_n\}$ and the set of corresponding weights is $Q = \{W, Z\} = \{\{w_k\}, \{z_i\}\}$ with $k = 1..m$, $i = 1..n$. Hence the BT-DYNSEN model for LEO satellite connections is presented by a graph G(P,R) with weight functions Q.

Furthermore, during the connection time only one sensor node (vertex) of a dynamic sensor field is chosen to connect with the sub-satellite point (center vertex) at one time [1]. A number of different nodes could be chosen based on the value of edge weights at different times. To manage the connections, the name of chosen node is kept in *Connection vector* [1] and the corresponding time is saved in *Time vector* [1] (as in Table 1).

Table 1 shows that in *connection 1*, center vertex s connects with v_1 at time t_1. In a similar way, in *connection 2* at time t_2 and *connection 3* at time t_3, v_4 and v_2 are chosen to connect with s respectively.

For instance, Figure 3 shows three graphs of the dynamic sensor field in three different connections with the center vertex s (a sub-satellite point) at different

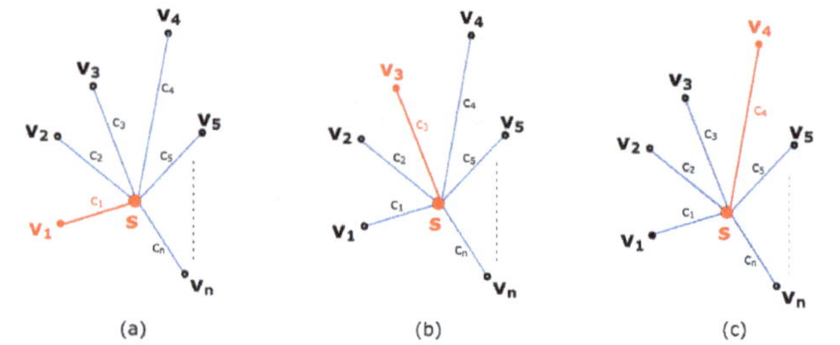

Figure 3. The BT-Graph of a dynamic sensor fiel in three di˙erent connections (solid red lines) at times t_1, t_2 and t_3.

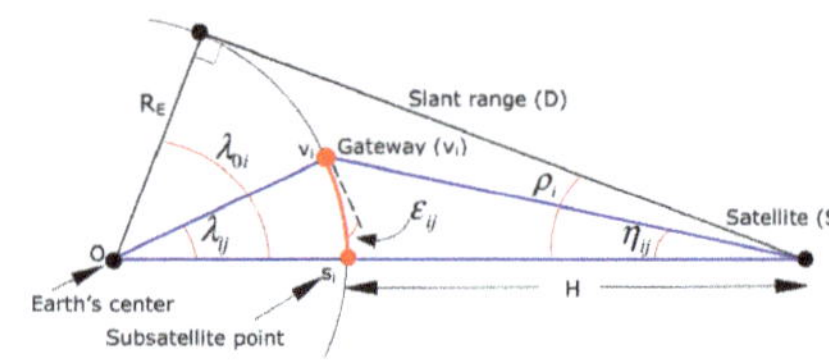

Figure 4. Gateway of sensor fiel geometry [12].

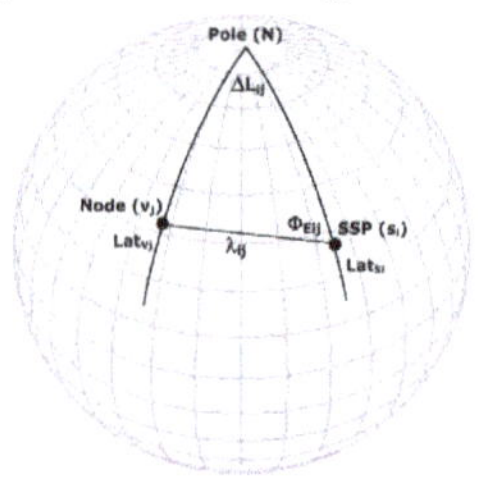

Figure 5. Relationship between sensor node of a DSF and sub-satellite point [12].

times t_1, t_2 and t_3. At time t_1 (Figure 3(a)), vertex v_1 is chosen and edge c_1 is established. Similarly, in Figure 3(b) and Figure 3(c) vertex v_3, v_4 are chosen for connections that leads to corresponding edges c_3, c_4 are established at time t_2 and t_3 respectively.

3.3. Compute the connection time

In this section, we describe the way to calculate connection time, t_{ij}, between a LEO satellite and a gateway of DSF, v_j [1] [12] [15]. Similarly, the calculation could be applied for all other nodes. Note that every node of the sensor field is assumed as a gateway for the connection with the satellite in calculating the values of connection time.

First, it is necessary to define the angles and related distances between satellite, a gateway on the ground and the Earth's center. The parameters are indicated on Figure 4. For angular radius of the spherical Earth, ρ_i, and the angular radius λ_{0_i} can be found from relations

$$\sin(\rho_i) = cos(\lambda_{0_i}) = \frac{R_E}{R_E + H} \tag{1}$$

$$\rho_i + \lambda_{0_i} = 90\ deg \tag{2}$$

where $R_E = 6378.14$ km is the Earth's radius and H is the altitude of the satellite above the Earth's surface.

With the coordinates of a sub-satellite point, s_i $(Long_{s_i}, Lat_{s_i})$ along each satellite's ground track and a sensor node, nodes v_j of the DSF $(Long_{v_j}, Lat_{v_j})$, and defining $\Delta L_{ij} = |Long_{s_i} - Long_{v_j}|$, the azimuth, $\Phi_{E_{ij}}$, measured eastward from north, and angular distance, λ_{ij}, from the sub-satellite point to the sensor node (see Figure 5) are given by

$$cos(\lambda_{ij}) = sin(Lat_{s_i})sin(Lat_{v_j}) + cos(Lat_{s_i})cos(Lat_{v_j})cos(\Delta L_{ij}) \tag{3}$$

$$cos(\Phi_{E_{ij}}) = ((sin(Lat_{v_i}) - cos(\lambda_{ij})sin(Lat_{s_i}))/(sin(\lambda_{ij})cos(Lat_{s_i})) \tag{4}$$

where $\Phi_{E_{ij}} < 180\ deg$ if v_i is east of s_i and $\Phi_{E_{ij}} > 180\ deg$ if v_i is west of s_i. Consider triangle Os_iv_j (in Figure 4), the distance, d_{ij}, between s_i and v_j can be found using the law of cosines:

$$d_{ij}^2 = R_E^2(1 - cos(\lambda_{ij})) \tag{5}$$

The *connection time*, t_{ij}, is given by

$$t_{ij} = (\frac{P}{180\ deg})\arccos(\frac{\cos(\lambda_{ij_{max}})}{\cos(\lambda_{ij_{min}})}) \tag{6}$$

where P is the *orbit period* in minutes. From Equations (5) and (6), the communication duration, t_{ij}, strongly depends on how close the nodes v_j is to the sub-satellite points s_i, the distances d_{ij}, along the ground track on any given orbit pass [12].

3.4. Constraint

The altitude of a LEO satellite must be in range from 275 km to 1400 km due to atmosphere drag and Van Allen radiation effects [12]. Besides, the experimental results were announced by High Altitude Society in United Kingdom that LoRa SemTech transceivers can communicate in distance up to 600 km in environment without any obstacle and 20-40 km in urban area [22]. Based on these factors, maximum communication range for all nodes in long range sensor fields is 40 km in this work. LEO's satellites are chosen in our experiments must have the orbit altitude less than 600 km. Furthermore, the satellite's relative speed over a fixed point on the Earth's surface must be around 7.5 - 8.0 km/sec [12]. The speed of the satellite is calculated by

$$v = \sqrt{\frac{Gm_E}{R_E + H}} \tag{7}$$

Equation 7 shows that the speed of the satellite in orbit is in inverse proportion of its altitude [12]. Where G is universal gravitational constant ($G = 6.67x10^{-11}$ Nm^2/kg^2) and m_E is the mass of the Earth ($m_E = 5.98x10^{24}$ kg). With orbit altitude of satellite in range 300-600 km, the speed of satellite on orbit must be in range 7.56-7.73 km/sec.

4. Optimization method

4.1. Verify the connectivity of DSF

In this work, top down construction algorithm is chosen in order to build balltree structure for verifying the network connectivity of the DSF. In this manner, the time complexity can be found in $O(nlog^2n)$ [6].

Algorithm 1 Verify the network connectivity of a DSF

```
procedure VERIFYCONNECTIVITY(D)
    if D.hasAPoint() then
        Create a leaf B is a point in D
        Return B
    else
        Let c is the dimension of greatest spread
        Let L, R are the set of points lying to 2 subsets
        Let r is a radius of ball
        Create B with two children:
            B.center ← c
            B.radius ← r
            B.leftChild ← balltreeConstruct(L)
            B.rightChild ← balltreeConstruct(R)
    end if
end procedure
```

Top down algorithm to construct the balltree is a recursive process from top to down. The process at each step is to choose the split dimension it and then split the set of values into two subsets. First, it is necessary to choose the point in the ball which is farthest from its center,p_1, and then choose a second point p_2 that is farthest from p_1. It is followed by assigning all points in the ball to closest one of two clusters corresponding to p_1 and p_2. Finally, the centroid of each cluster is calculated and the minimum cluster radius for enclosing all the points are determined. This process is stopped only when the leaf balls contain just two points. Note that D denotes current considered balltree structure and B denotes leaf node defined after each process.

4.2. Determine the sensorsets corresponding to each sub-satellite point

Defining a sensorset [1] based on BT-DYNSEN is to find k nearest sensor nodes (neighbors) under the satellite's coverage area at each sub-satellite. It is a depth-first traversal algorithm for traversing tree or graph data structures, starting with the root node. The value of Q

is updated during search process. At each considered node B, Q obtains k points which are nearest query q as following algorithm.

Algorithm 2 Determine a sensorset

```
procedure DETERMINESENSORSET
    if balltreeNode.isALeaf then
        if d < query.range then
            Write name of balltreeNode to file
        end if
    else
        if d < (query.range + radiusNode) then
            left ← balltreeNode.leftChild
            right ← balltreeNode.rightChild
            rnn: left, query
            rnn: right, query
        end if
    end if
end procedure
```

There are two different cases in *BT-DYNSEN* algorithm as follows. If current considered node is a leaf node and the distance from query point q to B is less than r ($d < r$), the obtained result is updated by adding B into Q. Otherwise, if B is not a leaf node and the distance from query point q to B is less than the total of r and the radius of B ($d < r + B.radius$), it is necessary to perform recursive algorithm for the two child nodes of a parent node B: left-child and right-child.

4.3. Select the gateways of DSF

If a DSF, V, has n nodes, there are $2^n - 1$ connection items, $G = \{g_l\}$, l=1..($2^n - 1$). It is necessary to find out a set of proper nodes which play as gateways of DSF to provide the longest length of time for the connection. To do this, the *association analysis algorithm* [18] is applied. In this way, a DSF is represented in a binary format, where each row corresponds to a connection item and each column corresponds to a node. A node value is one if the node appears in a connection item and zero otherwise. Weight of a connection item determines how often a connection item is applicable to a set of connection items. The weight of a connection item, $g_i \in G$, can be defined as $w(g_i) = |\{g_l \mid g_i \subseteq g_l, g_l \in G\}|$, where the symbol | . | denotes the number of elements in a set.

The algorithm for selecting the gateways of the DSF is briefly presented in following list.

For example, a DSF with 4 nodes, $V = \{v_1, v_2, v_3, v_4\}$ is shown in Figure 6. There are 4 sensorsets $A_1 = \{v_1\}$, $A_2 = \{v_1, v_2, v_3\}$, $A_3 = \{v_2, v_3, v_4\}$ and $A_4 = \{v_3, v_4\}$. The weights of connection items are presented in Table 2. The connection with the highest weight corresponds to the least number of times required to change the gateways. It leads to the connection duration time of the connection is longest. Because the weight of connection item (v_1, v_3) is highest, this connection is chosen.

Algorithm 3 Select the gateways of a DSF

```
Input: list of sensorsets
Output: gateways of DSF
C_k: candidate sensors (size k)
L_k: set of selected gateways (size k)
procedure SELECTGATEWAY
    L_1 ← {weight1 − sensorsets}
    for (k=2;Lk.count>0;k++) do
        C_{k+1} ← generate candidate sensors L_k
        for each transaction t ∈ data do
            Increment count of the candidate sensors in C_{k+1} that contained in t
        end for
        L_{k+1} ← candidate sensors in C_{k+1}
    end for
    Write counted frequent of all sensors L_k to file
end procedure
```

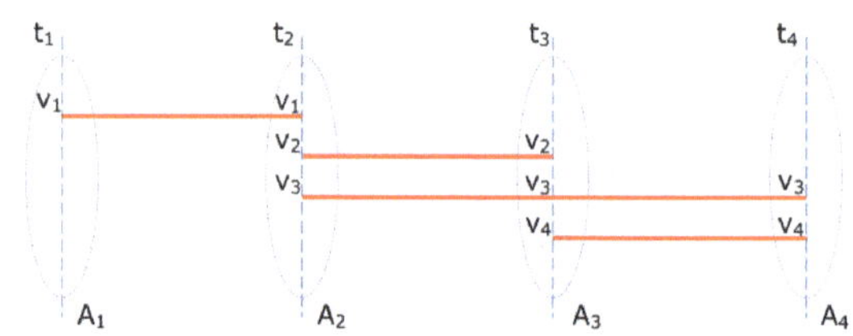

Figure 6. The connections between a 4-node DSF with a LEO satellite.

Table 2. The weights of connection items in a DSF with 4 nodes.

Connection items	Weights
(v_1, v_2, v_4)	1
(v_1, v_2, v_3)	1
(v_1, v_3)	3
(v_1, v_3, v_4)	1

4.4. Find shortest path for data dissemination

The problem of finding shortest path from each sensor node of DSF to the gateway can be solved by a graph search method. The algorithm is presented as Algorithm 4.

The algorithm proceeds in three following steps. First, the balltree graph-based model, G, is constructed. At the next step, the weight matrix M for edges connect each pair of vertices (sensor nodes) in V is computed based on their coordinates. For v_i, v_j are two different vertices in V, in case of $v_i \equiv v_j$ the edge weight is 0. If $v_i \not\equiv v_j$, the edge weight is given by $d(v_j, v_j)$ if $d(v_j, v_j) \leq r$, otherwise it is infinity, ∞. Finally, the shortest path from q to e in the weight matrix is figured out. Note

Algorithm 4 Find shortest path from each sensor node of DSF to the gateway

```
procedure FINDSHORTESTPATH
    Init matrix M
    S ← Start point
    T ← End point
    d is an array
    free is an array
    trace is as array
    d[S] ← 0
    while TRUE do
        u ← -1
        min ← INFINITE
        for v ← 0 to n-1 do
            if free[v] AND (d[v] > (d[u] + M[u,v])) then
                d[v] ← d[u] = arr[u][v];
                trace[v] ← u;
            end if
        end for
        if (u=-1) OR (u=T)) then
            Break
        end if
        free[u] ← false
        for (v ← 0 to n-1 do
            if free[v] AND (d[v]>(d[u]+M[u,v])) then
                d[v] ← d[u] = arr[u][v];
                trace[v] ← u;
            end if
        end for
    end while
end procedure
```

that q is starting point (a sensor node) and e is the destination point (the gateway of DSF).

4.5. Parallelization of RNN algorithmto determine sensorsets

As mentioned in section 4.2, range nearest neighbors (RNN) search method on BT-Graph can be used to determine the set of nodes in a sensorset associated with each position of satellite along ground track. In this section, a greedy algorithm is proposed to nearest neighbors search running on CUDA (namely RNN-CUDA as shown in Algorithm 5). The process of this greedy algorithm consists of two part. One part is parallel processes on GPU to compute distances from query the point to all nodes of a dynamic sensor field. Another part runs on CPU to sort the obtained distances in ascending order and then pick up k neighbor nodes associated with shortest distance values.

Algorithm 5 RNN-CUDA

```
▷ Host program executed on CPU
Load data into global memory
Assign the value of k
Copy data from CPU → GPU
▷ Parallel programs executed on GPUs
Compute distances from query point, q, to all nodes
  v ∈ V
Copy data back from GPU → CPU
▷ Host program executed on CPU
Sort the distance values as a ascending list
Pick up k nearest neighbor nodes associated to k
  first value of the list
Release GPU memory
```

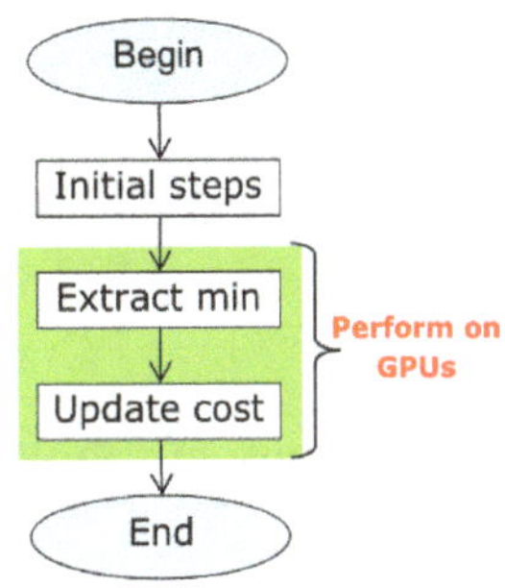

Figure 7. The flo chart of parallel Dijkstra algorithm for searching BT-Graph model.

4.6. Parallelization of search algorithmto fin the shortest path for data dissemination

To parellelize Dijkstra algorithm, we propose that each edge of BT-Graph is handled by a thread of CUDA. Figure 7 depicts the flow chart of parallel Dijkstra algorithm for searching BT-Graph model in which two procedures, *Extract min* and *Update cost*, are performed on NVIDIA GPUs. The parallelization of Dijkstra's algorithm is illustrated in Algorithm 6.

5. BT-DYNSEN implementation

We have developed the BT-DYNSEN tool in C/C++, that enables to model and optimize the connection time between a dynamic sensor field with a LEO satellite. GPredict [17] is used to provide the information about BEESAT-3 satellite's path. Besides, NetGen tool [21] is utilized to generate the abstract network of a 50-node dynamic sensor field from geographic data provided by Google maps service. The obtained result are the nodes of the DSF which should be configured as gateways for the best connection duration time and the shortest paths for data dissemination from each node to these gateways.

Algorithm 6 DijkstraCUDA

1: **Step 1:**
Initialize weight matrix (input vectors), start point and stop point

2: **Step 2:**
Allocate vectors in device memory corresponding to each edge of BT-Graph

3: **Step 3:**
Copy vectors from CPU (host) memory to GPU (device) memory

4: **Step 4:**
- Search for a free vertex u that has the least cost from the starting vertex S to u
+ If no vertex u is found, either it exists a path or not
+ Otherwise, from u we continue to consider the other free vertices
- Use CUDA to specify the thread for searching the paths between the vertex u and the others

5: **Step 5:**
Copy results from GPUs memory to CPU memory

6: **Step 6:**
Free device memory and host memory

Figure 8. Verify dynamicsensor networkconnectivity(a) A 50-node dynamicsensor network,(b) Balltree structureof the DSF, (c) Connectivityof the DSF.

6. Experiment

6.1. Experiment1: Verify networkconnectivity

The connectivity of the 50-node dynamic sensor network was verified by applying BT-DYNSEN as shown in Figre 8(a). Figure 8(b) depicts the balltree structure of this DSF in which there are 49 balls with radii from 10.124m to 321.165m. BT-Graph model then was utilized to ensure the full connectivity of all network nodes, the radius 30.497m was then chosen as shown in Figure 8(c).

6.2. Experiment2: Determinethe sensorsets

BT-DYNSEN tool was employed in determining sensorsets corresponding to sub-satellite points during visiting time. The map in Figure 9 shows the sensorset was determined with sub-satellites of BEESAT-3 at (latitude: 14.00, longitude: 102.48) in orbit 12794. Sub-satellite and satellite coverage were indicated by solid red square and red circle repectively. The sensorset consists of 19 sensor nodes which were under the satellite's coverage area (the shadow area). There are four sensorsets were created along the BEESAT-3's ground track in orbit 12794.

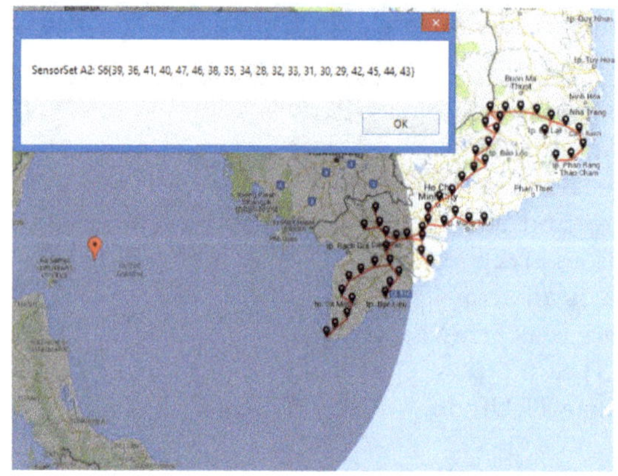

Figure 9. Determinethe sensorsets correspondingto each sub-satellite along the groundtrack of BEESAT-3 in orbit 12794.

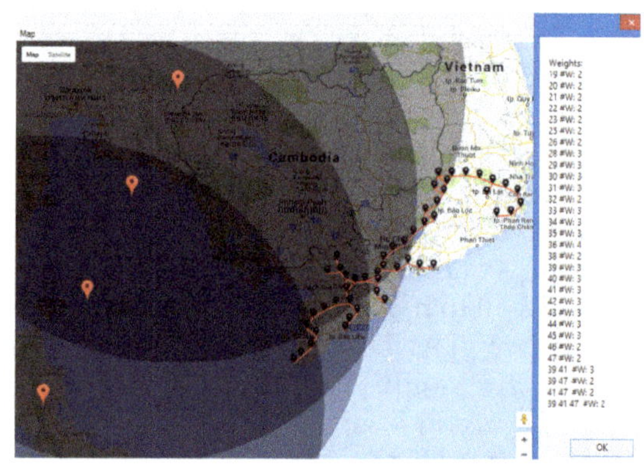

Figure 10. Select a set of nodes whichplay as gateways of the DSF accordingto the weights of connections.

6.3. Experiment3: Select the gateways

With each sensorset, a subset of connections is established. The weights of each connection in subset are then computed. A set of connection items is created by combining these subsets. The best connection is chosen based on the weights of connection items. For instance, in Figure 10 node v_{36} was chosen as the gateway of the 50-node DSF to connection with BEESAT-3 satellite in orbit 12794 because its weight is highest in sensor nodes.

6.4. Experiment4: Optimizedata dissemination

To ensure sensing data to be collected from all sensor nodes and then sent before the satellite leaving, it is necessary to find the shortest path from each

Figure 11. The shortest path for data dissemination from node v_{50} to gateway node v_{36}.

Table 3. Comparison on the execution time of RNN BT-Graph and RNN-CUDA.

	N=10000, Q=10000	N=100000, Q=100000
RNN BT-Graph	179.46	17898.13
RNN-CUDA	123.47	5440.29

sensor nodes to the gateway. The interconnection weights within the DSF are geographic distances between each pair of nodes that were carried out by applying BT-DYNSEN model. Figure 11, as an example, illustrates the chosen path (the bold blue line) for data dissemination from sensor node v_{50} to the gateway node v_{36}.

6.5. Experiment 5: Parallelization RNN algorithm for determining sensorsets

In order to make the comparison between the parallel algorithms and serial ones for searching BT-Graph, we have run the simulations on a computer which is equipped with a INTEL® Core i3-3220 3.3GHz processor, 4GB DDR3 SDRAM and a NVIDIA GeForce GTX660 graphic card. In our experiment, a comparison about BT-Graph's search speed is obtained by performing both sequential algorithm on CPU and parallel algorithm on GPUs. On each algorithm with the same input data, which are generated randomly by program, experimental program is executed in 10 times for getting the average value of execution time.

The results is illustrated in Table 3, where N is the amount of initial data points and Q is a number of query points. It shows that the search speed of RNN-CUDA be improved around 2.37 folds compared with that of RNN BT-Graph.

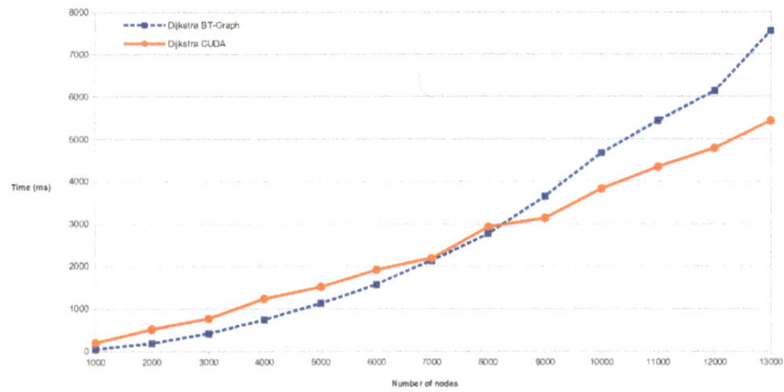

Figure 12. An execution time comparison of serial and parallel Dijkstra's algorithms.

6.6. Experiment 6: Parallelization Dijkstra algorithm for data dissemination

The purpose of this experiment is to compare search speed of traditional Dijkstra for BT-Graph and the corresponding speed of parallel Dijkstra (Dijkstra CUDA). The first step is to construct the weight matrix of BT-Graph model. Next, to obtain the average execution time, serial Dijkstra's algorithm and parallel one are run in turns in 10 times. These experimental results are illustrated in Figure 12.

From Figure 12, CPU can process efficiently for a small a number of vertices because it not waste extra time copying the data between CPU and GPU memories. Nevertheless, when a number of vertices are increased more than 8000, the duration time required for processing can be reduced significantly by using techniques of parallelism based on CUDA.

7. Conclusion

Based on BT-Graph, we have described a new approach in order to model and optimize the dynamic sensor field for LEO satellite connections. The distances between the sub-satellite points and each node of the sensor field is utilized as a key factor to find out the proper gateways for the longest connection time. The experimental results were obtained by applying several appreciate algorithms on BT-DYNSEN model to verify the connectivity of network, determine sensorsets at visiting time, choose set of gateway nodes and find shortest path for data dissemination in DSF. In order to improve the execution time, we implemented the corresponding parallel algorithms with CUDA on GPU, and analyzed its performance. The results show that the parallel algorithm on GPU is considerably superior to the serial one on CPU when BT-Graphs comprising more than 8000 vertices. Thus, our proposed graph-based model helps to increase the amount of time for data communications in long-range sensor field applications using satellite connections.

References

[1] Tuyen, P.T., Hoang, V.T., Hiep, X.H. and Pottier, B. (2015) Optimizing the Connection Time for LEO Satellite Based on Dynamic Sensor Field. In 4th *EAI International Conference on Context-Aware Systems and Applications (ICCASA15)*, Vungtau, Vietnam.

[2] Huong, H.L. and Hiep, X.H. (2014) Graph-based Model for Geographic Coordinate Search Based on Balltree Structure. In *Proceeding of the* 17th *International conference*, Daklak, Vietnam, 116-123.

[3] Ines, K., Pascale, M., Anis, L. and Saoucene, M. (2014) Survey of Deployment Algorithms in Wireless Sensor Networks: Coverage and Connectivity Issues and Challenges. In *International Journal of Autonomous and Adaptive Communications Systems (IJAACS)*, 24.

[4] Mohammad, R., Bijan, G. and Hassan, N. (2014) A Survey on Nearest Neighbor Search Methods. In *International Journal of Computer Applications*, 39-52. doi: 10.5120/16754-7073

[5] Panwar, D. and Neogi, S.G. (2013) Design of energy efficient routing algorithm for wireless sensor network (WSN) using Pascal graph. In *Computer Science & Information Technology*, 175-189. doi: 10.5121/csit.2013.3217

[6] Omohundro, S.M. (1989) Five Balltree Construction Algorithms. Technical Report.

[7] NVDIA, https://developer.nvidia.com/cuda-zone (accessed on 4 November 2015).

[8] GPU, http://www.nvidia.com/object/what-is-gpu-computing.html (accessed on 4 November 2015).

[9] Gunjan, S., Amrita, T., Dhirendra, P.S. (2013) New Approach for Graph Algorithms on GPU using CUDA. In *Internaltion Journal of Computer Application*, Volume 72, Issue 18, 38-42. doi: 10.5120/12645-9386

[10] Harish, P., Narayanan, P.J. (2007) Accelerating Large Graph Algorithm on the GPU using CUDA. In *Proceeding HiPC'07 Proceeding of the* 14th *international conference on High performance computing*, Vol. 4873, 197-208. doi: 10.1007/978-3-540-77220-0_21

[11] Lucas, P.Y., Long, N.H.V., Tuyen, P.T. and Pottier, B. (2015) Wireless sensor networks and satellite simulation. In 7th *EAI International Conference on Wireless and Satellite Systems*. doi: 10.1007/978-3-319-25479-1_14

[12] James, R.W. (2003) Space Mission Geometry. In Wiley, J.L. and textscJames, R.W. *Space Mission Analysis and Design*, 3rd *Edition* (Microcosm Press), chap. 5.

[13] Cakaj, Sh., Keim, W. and Malaric, K. (2007) Communication Duration with Low Earth Orbiting Satellites. In *Proceedings of IEEE, IASTED, 4th International Conference on Antennas, Radar and Wave Propagation*, 85-88. doi: 10.1002/sat.4600090403

[14] Dosiere, F., Zein, T., Maral, G., Boutes, J.P. (1993) A model for the handover traffic in low earth-orbiting (LEO) satellite networks for personal communications. In *International Journal of Satellite Communications*, 574-578. doi: 10.1002/sat.4600110304

[15] Capderou, M. (2014) Ground track of a satellite. In Capderou, M. [eds.] *Handbook of satellite orbits from Kepler to GPS* (Springer International Publishing), chap. 8, 301-338. doi: 10.1007/978-3-319-03416-4

[16] Walter, C., Kris, S., Nicolas, D. and Adam, D. (2008) Satellite based wireless sensor networks: global scale sensing with nano- and pico-satellites. In *Proceedings of the* 6th *ACM conference on Embedded network sensor systems (SenSys'08)*, 445-446. 10.1145/1460412.1460495

[17] Alexandru Csete, GPredict project, http://gpredict.oz9aec.net/ (accessed on 4 November 2015).

[18] Pang-Ning, T., Michael, S., Vipin, K. (2005) Association Analysis: Basic Concepts and Algorithms. In *Introduction to Data Mining* (Addison-Wesley Longman Publishing Co., Inc. Boston, MA, USA), chap. 6.

[19] Celandroni, N. *et al.* (2013) *A survey of architectures and scenarios in satellite-based wireless sensor networks: system design aspects* (International Journal of Satellite Communications and Networking), vol. 31, 1-38. doi: 10.1002/sat.1019

[20] Dong, J., Chen, Q. and Niu, Z. (2007) Random graph theory based connectivity analysis in wireless sensor networks with Rayleigh fading channels. In *Asia-Pacific Conference on Communications*, 123-126. doi: 10.1109/APCC.2007.4433515

[21] Pottier, B. and Lucas, P.Y. (2015) Dynamic networks-NetGen. Technical report, Universite' de Bretagne Occidentale, France.

[22] UK High Altitude Society, http://www.instructables.com/id/Introducing-LoRa-/step19/LoRa-receiver-links/ (accessed on 4 November 2015).

[23] Berlin Experimental and Educational Satellite-2 and -3. https://directory.eoportal.org/web/eoportal/satellite-missions/b/beesat-2-3 (accessed on 4 November 2015).

[24] Ting, Y. and ChunJian, K. (2012) An energy-efficient and fault-tolerant convergecast protocol in wireless sensor networks. In *International Journal of Distributed Sensor Networks*, 8 pages.

[25] Camtepe, S., Yener, B. and Yung, M. (2006) Expander Graph based Key Distribution Mechanisms in Wireless Sensor Networks. In *IEEE International Conference on Communications*, 2262-2267. doi: 10.1109/ICC.2006.255107

23

Improvement of natural image search engines results by emotional filtering

Patrice Denis[1], Vincent Courboulay[1,*], Arnaud Revel[1],
Syntyche Gbèhounou[2], François Lecellier[2], Christine Fernandez-Maloigne[2]

[1] Laboratoire Informatique, Image et Interaction (L3i)
Pôle Sciences et Technologie - Université de La Rochelle
Avenue Michel Crépeau - 17042 La Rochelle Cedex 1 - France
[2] Département SIC du Laboratoire XLIM - UMR CNRS 7252
Université de Poitiers - Bât SP2MI Téléport 2 - 11 Bd Marie et Pierre Curie
BP 30179 - 86962 Futuroscope Chasseneuil Cedex - France

Abstract

With the Internet 2.0 era, managing user emotions is a problem that more and more actors are interested in. Historically, the first notions of emotion sharing were expressed and defined with emoticons. They allowed users to show their emotional status to others in an impersonal and emotionless digital world. Now, in the Internet of social media, every day users share lots of content with each other on Facebook, Twitter, Google+ and so on. Several new popular web sites like FlickR, Picassa, Pinterest, Instagram or DeviantArt are now specifically based on sharing image content as well as personal emotional status. This kind of information is economically very valuable as it can for instance help commercial companies sell more efficiently. In fact, with this king of emotional information, business can made where companies will better target their customers needs, and/or even sell them more products.

Research has been and is still interested in the mining of emotional information from user data since then. In this paper, we focus on the impact of emotions from images that have been collected from search image engines. More specifically our proposition is the creation of a filtering layer applied on the results of such image search engines. Our peculiarity relies in the fact that it is the first attempt from our knowledge to filter image search engines results with an emotional filtering approach.

Keywords: creative/affective experiences, creative technologies, creative user experience, emotions, image search engine, natural images, image retrieval, image web browsing, bottom-up saliency, regions of interest

1. Introduction

Needs from users of the Web are evolving with new technologies. Nowadays, people are accustomed to participate and give their opinions on websites or social networks. For example, they share everyday emoticons, *likes* on Facebook, *+1's* on Google+, short messages on Twitter. They even share their professional networks connections on LinkedIn or Viadeo. The *Internet of Things* is a technology that now allows to interact with computers not only with keyboard and mouse but in new user-oriented devices like connected watches, pedometers, refrigerators, electronic scales or Google glasses for instance.

Web search engines users are often lost in the huge amount of answers they got from anonymous requests and want specific and personalized replies to their own needs. Semantic Web, sometimes called Web 3.0, is evolving to focus more and more on structuring its data to answer more precisely and personally to these users.

In this context, emotional status of users is now taken into account by recent works. Business industries are in

*Corresponding author. Email: vcourbou@univ-lr.fr

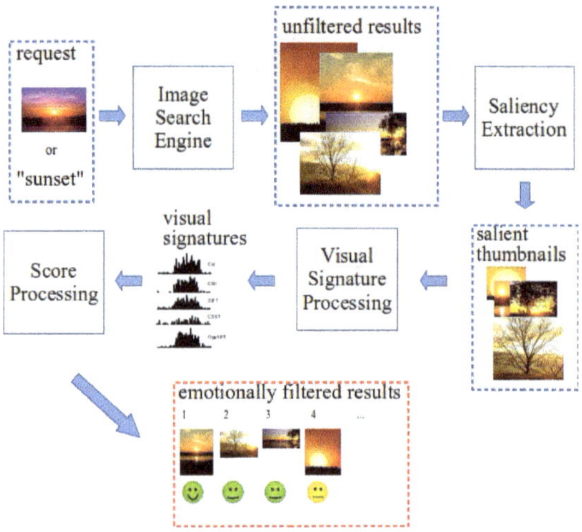

Figure 1. Global emotional filtering model.

because they think they could increase their customers market delivering better customized products arousing positive emotions. This paper proposes to contribute to these researches introducing a new layer to image search engines filters according them to add an image emotion status. We do not known any search engine that has addressed these issues the way we have.

We first introduce in Section 2 background on image search engines (2.1). Emotion definitions and modelling are given in 2.2. We discuss in Section 2.2 how emotions have been extracted from digital documents in the past. More details are given specifically on texts and images. 3 is devoted to the description of our proposed system. We first present Support Vector Machines technology (SVMs), used for our classifiers. Then, an image database, called SENSE, is presented and we justify its usage to train our system. Next, we expose a bottom-up saliency model we used to focus on the most important emotional information of each image. Then we detail the whole processing string our system. We show some of its results on our Internet crowded image database of cooking recipes and compare it to other systems. Last section, (4), is kept for conclusions and perspectives.

2. Background and related works

2.1. Image search engines filtering

Image research engines often allow filtering their results. This is usually done by extraction and comparison of image features. Those can be deduced from bottom-up or top-down approaches [15]. Bottom-up methods extract low level features on image data, while top-down methods are task-driven ones using supervised learning. Common filtering techniques get information from direct access within the image file metadata. Examples are bottom-up features like the size; dimension proportion ratio: height, width; style: photo, drawing. Top-down classical features extraction from metadata includes: characters, faces, portraits with head and shoulders...The number of indexed images where features were extracted is also a criteria to compare search engines filters. Table 1 shows a comparison on common filters.

Here is a non exhautive list of several famous and not so famous image search engines:

Google images[1]; *Bing images (Microsoft)*[2]; *TinEye*[3]; *Picsearch*[4]; *oSkope visual search*[5]; *Search22* [6].

The following section is about emotion modelling and extraction within digital documents.

2.2. Emotion and digital images

Human emotions are answers of human beings to their environment. This environment is interpreted by their brains through several sensors: eyes, ears, nose, mouth, skin...People use their emotions to interact with other humans. They rely on their empathy to interprets these emotions correctly.

Emotion modelling. Many definitions of emotion have been published and differ from different schools of psychology. Refering to computational approach of emotion recognition, two theories are considered [25]:

- *Basic emotion theories* [39, 45] establishing the existence of basic emotions. These basic emotions are also referred as fundamental or primary.
- *Theories of evaluation* defining emotion as a set of appraisal states that occur when a human being is faced with an external stimulus.

These two theories have defined the two more used emotion classification:

1. **Discrete approach** [10, 37]: emotional process can be explained with a set of basic or fundamental emotions, innate and common to all human (sadness, anger, happiness, disgust, fear, ...). There is no consensus about the nature and number of these fundamental emotions. This modelling is usually preferred in emotion extraction based on facial expressions. An example of discret classification of emotion is given in the Figure 2(a); it represents the Plutchik's wheel of emotions [37]. The author defines eight basic emotions; combinations of primary emotions define secondary ones.

[1]http://www.google.fr/imghp
[2]http://www.bing.com/?scope=images
[3]http://www.tineye.com
[4]http://www.picsearch.com
[5]http://www.oskope.com
[6]http://www.search-22.com/downloads/images.php

Filter	date	size	orientation	color	file format	licence	style	faces	image query	textual query
Google image	x	x		x	x	x	x	x	x	x
Bing image	x	x	x	x			x	x		x
TinEye	x	x				x			x	
Picsearch		x	x	x			x	x		x
Exalead		x	x	x	x	x	x	x		x

Table 1. Comparison of common filters proposed by image search engines

2. **Dimensional approach**: emotions are considered as the result of a fixed number of concepts represented in a dimensional space. The dimensions can be pleasure, arousal or power for instance. They vary depending to the needs of the model. Russel's dimensional model, represented in the Figure 2(b), is the most used with the dimensions valence and arousal:

   • ***The valence*** corresponds to the way a person feels when looking at a picture. This dimension varies from negative to positive and allows to distinguish between negative emotions and pleasant ones.

   • ***The arousal*** represents the activation level of the human body.

   The advantage of these models is to define a large number of emotions. Despite this, some emotions can be confused (such as fear and anger in the circumplex of Russel) or unrepresented (among others surprise in Russel's model).

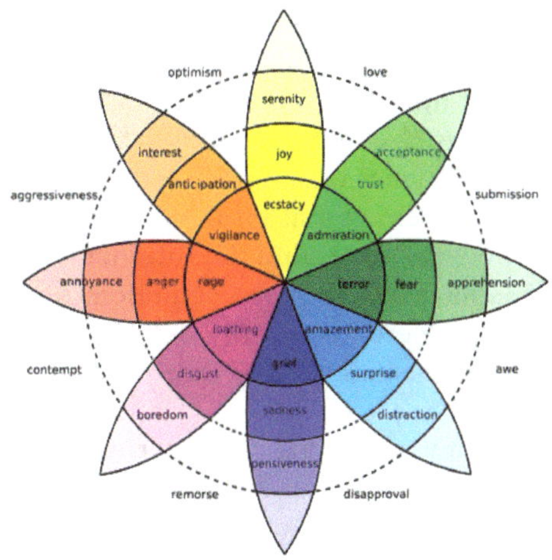

(a) Plutchik's wheel of emotions

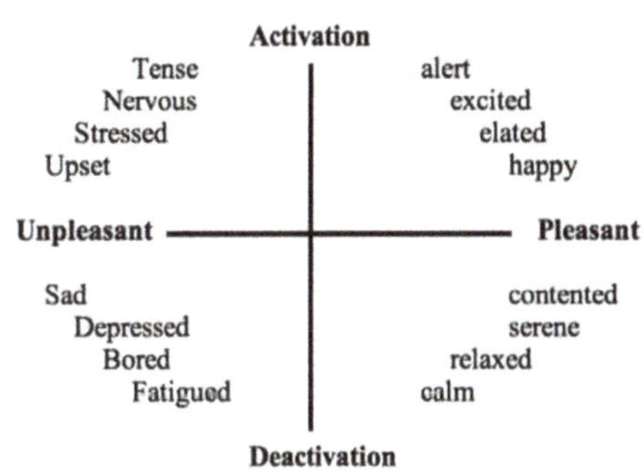

(b) Rusell's dimensional model

Figure 2. Examples of discrete and dimensional emotion classification.

Emotion extraction. As a reaction to our environment, emotions are linked to our five senses. But let's face that only seight and hearing are used in our digital world of multimedia documents. Many researches have been made to achieve emotion extraction by computers and improve interaction between humans and machines. The holy grail to achieve being able to bring human-empathy to computers aptitudes. Though, the multiplicity and diversity of digital documents is still a major research challenge for computers to correctly gather and interpret emotions. Moreover, extraction of emotions can either be targeted to the content or to the user of the document. In other words, the goal may be to extract emotions intrinsically embedded into a document or those that will arouse on humans reading it.

From audio To decode emotions from audio files, focus is directed on our hearing aptitudes. Everyone knows that music or voice provoke emotions on people. For instance, one can often tell the mood of another person by listening her even in a foreign language. Competitive challenges like the avec [7] or the isca-speech [8] for instance are gathering new contributors every year [22, 31].

Table 2 summarizes these emotion extraction approaches on audio documents.

The majority of digital documents are captured by our eyes. Most contributions in the literature are then focused on visual documents as we do in the following.

[7]International Audio/Visual Emotion Challenge and Workshop ; http://sspnet.eu/avec2014/

[8]Interspeech Conferences of the International Speech Communication Association; http://www.isca-speech.org/iscaweb/

Method	features	extraction/transmission	context/application
Krishnan	speech	speaker's emotional state	recognition system
Shing	audio	audio emotion	recognition system

Table 2. Extraction approaches on audio documents

From text documents The first scrutinized documents to extract emotions were digital textual documents. Text is linked to language. If computers could interpret texts, they would be able to interpret our emotions expressed. This textual information, is still combed to decrypt human emotions. For instance, Salway and Graham used audio description text streams to extract characters emotions in movies [41]. Kamvar and Harris extracted emotional information of textual contents from bloggers all around the Internet [20].

With social media, needs to express emotions brought up emoticons. They were historically defined by small groups of characters, usually two or three, representing a facial expression (:-) :-| :-((^_^) and (;_;) are examples). Now they are single character encoded (☺ or ☹). Emoticons were used to dig up emotions from text documents [38, 46]. Hu et al. defined different emotional signals extracted from social media datasets (Twitter and Facebook short messages) to built an unsupervised sentiment analysis system [17]. Lin et al. proposed an emotion classification system based on five textual feature types [24]. They pointed out that authors and observers points of view on emotions are often not the same as in [21].

Table 3 summarizes these emotion extraction approaches on textual documents.

The following paragraph focuses on emotion extraction from image documents.

From image documents Image emotional extraction is a challenge because searchers are still learning how to link computer language and human being interpretations of images. The first approaches to compute image emotions were to find associated text. That way, previous works done on text (as in Section 2.2) helped extract emotions on images.

Emotion search engines or interface based on textual metadata and users tagging were proposed in [11, 18, 21, 52].

A large part of the literature is devoted to the links between emotions and colours [2, 6, 7, 27, 32–34, 49]. In fact, a consensus states the existence of a link between colours and particular emotions. As stated by Ou et al. [32], colours play an important role in decision-making, evoking different emotional feelings. In a serie of publications, Ou et al. [32–34] studied the relationship between emotions, preferences and colours. They established a model of emotions associated with colours from psychophysical experiments.

Emotions are also extracted from facial features (such as eyebrows, lips) using faces contained in images [35]. It seems to be the easiest way to predict emotions, since facial expressions are common to human to express basic emotional feelings (happy, fear, sadness).

More recently some authors considered the emotion recognition as a CBIR task [28, 44, 50]. The underlying idea is to use the traditional techniques of image retrieval to extract the emotional impact. Extraction of traditional image features (colours, textures, shapes) combined with a classification system allows the authors to predict the emotional impact of images after a learning step. It is often necessery to add complementary information. For example, Wang and Yu [48] used the semantic description of colours to associate an emotional semantic to an image. Liu et al. [25] stated that oblique lines could be associated with dynamism and action; horizontal and vertical ones with calm and relaxation.

Although research has been made to characterize still images emotions with low level features extraction as said above, we do not know any emotional image search engine focused only on low level extraction from image as we developed in this paper. However some studies have been made on image search engines.

Table 4 summarizes these emotion extraction approaches on images documents.

Most of these former presented papers use text metadata mining to extract emotional information about images in their search engines. A novelty of our paper is that we do not use emotional associated text, or as Solli emotional Bags of Emotions [43], but we pull out emotion in images from bags of visual words [4] based on low level descriptors extracted from pixel level processing.

3. Proposed system

Our proposed method starts by a Query By Image Content (QBIC) and defines an emotional filter system based on low level images features. This filter can then be applied on image search engines results in Web technologies for instance. For that, we processed an emotional score for each image in a set acquired by QBIC. This score is gained by a fusion of five classifier scores. The classifiers are based on Support Vector Machines (SVMs) [9] described in Subsection 3.1 and powered by libSVM [8]. Classifiers are trained on a low semantic database called SENSE which is described in

Method	features	extraction/transmission	context/application
Salway	audio descrition text	characters emotion	movies
Kamvar	textual content	emotion info	bloggers
Read, Urabe	emoticons	user emotion	machine learning
Hu	social media datasets	emotional signals	sentiment analysis
Lin	text	reader's emotion	classification

Table 3. Emotion extraction approaches on textual documents

Method	features	extraction/transmission	context/application
Knautz	textual metadata, tags	felt and/or shown emotion	animals DB/emotion search engine
Fonseca	textual metadata, tags	observer's perpective	image search interface
Yanulevskaya	texture descriptors, text	emotion	emotion categorization system
Zhang	metadata	image attractiveness	image search engine interface
Huisman	visual appeal	visual content from web pages	emotion evaluating interface

Table 4. Emotion extraction approaches on images documents

Subsection 3.2. The description and justification of the usage of a saliency model can be read in Subsection 3.3. Eventually, the whole stream of our emotional filtering system is described in details in the last Subsection 3.4.

3.1. Support Vector Machines

Support Vector Machines (SVMs), introduced as Support Vector Networks in [9] are supervised learning models used for classification and regression analysis.

We briefly introduce the basics of SVMs to explain how we used them in our approach. When one want to perform classification, he has to get data first. A model is learnt splitting data into training and testing sets. Each sample of the training set is described by a feature vector and labeled to the class it belongs to. The goal of SVMs is to produce a model, based on the training set which predicts the target class or label of every sample of the test set given only the its attributes (feature vector). To achieve that, samples projected in a dimensional space are separated by a linear hyperplane.

SVMs classification process is illustrated by Figure 3. Let $(\mathbf{x}_i, y_i), i = 1, ..., m$ where $\mathbf{x}_i \in \mathbb{R}^n$, n the dimension of the feature vector and $\mathbf{y} \in \{1, -1\}^m$ be a training set of m sample-label pairs. In both sckeches, the separating hyperplane, defined by equation $\mathbf{w}.\mathbf{x} + b = \mathbf{0}$, is represented by the black line. For every sample, $r = \frac{|\mathbf{w}.\mathbf{x}+b|}{\|\mathbf{w}\|}$ is the distance to the separating hyperplane. The support vectors (rounded points) are the closest points to the separating hyperplane. d is the distance between the two classes (red and blue). It is also called the margin. The goal of SVMs is to maximize this distance. In the case of hard margin SVMs (Figure 3.a), SVM finds a linear separating hyperplane with the maximal margin. When using soft margin (Figure 3.b), an error term ξ is introduced to take classification errors and noise into account. $C > 0$ is the penalty parameter. It is a constant used to control the balance between the number of classification errors and the margin width.

The support vector machines (SVMs) [5, 9] require the solution of the following optimization problem:

$$\min_{\mathbf{w},b,\xi} \frac{1}{2}\mathbf{w}^T\mathbf{w} + C\sum_{i=1}^{m} \xi_i$$

$$\text{subject to } y_i(\mathbf{w}^T\Phi(\mathbf{x}_i) + b) \geq 1 - \xi_i,$$

$$\xi_i \geq 0.$$

Here training vectors $\mathbf{x}_i$ are mapped into a higher (maybe infinite) dimensional space by the function Φ.

Furthermore, $k(\mathbf{x}_i, \mathbf{x}_j) \equiv \Phi(\mathbf{x}_i)^T\Phi(\mathbf{x}_j)$ is called the kernel function. Though new kernels are often being proposed by researchers, the most known and most commonly used in the literature are the following:

- linear: $k(\mathbf{x}_i, \mathbf{x}_j) = \mathbf{x}_i^T\mathbf{x}_j$;
- polynomial: $k(\mathbf{x}_i, \mathbf{x}_j) = \left(\gamma\mathbf{x}_i^T\mathbf{x}_j + r\right)^d, \gamma > 0$;
- RBF[9]: $k(\mathbf{x}_i, \mathbf{x}_j) = \exp^{-\gamma\|\mathbf{x}_i - \mathbf{x}_j\|^2}, \gamma > 0$;
- sigmoid: $k(\mathbf{x}_i, \mathbf{x}_j) = \tanh(\gamma\mathbf{x}_i^T\mathbf{x}_j + r)$.

$\gamma \in \mathbb{R}$ being a kernel parameter.

In addition to performing linear classification, SVMs can efficiently perform a non-linear classification using what is called the kernel trick, implicitly mapping their inputs into high-dimensional feature spaces. This property of the SVMs is not described furthermore here as we used linear classification.

3.2. SENSE database

Several image databases have been used to study emotions [28, 44, 50]. The most known and used

[9] RBF = radial basis function

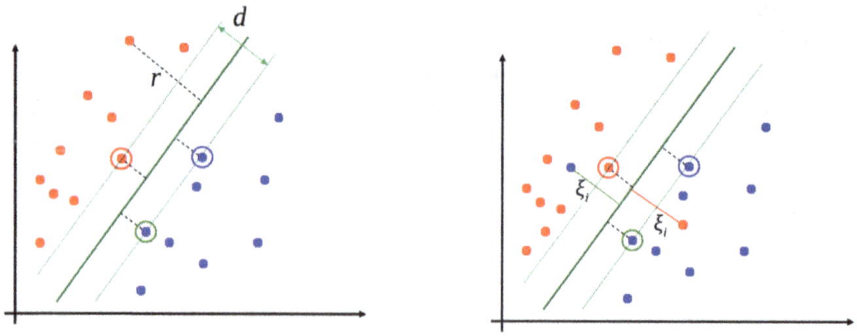

(a) hard margin SVM:
r is the distance to the hyperplane;
d is the margin

(b) soft margin SVM:
ξ is an error term introduced to take classification errors and noise into account

Figure 3. SVMs' principle: The two classes (red and blue points) are separated by the SVM hyperplane (black line). Support vectors (rounded points) are the closest samples to the hyperplane.

is the International Affective Picture System (IAPS) [10] [23]. As there were no universal emotional image database focused on extraction of emotional impact, Gbèhounou et al. [13] proposed one. This image set is named SENSE for Studies of Emotion on Natural image databaS. It is composed of images as diversified as possible and with low semantic power. That is to say that the images do not shock or force a strong emotional answer during subjective evaluation in opposition to some IAPS images for example. Each image has been evaluated by users during very short period of time: eight seconds. This allows to look for primary emotion impact of the presented images. In fact, Schyns et al. demonstrated that emotions on facial expressions [42] are transmitted during the very first 200*ms* after the reception of the observer signal. During this very short period time, semantic interpretation can not be done by the user. On the contrary, image from the IAPS, embed high semantic power. High semantic can introduce a bias on the evaluation of multiple images as strong emotions remains longer that law emotions in the mind of evaluators [40]. The authors focused to include only low semantic images in their database in order to minimize or to remove that kind of bias during evaluation. The goal was to evaluate each image for its content and not for the remnant of the previous image emotion evaluation.

The database is free to use for research and contains 350 images of landscapes, characters, animals, food, drink, historic and tourist buildings images as shown in Figure 4. Furthermore, the database also contains very few images with human faces (4.86%). This last point reflects the desire to limit the interpretation of facial expressions which embed bigger emotional reactions.

[10]from the Center for Emotion and Attention (CSEA) at the University of Florida

The SENSE database was then evaluated in different subjective experiments. Twenty-five subjects participated in the experiments, 28% women and 72% men. Half were aged 18 to 24 years. They took part on voluntary basis and did not receive a financial reward. The test strategy is closely linked to the model of emotions the authors have chosen. They decided to work with a dimensional model which dimensions are:

- The nature of the emotion,
- The power of the emotion.

This emotion modelling is equivalent to Valence/Arousal model in which, the valence allows to distinguish positive and negative emotions and the arousal that varies from low to high defines the arousal body. This parameter is to describe the intensity of the emotion associated with the nature of the chosen emotion. During the different tests, observers evaluated the nature and the power of the emotion aroused by the image. As shown on Figure 5, for the nature of the emotion, the participants had choice between "Negative", "Neutral" and "Positive". The power was chosen from low to high. During the tests, some presented images have the same content but are modified with some transformations. Rotations and changes in colour balance are applied. Even if they change the natural aspects of some pictures, they allowed to see the results of some invariants features used to recognise the emotion associated with the image. Only 2.29% images represent these transformed "non natural" content.

The next subsection will describe why we chose to process our features extraction algorithms on salient thumbnails instead of full images.

3.3. Bottom-up saliency model

During the emotional evaluation of images of SENSE, the authors realized that a number of pictures were

Figure 4. Images examples from the SENSE database

Figure 5. Emotional evaluation of SENSE images

not clearly classified by all users. This result is not surprising because of the construction of the image database. In fact, images on the database are selected so they do not cause strong emotional answers to observers. Therefore it seems logical that a number of them are not easily emotionally classifiable. A new experiment was then mounted: evaluate the most salient areas for each SENSE image. Maybe focusing on the most salient zones would allow better emotional classification? This experiement was made in [13] resulting in a novel SENSE2 database. For each SENSE image, its most salient region counterpart was added to SENSE2. SENSE2 analysis showed a correlation of emotion evaluations from each SENSE image and their salient counterpart in SENSE2.

In our proposed system, as we will see in Section 3.4, we extract emotions on images from CBIR techniques. We designed our system to work on low semantic images as we think that images answers from image search engines will not embed strong emotional impacts on users. This is why we choose to process features from image thumbnails acquired by a saliency model. The saliency model we use was developed in our laboratory from previous works by Perreira et al.[36] and is based on a computational model of attention.

This was implemented to improve the performances of our system:

- working only on smaller images enhances time processing;
- focusing on most significant regions performs better emotional classifications.

Many researchers have worked on visual attention, one can report to [3] for a good review. A classiquely used model of visual attention is Laurent Itti's [19]. The first part of its architecture relies on the extraction of three conspicuity maps based on low level characteristics computation, that's correspond to the production of information on retina. These three conspicuity maps are color, intensity and orientation.

The second part of Itti's architecture proposes a medium level system which allows merging conspicuity maps and then simulates a visual attention path on the observed scene. The focus is determined by a *"winner-takes-all"* and an *"inhibition of return"* algorithms (Figure 6).

Perreira et al. proposed to substitute this second part by their optimal competitive architecture: a preys / predators system. This optimal criteria, preys / predators equations are particularly well adapted for such a task:

- preys / predators systems are dynamic, they include intrinsically time evolution of their activities. Thus, the visual focus of attention, seen as a predator, can evolve dynamically;
- without any objective (top-down information or pregnancy), choosing a method for conspicuity maps fusion is hard. A solution consists in developing a competition between conspicuity maps and waiting for a natural balance in the preys / predators system, reflecting the competition between emergence and inhibition of elements that engage or not our attention;
- discrete dynamic systems can have a chaotic behavior. Despite the fact that this property is not often interesting, it is an important one. Actually, it allows the

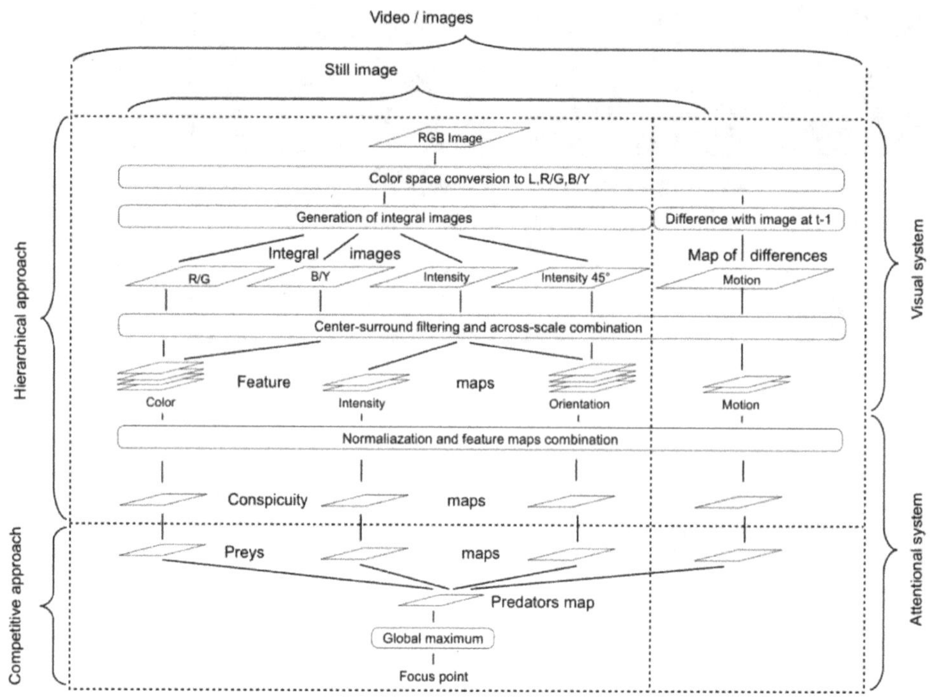

Figure 6. Architecture of the Perreira computational model of attention

emergence of original paths and exploration of visual scene, even in non salient areas, reflecting something like curiosity.

General architecture of the Perreira model is represented in Figure 7

Starting from the "basic" version of preys / predators equations, Perreira et al. enriched processing in several ways:

• the number of parameters can be reduced by replacing s' by s. Indeed, mortality rates differences between preys and predators can be modeled by an adjustment of factors b and m_I

• the original model represents the evolution of a single quantity of preys and predators over time. It can be spatially extended in order to be applied to 2D maps where each point represents the amount of preys or predators at a given place and time. Preys and predators can then "move" on this map using a classical diffusion rule, proportional to their Laplacian Δ_C and a diffusion factor f.

• natural mortality of preys in the absence of predation is not taken into account. If the model only changes temporally, mortality is negligible when compared to predation. However, when the model is applied to a 2D map (which is the case in our system), some areas of the map may not contain any predator. Natural mortality of prey can no longer be considered negligible. A new mortality term $-m_c$ need to be added to the model.

This yields to the following set of equations, modeling the evolution of preys and predators populations on a two dimensional map:

$$\begin{cases} \frac{dC_{x,y}}{dt} &= bC_{x,y} + f\ \Delta_{C_{x,y}} - m_C C_{x,y} - sC_{x,y}I_{x,y} \\ \frac{dI_{x,y}}{dt} &= sC_{x,y}I_{x,y} + sf\ \Delta_{P_{x,y}} - m_I I_{x,y} \end{cases}$$

A last phenomenon can be added to this model: a positive feedback, proportional to C^2 or I_2 and controlled by a factor w. This feedback models the fact that (provided unlimited resources) the more numerous a population is, the better it is able to grow (more efficient hunting, higher encounter rater favoring reproduction, etc.). The final preys / predators system is then:

$$\begin{cases} \frac{dC_{x,y}}{dt} &= b(C_{x,y} + w(C_{x,y})^2) + f\ \Delta_{C_{x,y}} - m_C C_{x,y} - sC_{x,y}I_{x,y} \\ \frac{dI_{x,y}}{dt} &= s(C_{x,y}I_{x,y} + w(I_{x,y})^2) + sf\ \Delta_{P_{x,y}} - m_I I_{x,y} \end{cases}$$

In order to simulate the evolution of the focus of attention, Perreira et al. proposed a preys / predators system (as described above) with the following features:

• the system is comprised of four types of preys and one type of predators;

• these four types of preys represent the spatial distribution of the curiosity generated by four types of conspicuity maps;

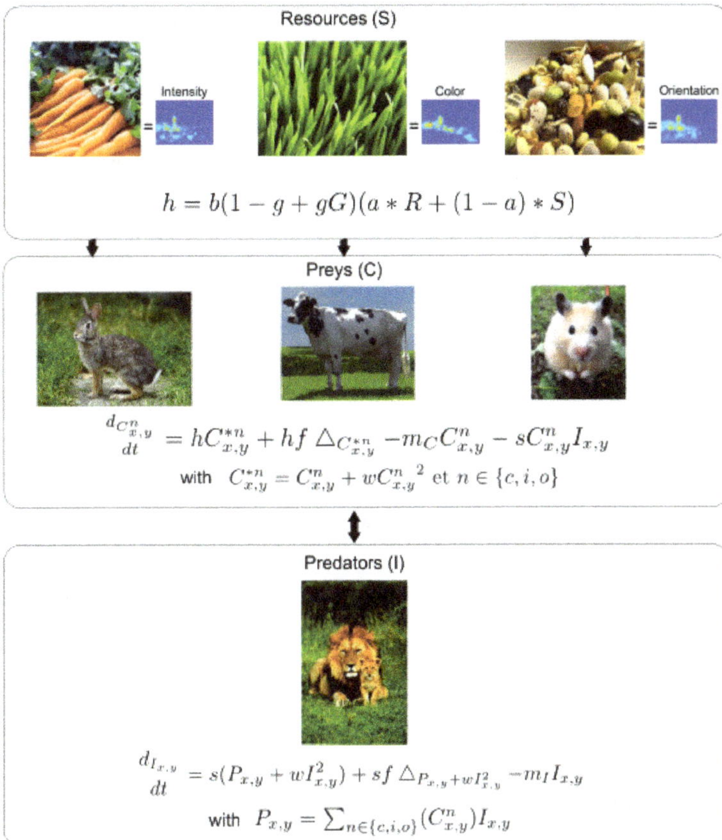

Figure 7. Competitive preys / predators attention model. Singularity maps are the resources that feed a set of preys which are themselves eat by predators. The maximum of the predators map represents the location of the current focus of attention.

- the predators represent the interest generated by the consumption of curiosity (preys) associated to the different conspicuity maps;
- the global maximum of the predators maps (interest) represents the focus of attention at time t.

The authors showed that despite the non deterministic behavior of preys / predators equations, the system exhibits interesting properties of stability, reproducibility and reactiveness while allowing a fast and efficient exploration of the scene.

With this visual saliency model, we can focus our emotion extraction on the minimal local region of an image where the its emotion impact should be the most meaningful. In other words, this enables us to summarize every image emotion by its most salient thumbnail counterpart and enhances as well feature extraction processing time as thumbnails dimensions are smaller.

The next subsection represents the core of our system, it details how we constructed our emotional filtering structure layer by layer.

3.4. Emotional filtering system structure

Our emotional filter is based on an emotional score processed from a machine learning approach. The idea is to process a signature for a request image and to inject it on trained classifiers in order to output an emotional score. The global system is illustrated by Figure 1 and Figure 8.

First, we collect an input set of images by a textual or image request from any image search engine. Then we extract the main salient region for each image of the set using the computational model of attention [36] formerly described in Section 3.3. This additional processing is justified by the fact that the emotional impact of an image is correlated to its main salient region [13]. Moreover, as thumbnails, salient regions take less weight and then less processing time.

Next, each salient image (thumbnail) is processed by a Harris Laplace key-point detector[16, 29]. A dynamic algorithm is performed to ensure a minimum amount of detected key-points to describe. The size of the thumbnail to process is sometimes enlarged to ensure a minimal amount of detected key-points.

Each key-point is then described using five low level descriptors:

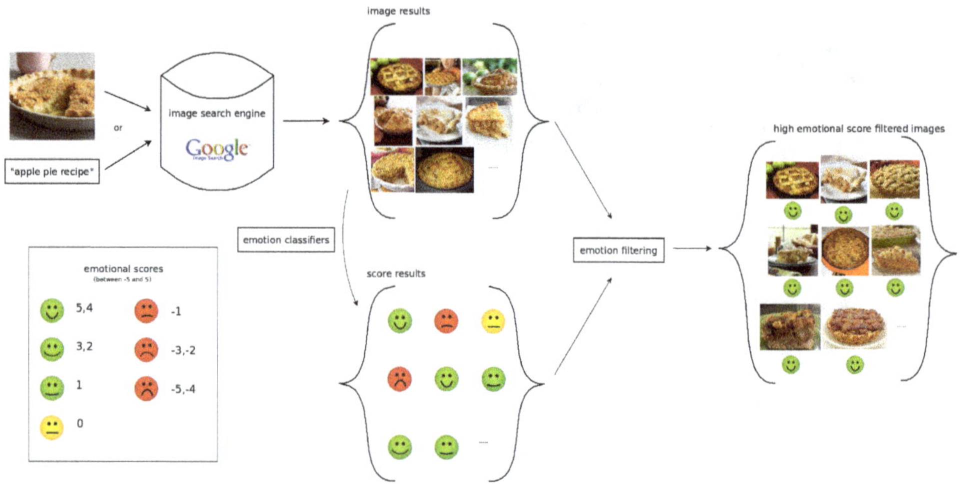

Figure 8. Emotional filtering model. In this example, only high emotional score images are displayed as a result of the filter

- color moments (CM) [51],
- color moments invariant (CMI) [30],
- scale invariant feature transform (SIFT) [26],
- color SIFT (CSIFT) [1]
- opponent SIFT (OppSIFT) [47].

These descriptors, given in [12], are justified by their properties of invariance over geometric transformations such as translations, rotations and image scaling.

Then a k-means reduction algorithm is applied on each vector to reduce the dimensionality of feature vectors. An image CBIR's Bag of Words (BoW) method [4] is used to product visual words from these feature vectors as it was for training of the emotional classifiers in [12]. Training process is illustrated in Figure 9. Note that emotions are trained offline on the SENSE database.

So we classify each image with these five classifiers by the same training sets used with the SENSE database in [12] using Support Vector Machine classifiers (SVMs) implemented from libSVM library with linear kernels [8]. Each classifier outputs a real score between -1 and +1. -1 stands for a negative emotion impact whereas +1 stands for a positive emotion impact. All the scores are merged by a simple addition into a final emotional score. This is illustrated by Figure 10.

3.5. Results and discussions

A 100,000 Internet cooking recipes images database was created by a partner lab in our project. Figure 11 shows image examples of its content. That was a good opportunity to test our model on a large scale. This database is full of a large diversity in the shapes and looks of the displayed dishes. We so presumed that its content do not imply hight emotional impact on viewers. This is then why we choose to use the SENSE database as a trainer for our application.

Our system can output different views. First, images can be displayed by processed emotional score order; second, system can output the set of most positive emotional images to fulfill a better emotional impact on viewer when applied on research engine's results; we can as well show negative emotional score images only.

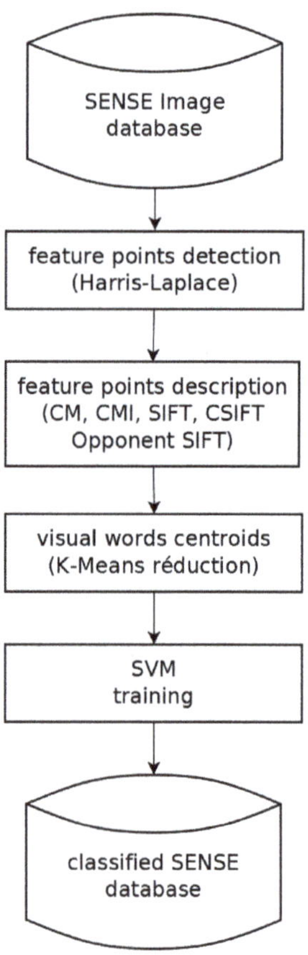

Figure 9. Emotional classifiers training

Figure 8 shows some results of our system on this cooking database. We can see how our system is filtering images and that the result lacks of any

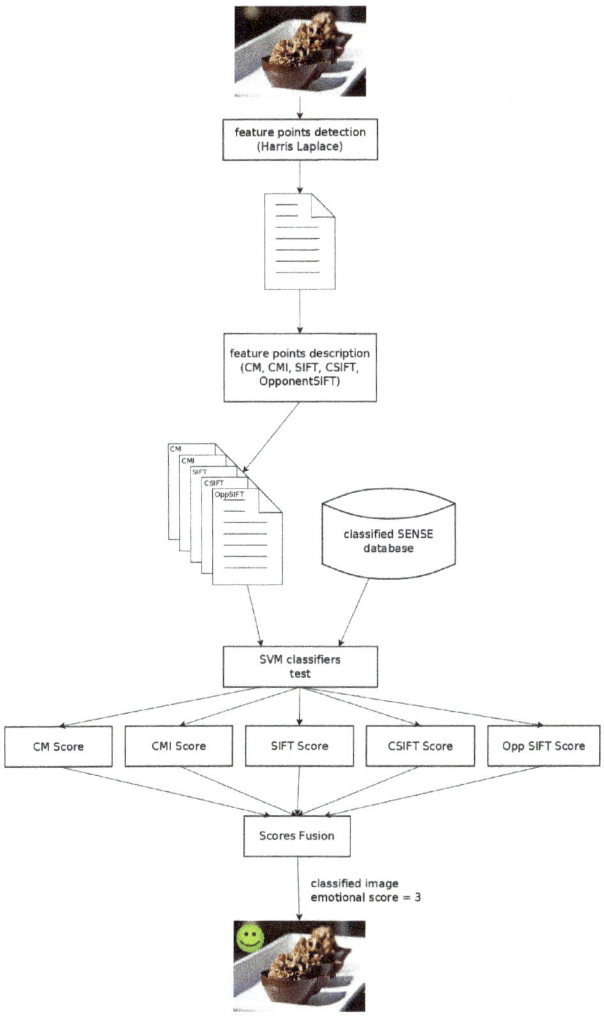

Figure 10. Emotional score processing

emotional validation. But again, the main novelty of our proposition was to apply for the first time an emotional filtering layer on the results of image search engines. This validation process will be investigated on future works. Another contribution was to use low-level features from the most salient regions only to describe image emotions.

In [14], authors used a hybrid approach to define a notion of attractiveness in web image search. This is one of the nearest works we could find from our proposed system. They used EXIF metadata associated with trained high level visual features (quality assessment, aesthetic prediction, affective classification) to define their attractiveness ranks. They also confess that their results are still widely improvable. To improve them, they add images EXIF textual metadata to their model. They also look into the image source web pages for other textual information to help in the classification process.

Our approach here is different, we want to extract the emotion information only on the image content. We are not at all interested with additional information to enhance the performance of classification as it is usually done in other approaches. Also our system scores image sets where theirs scores Web pages where images can be found.

4. Conclusions and Perspectives

In this paper, we focused on the impact of emotions from images collected from research image engines. We proposed for the first time to add an emotional filtering system to image search engines results.

With this new filtering technique, web users will enhance their experience on image searches.

We saw that our solution still needs a validation phase. We thought of several centered on human evaluation:

- a standalone application where evaluators will validate the engine classification. Image search engine result sets will be labeled by people on an emotional perspective. Comparison between labeled grountruth samples and the system's results can then be reached;
- evaluators could compare their emotional feelings from watching direct search engines result sets with respect to emotionally filtered ones;
- users needs for individual contents may be achieved using backpropagation labelling to train classifiers on personalized emotional result sets. iv. a user interface to re-rank (drag and drop) images ordered by the emotional filtering system will take their prefered ranking positions as new labeled training sets.
- enhancement of user personalized results by deletion of images from his displayed results. This will again help the classifier training with backpropagation techniques.

New technologies such as eye-tracking can help analysing user behaviour when observing results from search engines. We already started experiments in our lab. Given two images displayed, users are asked to focus on their prefered one. The analysis of eye-tracking data allows the computer to determine the prefered one. Future works may reveal eye-tracking path patterns linked to emotion impact factors when looking to images.

Big data technologies will also be necessary to perform customizable results for users. Behavior patterns are already extracted from your navigation over the Internet. Merging Internet extracted emotional information to emotional image descriptors may also allow huge enhancements in the efficiency of image search engine filters results.

If all were implemented, Internet could eventually be focused on users and not only on data.

Acknowledgement. This work was supported by the ANR Visiir project, grant ANR-13-CORD-0009 of the French Agence Nationale de la Recherche.

Figure 11. Images examples from the our cooking recipes database

References

[1] A. E. Abdel-Hakim and A. A. Farag. Csift: A sift descriptor with color invariant characteristics. In *Comp. Society Conf. on Comp. Vis. and Pat. Rec.*, volume 2, pages 1978–1983. IEEE, 2006.

[2] L. Beke, G. Kutas, Y. Kwak, G. Y. Sung, D. Park, and P. Bodrogi. Color preference of aged observers compared to young observers. *Color Research & Application*, 33(5):381–394, 2008.

[3] A. Borji and L. Itti. State-of-the-art in visual attention modeling. *Trans. on Pat. Analysis and Mach. Intelligence*, 35(1):185–207, 2013.

[4] A. Bosch, X. Muñoz, and R. Martí. Which is the best way to organize/classify images by content? *Image and vision computing*, 25(6):778–791, 2007.

[5] B. E. Boser, I. M. Guyon, and V. N. Vapnik. A training algorithm for optimal margin classifiers. In *Proc. of the 5th annual workshop on Computational learning theory*, pages 144–152. ACM, 1992.

[6] C. Boyatziz and R. Varghese. Children's emotional associations with colors. *The Journ. of Genetic Psychology*, 155:77–85, 1993.

[7] M. M. Bradley, M. Codispoti, D. Sabatinelli, and P. J. Lang. Emotion and motivation ii: Sex differences in picture processing. *Emotion*, 1(3):300–319, 2001.

[8] C.-C. Chang and C.-J. Lin. Libsvm: A library for support vector machines. *ACM Trans. Intell. Syst. Technol.*, 2(3):27:1–27:27, May 2011.

[9] C. Cortes and V. Vapnik. Support-vector networks. *Machine learning*, 20(3):273–297, 1995.

[10] P. Ekman. Facial expressions of emotions. *Psychological science*, 3(1):34–38, 1992.

[11] D. Fonseca, O. García, J. Duran, M. Pifarré, and E. Villegas. An image-centred "search and indexation system" based in user's data and perceived emotion. In *Proc. of the 3rd ACM Int. Workshop on Human-centered Computing*, HCC '08, pages 27–34, New York, NY, USA, 2008. ACM.

[12] S. Gbèhounou, F. Lecellier, and C. Fernandez-Maloigne. Extraction of emotional impact in colour images. In *6th European CGIV Conference*, pages 314–319, Pays-Bas, May 2012.

[13] S. Gbèhounou, F. Lecellier, C. Fernandez-Maloigne, and V. Courboulay. Can salient interest regions resume emotional impact of an image? In R. Wilson, E. Hancock, A. Bors, and W. Smith, editors, *Computer Analysis of Images and Patterns*, volume 8047 of *Lecture Notes in Comp. Science*, pages 515–522. Springer Berlin Heidelberg, 2013.

[14] B. Geng, L. Yang, C. Xu, X.-S. Hua, and S. Li. The role of attractiveness in web image search. In *Proc. of the 19th ACM Int. Conf. on Multimedia*, MM '11, pages 63–72, New York, NY, USA, 2011. ACM.

[15] J. S. Hare, P. A. Sinclair, P. H. Lewis, K. Martinez, P. G. Enser, and C. J. Sandom. Bridging the semantic gap in multimedia information retrieval: Top-down and bottom-up approaches. In *In 3rd European Semantic Web Conf. (ESWC-06), LNCS 4011*, 2006.

[16] C. Harris and M. Stephens. A combined corner and edge detector. In *In Proc. of 4th Alvey Vision Conf.*, pages 147–151, 1988.

[17] X. Hu, J. Tang, H. Gao, and H. Liu. Unsupervised sentiment analysis with emotional signals. In *Proc. of the 22nd Int. Conf. on WWW*, WWW '13, pages 607–618, Republic and Canton of Geneva, Switzerland, 2013. Int. WWW Conf. Steering Committee.

[18] G. Huisman, M. van Hout, E. van Dijk, T. van der Geest, and D. Heylen. Lemtool: Measuring emotions in visual interfaces. In *Proc. of the SIGCHI Conf. on Human Factors in Computing Systems*, CHI '13, pages 351–360, New York, NY, USA, 2013. ACM.

[19] L. Itti, C. Koch, E. Niebur, et al. A model of saliency-based visual attention for rapid scene analysis. *IEEE Trans. on Pat. Analysis and Mach. Intelligence*, 20(11):1254–1259, 1998.

[20] S. D. Kamvar and J. Harris. We feel fine and searching the emotional web. In *Proc. of the 4th Int. Conf. on Web Search and Web Data Mining (WSDM)*, pages 117–126. ACM, 2011.

[21] K. Knautz, T. Siebenlist, and W. G. Stock. Memose: Search engine for emotions in multimedia documents. In *Proc. of the 33rd Int. ACM SIGIR Conf. on Research and Development in Information Retrieval*, SIGIR '10, pages 791–792, New York, NY, USA, 2010. ACM.

[22] A. Krishnan and M. Fernandez. System and method for recognizing emotional state from a speech signal, Nov. 2013. US Patent 8,595,005.

[23] P. J. Lang, M. M. Bradley, B. N. Cuthbert, et al. International affective picture system (iaps): Affective ratings of pictures and instruction manual. Technical Report A-8, Univ. of Florida, Gainesville, Florida, 2008.

[24] K. H.-Y. Lin, C. Yang, and H.-H. Chen. Emotion classification of online news articles from the reader's perspective. In *Proc. of the Int. Conf. on Web Intelligence and Intelligent Agent Technology - Volume 01*, WI-IAT '08, pages 220–226, Washington, DC, USA, 2008. IEEE Computer Society.

[25] N. Liu, E. Dellandréa, B. Tellez, and L. Chen. Evaluation of features and combination approaches for the classification of emotional semantics in images. In *Int. Conf. on Computer Vision Theory and Applications*, 2011.

[26] D. G. Lowe. Object recognition from local scale-invariant features. In *Proc. of the 7th Int. Conf. on Comp. Vision*, volume 2, pages 1150–1157. Ieee, 1999.

[27] M. P. Lucassen, T. Gevers, and A. Gijsenij. Adding texture to color: quantitative analysis of color emotions. In *Proc. of CGIV*, 2010.

[28] J. Machajdik and A. Hanbury. Affective image classification using features inspired by psychology and art theory. In *Proc. of the Int. Conf. on Multimedia*, pages 83–92. ACM, 2010.

[29] K. Mikolajczyk and C. Schmid. Scale & affine invariant interest point detectors. *Int. Journal of Comp. Vision*, 60:63–86, 2004.

[30] F. Mindru, T. Tuytelaars, L. V. Gool, and T. Moons. Moment invariants for recognition under changing viewpoint and illumination. *Comp. Vision and Image Understanding*, 94(1):3–27, 2004.

[31] C. S. Ooi, K. P. Seng, A. Li-Minn, and L. W. Chew. A new approach of audio emotion recognition. *Expert Systems with Applications*, 41(13):5858 – 5869, 2014.

[32] L. C. Ou, M. R. Luo, A. Woodcock, and A. Wright. A study of colour emotion and colour preference. part i: Colour emotions for single colours. *Color Research & Application*, 29(3):232–240, 2004.

[33] L. C. Ou, M. R. Luo, A. Woodcock, and A. Wright. A study of colour emotion and colour preference. part ii: Colour emotions for two-colour combinations. *Color Research & Application*, 29(4):292–298, 2004.

[34] L. C. Ou, M. R. Luo, A. Woodcock, and A. Wright. A study of colour emotion and colour preference. part iii: Colour preference modeling. *Color Research & Application*, 29(5):381–389, 2004.

[35] M. Paleari and B. Huet. Toward emotion indexing of multimedia excerpts. *Proc. on Content-Based Multimedia Indexing, Int. Workshop*, pages 425–432, 2008.

[36] M. Perreira Da Silva and V. Courboulay. Implementation and evaluation of a computational model of attention for computer vision. In *Developing and Applying Biologically-Inspired Vision Systems: Interdisciplinary Concepts*, pages 273–306. Hershey, Pennsylvania: IGI Global, Aug. 2012.

[37] R. Plutchik and H. R. Conte. *Circumplex Models of Personality and Emotions*. American Psychological Association, 1997.

[38] J. Read. Using emoticons to reduce dependency in machine learning techniques for sentiment classification. In *Proc. of the ACL Student Research Workshop*, ACLstudent '05, pages 43–48, Stroudsburg, PA, USA, 2005. Association for Computational Linguistics.

[39] S. Rigoulot. *Impact comportemental et électrophysiologique de l'information émotionnelle en vision périphérique*. PhD thesis, Univ. du Droit et de la Santé - Lille II, Dec. 2008.

[40] B. Rimé. Émotion et mémoire: La rémanence des expériences émotionnelles. *Cognition et Émotions*, pages 79–96, 2004.

[41] A. Salway and M. Graham. Extracting information about emotions in films. In *Proc. of the 11th ACM Int. Conf on Multimedia*, MULTIMEDIA '03, pages 299–302, New York, NY, USA, 2003. ACM.

[42] P. G. Schyns, L. S. Petro, and M. L. Smith. Transmission of facial expressions of emotion co-evolved with their efficient decoding in the brain: behavioral and brain evidence. *Plos one*, 4(5):e5625, 2009.

[43] M. Solli and R. Lenz. Color based bags-of-emotions. In *Computer Analysis of Images and Patterns*, volume 5702 of *Lecture Notes in Comp. Science*, pages 573–580. Springer Berlin Heidelberg, 2009.

[44] M. Solli and R. Lenz. Emotion related structures in large image databases. In *Proc. of the ACM Int. Conf. on Image and Video Retrieval*, CIVR '10, pages 398–405, New York, NY, USA, 2010. ACM.

[45] I. Tayari, N. Le Thanh, and C. B. Amar. Modélisation des états émotionnels par un espace vectoriel multidimensionnel. Technical report, Technical report, Laboratoire Informatique, Signaux et Systèmes de Sophia Antipolis, 2009.

[46] Y. Urabe, R. Rzepka, and K. Araki. Emoticon recommendation system for effective communication. In *Proc of the Int. Conf. on ASONAM*, pages 1460–1461, New York, USA, 2013. ACM.

[47] K. E. Van De Sande, T. Gevers, and C. G. Snoek. Evaluating color descriptors for object and scene recognition. *Trans. on Pat. Analysis and Mach. Intelligence*, 32(9):1582–1596, 2010.

[48] W. Wang and Y. Yu. Image emotional semantic query based on color semantic description. In *Proc of the 4th Int. Conf. on Machine Learning and Cybernectics*, volume 7, pages 4571–4576, 2005.

[49] K. Wei, B. He, T. Zhang, and W. He. *Image Emotional Classification Based on Color Semantic Description*, volume 5139 of *Lecture Notes in Comp. Science*, pages 485–491. Springer Berlin / Heidelberg, 2008.

[50] V. Yanulevskaya, J. Van Gemert, K. Roth, A.-K. Herbold, N. Sebe, and J.-M. Geusebroek. Emotional valence categorization using holistic image features. In *15th ICIP*, pages 101–104. IEEE, 2008.

[51] H. Yu, M. Li, H.-J. Zhang, and J. Feng. Color texture moments for content-based image retrieval. In *ICIP*, volume 3, pages 929–932. IEEE, 2002.

[52] L. Zhang, L. Chen, F. Jing, K. Deng, and W.-Y. Ma. Enjoyphoto: a vertical image search engine for enjoying high-quality photos. In *Proc. of the 14th annual ACM Int. Conf. on Multimedia*, pages 367–376, 2006.

Permissions

All chapters in this book were first published by EAI European Alliance for Innovation; hereby published with permission under the Creative Commons Attribution License or equivalent. Every chapter published in this book has been scrutinized by our experts. Their significance has been extensively debated. The topics covered herein carry significant findings which will fuel the growth of the discipline. They may even be implemented as practical applications or may be referred to as a beginning point for another development.

The contributors of this book come from diverse backgrounds, making this book a truly international effort. This book will bring forth new frontiers with its revolutionizing research information and detailed analysis of the nascent developments around the world.

We would like to thank all the contributing authors for lending their expertise to make the book truly unique. They have played a crucial role in the development of this book. Without their invaluable contributions this book wouldn't have been possible. They have made vital efforts to compile up to date information on the varied aspects of this subject to make this book a valuable addition to the collection of many professionals and students.

This book was conceptualized with the vision of imparting up-to-date information and advanced data in this field. To ensure the same, a matchless editorial board was set up. Every individual on the board went through rigorous rounds of assessment to prove their worth. After which they invested a large part of their time researching and compiling the most relevant data for our readers.

The editorial board has been involved in producing this book since its inception. They have spent rigorous hours researching and exploring the diverse topics which have resulted in the successful publishing of this book. They have passed on their knowledge of decades through this book. To expedite this challenging task, the publisher supported the team at every step. A small team of assistant editors was also appointed to further simplify the editing procedure and attain best results for the readers.

Apart from the editorial board, the designing team has also invested a significant amount of their time in understanding the subject and creating the most relevant covers. They scrutinized every image to scout for the most suitable representation of the subject and create an appropriate cover for the book.

The publishing team has been an ardent support to the editorial, designing and production team. Their endless efforts to recruit the best for this project, has resulted in the accomplishment of this book. They are a veteran in the field of academics and their pool of knowledge is as vast as their experience in printing. Their expertise and guidance has proved useful at every step. Their uncompromising quality standards have made this book an exceptional effort. Their encouragement from time to time has been an inspiration for everyone.

The publisher and the editorial board hope that this book will prove to be a valuable piece of knowledge for researchers, students, practitioners and scholars across the globe.

List of Contributors

Luis G. Montané-Jiménez, Edgard Benítez-Guerrero and Carmen Mezura-Godoy
Facultad de Estadística e Informática, Universidad Veracruzana, Xalapa, Mexico 91020

Nesrine Khabou
ReDCAD laboratory, University of Sfax, National School of Engineers of Sfax, Tunisia

Ismael Bouassida Rodriguez
ReDCAD laboratory, University of Sfax, National School of Engineers of Sfax, Tunisia

Univ de Toulouse, LAAS, Toulouse, France

Mohamed Jmaiel
ReDCAD laboratory, University of Sfax, National School of Engineers of Sfax, Tunisia

Research Center for Computer Science, Multimedia and Digital Data Processing of Sfax, Sfax, Tunisia

Katayoon Etemad, Dominikus Baur, John Brosz, Sheelagh Carpendale and Faramarz F.Samavati
University of Calgary

Maurizio Mancini, Gualtiero Volpe, Giovanna Varni and Antonio Camurri
Casa Paganini - InfoMus, DIBRIS, University of Genoa, Italy

I. Renna, F. Catteau, C. Vincent and D. Boutet
Structures Formelles du Langage, UMR 7023 (CNRS / Université Paris 8)

S. Delacroix
Laboratoire d'Analyse du Mouvement, Institut National de Podologie, Paris

Phuong T. Nguyen
Research and Development Center, Duy Tan University, Da Nang, Vietnam

Volkmar Schau and Wilhelm R. Rossak
Department of Computer Science, Friedrich-Schiller-Univerität Jena, Ernst-Abbe-Platz 2-4, D-07743 Jena, Germany

Nguyen Thanh Tung
International School, Hanoi Vietnam National University , Vietnam

Fabrizio Nunnari
Deutsches Forschungszentrum für Künstliche Intelligenz (DFKI), CAMPUS D3.2, 66123, Saarbrücken, Germany

Alexis Heloir
Deutsches Forschungszentrum für Künstliche Intelligenz (DFKI), CAMPUS D3.2, 66123, Saarbrücken, Germany

LAMIH-UMR CNRS 8201, Le Mont Houy, Univ. Valenciennes, France

Phan Cong Vinh
Nguyen Tat Thanh University, Ho Chi Minh City, Vietnam

Sidi Ahmed Mahmoudi
University of Mons, Faculty of Engineering, Computer Science Department. Place du Parc, 20. 7000, Mons, Belgium

Minoru Nakayama and Naoya Takahashi
Human System Science, Tokyo Institute of Technology Ookayama, Meguro, Tokyo, Japan

Daniele Grassi and Anas Bouhtouch
DM Digital, Modena, Italy

Giacomo Barigazzi
DM Digital, Modena, Italy

Università di Modena e Reggio Emilia, Modena, Italy

Giacomo Cabri
Università di Modena e Reggio Emilia, Modena, Italy

Andrea Sanna, Fabrizio Lamberti, Gianluca Paravati, Gilles Carlevaris and Paolo Montuschi
Dipartimento di Automatica e Informatica, Politecnico di Torino, Torino 10129, Italy

Carla Passiatore and Pietro Camarda
DEI, Politecnico di Bari, Bari, Italy

Huong Hoang Luong
CUSC-CTU, Cantho, Vietnam

Thanh Hai Nguyen
Ministry of Education and Training, Vietnam

Hung Huu Huynh
The University of Danang, Vietnam

Hiep Xuan Huynh
DREAM-CTU/IRD, CICT-CTU, Cantho, Vietnam

Phan Cong Vinh
Nguyen Tat Thanh University, 300A Nguyen Tat Thanh street, Ward 13, District 4, Ho Chi Minh City, Vietnam

S. W. Loke
Department of Computer Science and Information Technology, La Trobe University, Kingsbury Drive, VIC 3086, Australia

Kristin Carlson, Thecla Schiphorst and Steve DiPaola
The School of Interactive Arts + Technology, Simon Fraser University, Surrey, British Columbia, Canada

Bao Hoai Lam and Bernard Pottier
Université de Bretagne Occidentale, LAB-STICC, UMR CNRS 6285, France

Hiep Xuan Huynh
DREAM Team/UMI 209 UMMISCO-IRD, Cantho University, Vietnam

Hoang Van Tran and Hiep Xuan Huynh
Department of Software Engineering, College of Information & Communication Technology, Cantho University, Cantho city, Vietnam

Vinh Cong Phan
Department of Computer Engineering, Faculty of Information Technology, Nguyen Tat Thanh University, Hochiminh city, Vietnam

Bernard Pottier
Département Informatique, Université de Bretagne Occidentale, Brest, France

Arie Pratama Sutiono, Rido Ramadan and Ayu Purwarianti
Bandung Institute of Technology, Bandung, Indonesia

Peetikorn J arukasetporn
Chulalongkorn University, Bangkok, Thailand

J unki Takeuchi and Hiroyuki Iida
Japan Advanced Insitute of Science and Technology, Ishikawa, Japan

Tuyen Phong Truong and Bernard Pottier
Lab-STICC, UMR CNRS 6285, Université de Bretagne Occidentale, France

Huong Hoang Luong and Hiep Xuan Huynh
Can Tho University, Vietnam

Vinh Cong Phan
Nguyen Tat Thanh University, Vietnam

Patrice Denis, Vincent Courboulay and Arnaud Revel
Laboratoire Informatique, Image et Interaction (L3i) Pôle Sciences et Technologie - Université de La Rochelle Avenue Michel Crépeau - 17042 La Rochelle Cedex 1 - France

Syntyche Gbèhounou, François Lecellier and Christine Fernandez-Maloigne
Département SIC du Laboratoire XLIM - UMR CNRS 7252 Université de Poitiers - Bât SP2MI Téléport 2 - 11 Bd Marie et Pierre Curie BP 30179 - 86962 Futuroscope Chasseneuil Cedex – France

Index

www.ingramcontent.com/pod-product-compliance
Lightning Source LLC
LaVergne TN
LVHW082202200926
848484LV00008B/2200
* 9 7 8 1 6 8 2 8 5 3 4 2 9 *